Abiotic Stress Management in Fruit Crops

Abiotic Stress Management in Fruit Crops

Manju
College of Horticulture, Veer Chandra Singh Garhwali Uttarakhand University of Horticulture and Forestry, Bharsar, Uttarakhand, India

Sarita Agrawal
College of Agriculture and Research Station, Raigarh, Indira Gandhi Krishi Vishwavidyalaya (IGKV), Raipur, Chhattisgarh, India

Md Masidur Alam
Department of Botany, Malda College, University of Gour Banga, Malda, West Bengal, India

Muzafar Mir
Division of Fruit Science, Sher-e-Kashmir University of Agricultural Sciences & Technology, Jammu (J & K), India

Kh. Chandrakumar Singh
Department of Natural Resource Management, VCSGUUHF Bharsar, U.K., India

Aman Sharma
Department of Floriculture & Landscaping, PAU, Ludhiana, India

-EPH-
Elite Publishing House

CRC Press
Taylor & Francis Group
Boca Raton London New York

CRC Press is an imprint of the Taylor & Francis Group, an **informa** business

First published 2026
by CRC Press
4 Park Square, Milton Park, Abingdon, Oxon, OX14 4RN

and by CRC Press
2385 NW Executive Center Drive, Suite 320, Boca Raton FL 33431

CRC Press is an imprint of Informa UK Limited

British Library Cataloguing-in-Publication Data
A catalogue record for this book is available from the British Library

Print edition not for sale in India.

ISBN: 9781041205531 (hbk)
ISBN: 9781041206484 (pbk)
ISBN: 9781003717409 (ebk)

DOI: 10.1201/9781003717409

Typeset in A Caslon Pro
by Elite Publishing House, Delhi

-EPH-

Contents

Preface

In recent years, the agricultural sector has faced unprecedented challenges due to the growing prevalence and intensity of abiotic stresses. Climate change, soil degradation, water scarcity, and increased environmental pollution have imposed considerable strain on fruit crops, directly impacting productivity, quality, and sustainability in horticultural practices. As our global population continues to rise, the demand for consistent and resilient fruit production intensifies, underscoring the urgent need for research-driven solutions and advanced management practices that can help mitigate these environmental stresses. This textbook, **Abiotic Stress Management in Fruit Crops**, is crafted to serve as an indispensable guide for students, researchers, practitioners, and professionals in the field of Horticultural Sciences. It combines scientific principles with practical strategies to address the dynamic and multifaceted challenges of managing abiotic stress in fruit crops, aiming to bridge the gap between theoretical understanding and practical application.

Authors

About the Authors

Dr. Manju is a renowned academician, with nearly 20 years of illustrious teaching experience to the UG, PG, and PhD students in the area of Horticulture. She has contributed vastly to the research activities related to the Multiplication of cultivated as well as wild fruit species, Sustainable production of horticultural crops, and floral biological studies of cultivated as well as wild fruit crops. She has vast administrative experience too. Dr. Manju has a Doctorate in Horticulture (Fruit Science) and M.Sc. In Horticulture (Fruit Science) from Hemwati Nandan Garhwal University, Srinagar Garhwal Uttarakhand and has extensively contributed to the field of Horticulture (Fruit Science) for the past 22 years. She is an Assistant Professor of Horticulture at the College of Horticulture Veer Chandra Singh Garhwali Uttarakhand University of Horticulture and Forestry Bharsar, Uttarakhand. She has contributed to research as a research guide and through working on different development and research projects, bringing out more than 35 publications. She has been a recipient of the Best Horticulturist Award 2021 by the Agricultural and Environmental Technology Development Society, Education Excellence Award 2022 by the Environment and Social Development Association, and the Excellence of Research Award 2024 Given by the Indian Society of Agriculture Science and Technology for her contribution in the field of Horticulture.

Dr. Sarita Agrawal, Assistant Professor at College of Agriculture and Research Station, Raigarh, Indira Gandhi Krishi Vishwavidyalaya (IGKV), Raipur, Chhattisgarh. She completed her Ph.D degree in the Department of Horticulture, College of Agriculture, IGKV, Raipur, C.G. during the year 2013. She has a University Gold Medalist in Ph.D. degree. She has 10 years of experience in teaching and research. She has written two book chapters, 20 research papers, and more than 20 popular articles. Dr. Agrawal has received three awards from different societies for her outstanding contributions to the relevant discipline. She is a life member of the Agricultural and Environmental Technology Development Society (AETDS), U.S. Nagar, UK, India, and Agro-Environmental Development Society (AEDS), Rampur, UP, India.

Md Masidur Alam is an Assistant Professor in the Department of Botany at Malda College, affiliated to the University of Gour Banga, India. He has twenty years of research and eight years of teaching experience. He was awarded PhD degree by Aligarh Muslim University in the year 2013. His research area focuses on the influence of various plant growth regulators

(PGRs) on medicinal plants under normal and abiotic stress conditions. He published 15 research papers in various national and international journals. He also contributed his research work in five book chapters. In the meantime, his team surveyed the impact of arsenic contamination in local crops and published a book chapter in an Agriculture book. He presented so many papers at national and international seminars and conferences. Recently (2024) he was awarded with 'Most Innovative Teacher' award by Aranya Bangla Foundation. At present his team started work on local rice varieties growing under arsenic-contaminated irrigation water and soil to measure the tolerance of varieties.

Dr. Muzafar Mir, working as Sr. Scientist in the Division of Fruit Science, Sher-e-Kashmir University of Agricultural Sciences & Technology of Jammu. He has a Doctorate in Fruit Science from Dr. YS Parmar University of Horticulture & Forestry, Solan (H.P), and qualified for the ICAR-NET in Fruit Science. He has 45 Research publications in reputed National and International Journals, besides two books and three book chapters to his credit. He has more than eleven years of experience in the field of Teaching, Research, and Extension. Presently he is working on germplasm conservation and production technology for fruit crops, particularly in Pome and Nut crops. He has developed and released three varieties of fruit crops two in walnut and one in lemon. In addition to that four other varieties two each in walnut and pecan nut are in the pipeline for release. Currently, his research focuses on the exploration of indigenous germplasm through smart breeding approaches in nut crops. Dr. Mir is a life member of 08 professional societies.

Dr. Kh. Chandrakumar Singh graduated in Agriculture (2009) from HNBGU, Uttarakhand, and achieved both master's (2011) and doctoral (2015) degrees in Agricultural Chemistry and Soil Science from Bidhan Chandra Krishi Viswavidyalaya, Mohanpur. Dr. Singh qualified ICAR-NET conducted by ASRB in 2013. Dr. Singh served KVK – North Sikkim as a Subject Matter Specialist (SMS) and currently serving as an Assistant Professor at VCSGUUHF Bharsar, Uttarakhand. Dr. Singh is actively involved in teaching and research in the field of Soil Science. So far Dr. Singh has guided 13 students in their master's degree programs and published 8 Books and many research papers.

Dr. Aman Sharma joined the Regional Research Station, Faridkot in All India Coordinated Research Project (cotton) on 6th Dec 2013 at Punjab Agricultural University Ludhiana. He has 11 years of experience in Research and Extension and has contributed in the Fields of Cotton Pathology, Extension, and Floriculture Pathology. Dr. Aman Sharma Completed his Ph. D from Chaudhary Charan Singh Haryana Agricultural University, Hisar. He is presently working as a Plant Pathologist in the Department of Floriculture and Landscaping, PAU, Ludhiana. He has about 8 Research papers and has evaluated fifteen cotton varieties against the Cotton leaf curl virus. He has also evaluated three varieties of Floriculture crops against different diseases. He is associated with the publication of Three research and two extension booklets. He has been a CO-PI of three ICAR-funded projects and PI of one DBT–funded project.

1. Weather-based prediction model for cotton leaf curl virus disease and its vector *Bemisia tabaci* in context to climate change (UGC 2013- 2017) (Co-P1)"
2. E-Kapas and Technology documentation (ICAR-TMC, 2013-17) (Co-P1)
3. High-density planting system and insecticide resistance management in cotton (ICAR (2014)), (Co-P1)
4. P1 of project Funded by Department of Biotechnology "Popularization of Foldscope for agriculture pest diagnostics in Punjab (DBT), (2018-2020)

Chapter - 1

Basic Aspects and Principles of Plant Stress

Stress: Definition and Classification

1. Introduction to Stress in Plants

The term "stress" in plant science denotes any external condition that disrupts a plant's metabolic balance, growth, and overall functionality. In the context of agriculture and horticulture, understanding stress factors is critical, as plants are often exposed to diverse environmental conditions that may inhibit or even halt growth and yield. Unlike animals, plants cannot move away from unfavorable conditions and thus have evolved complex physiological and biochemical mechanisms to cope with or mitigate stress impacts. The impact of environmental stress on plants is especially significant in fruit crops, which are highly sensitive to external environmental changes due to their specific physiological needs and stages of fruit development. For fruit crops, stress factors can lead to reductions in yield, quality, and economic returns, making the study of plant stress a vital area within horticultural sciences.

The specific types of stresses that affect fruit crops, as well as the physiological mechanisms plants use to perceive and respond to these stresses, is essential for developing sustainable practices in crop management. The physiological and biochemical responses that fruit crops exhibit to cope with stress vary depending on the nature, duration, and intensity of the stressor. These responses are adaptive mechanisms that allow plants to maintain internal balance and conserve resources, despite adverse conditions. In agriculture, learning about these responses and how different crops react to stress enables farmers, researchers, and practitioners to make informed management decisions that can enhance crop resilience, promote better yields, and minimize economic losses. By comprehensively studying stress and its effects on fruit crops, horticulturalists can develop targeted approaches to reduce the adverse impacts of environmental stressors, thereby supporting food security and sustainable agricultural practices.

Explanation of Stress as an Environmental Factor Impacting Plant Productivity

Stress is a condition or event that disrupts normal plant growth, development, and productivity. The concept of stress in plants encompasses a wide range of factors, both biotic (living) and abiotic (non-living). Abiotic stresses are those that arise from non-living factors such as water scarcity, temperature extremes, radiation, wind, soil salinity, and other physical or chemical changes in the environment. These stresses affect plant productivity by impairing physiological and biochemical processes, ultimately

DOI: 10.1201/9781003717409-1

reducing growth and crop yield. Unlike animals, plants cannot escape from unfavorable conditions, meaning they must instead adjust their internal systems to cope with adverse environments. This adaptation process, however, often comes at the cost of reduced energy availability for growth and reproduction, thereby limiting productivity and, in severe cases, leading to plant death.

Environmental stresses affect plant productivity in several ways, each of which involves complex interactions between the stressor, the plant's physiology, and its surrounding environment. For instance, drought stress reduces water availability for metabolic processes, leading to wilting, stunted growth, and decreased photosynthesis. Temperature extremes, such as high temperatures, can disrupt enzyme functions and damage cellular structures, while low temperatures can lead to freezing injury and cell death. Salinity and nutrient toxicity disrupt the osmotic balance, making it difficult for plants to absorb water and essential nutrients. These stress conditions trigger specific physiological and biochemical responses within the plant, such as stomatal closure, accumulation of osmolytes, and activation of stress-related genes, all of which help the plant survive in adverse conditions but often reduce its growth rate and yield potential.

Fruit crops are especially vulnerable to stress due to their high water, nutrient, and light requirements for optimal growth and fruit development. Stress can not only affect the quantity of fruit produced but also its quality, impacting characteristics like size, flavor, color, and shelf life. For example, water stress during critical stages of fruit development can result in smaller fruit sizes and lower sugar content, while excessive heat can cause sunscald on the fruit surface, reducing marketability. Soil-related stresses, such as high salinity or alkalinity, impede nutrient uptake, resulting in nutrient deficiencies that compromise growth and yield. The cumulative impact of these stress factors highlights the importance of managing and mitigating stress to protect fruit crops and maintain productivity.

Importance of Understanding Stress for Sustainable Fruit Crop Management

Sustainable agriculture aims to produce high-quality crops in a manner that preserves environmental resources, supports economic viability, and minimizes the ecological impact of farming practices. In fruit crop production, managing stress is a fundamental component of achieving sustainability, as it enables farmers to optimize resource use, reduce crop losses, and maintain productivity even under challenging environmental conditions. As global demand for fruit products increases, driven by population growth and consumer preference for fresh produce, the need for sustainable production practices that mitigate stress becomes more urgent. By understanding how stress affects fruit crops and identifying effective management strategies, growers can achieve consistent yields while reducing their reliance on external inputs like water, fertilizers, and pesticides.

One of the key challenges in sustainable fruit crop management is water scarcity, a common stress factor that poses significant risks to agriculture worldwide. Drought stress can severely impact fruit crops by limiting water availability, leading to wilting, reduced growth, and decreased fruit production. In regions prone to water scarcity, understanding the water requirements of specific fruit crops and implementing efficient irrigation techniques can help conserve water resources while ensuring that crops receive sufficient moisture. Practices such as drip irrigation, mulching, and soil moisture monitoring allow for precise water management, reducing water use without compromising crop health. Additionally, selecting drought-tolerant crop varieties and adopting practices that improve soil moisture retention, such as organic matter incorporation, can further enhance resilience against water-related stress.

Temperature stress is another critical factor influencing fruit crop productivity, as extreme heat or cold can disrupt growth processes and reduce yield. High temperatures accelerate evaporation, leading to increased water stress, while also causing thermal damage to leaves and fruits. Cold stress, on the other hand, can cause chilling injury, frost damage, and even cell death in fruit crops that are sensitive to low temperatures. Understanding the temperature thresholds for different fruit crops allows growers to choose appropriate planting dates, select varieties suited to local climates, and implement protective measures like shading or frost covers to mitigate temperature effects. Moreover, advances in breeding for heat and cold tolerance have led to the development of crop varieties that can withstand temperature extremes, enhancing the resilience of fruit production systems in variable climates.

Soil management is also essential for sustainable agriculture, as soil-related stresses like salinity, alkalinity, and nutrient toxicity can impede plant growth and reduce productivity. Salinity stress, which occurs when high levels of soluble salts accumulate in the soil, restricts water uptake by roots and leads to osmotic stress. Alkalinity, or high soil pH, limits the availability of certain nutrients, causing deficiencies that impair growth. Understanding the causes and effects of soil stress enables growers to implement soil management practices that improve soil quality and reduce stress impacts. For example, applying organic amendments, such as compost or biochar, can enhance soil structure and nutrient retention, while gypsum or sulfur can help neutralize excess alkalinity. Soil testing and regular monitoring allow for informed decisions about nutrient management and soil amendments, promoting healthy crop growth and reducing reliance on synthetic fertilizers.

Fruit quality is another aspect of production that is closely linked to stress management, as abiotic stresses often impact the taste, texture, color, and nutritional value of fruit. Consumers have specific preferences for fruit quality attributes, and stress-induced changes can affect marketability and economic returns. For instance, excessive sunlight and high temperatures can cause sunburn on fruit surfaces, altering color and reducing visual appeal, while water stress during ripening stages may reduce fruit size and sweetness. By understanding the relationship between stress and fruit quality, growers can adopt practices that enhance desirable traits while minimizing the negative effects of stress. In some cases, controlled exposure to mild stress, such as regulated deficit irrigation, may even enhance certain quality attributes, such as sugar concentration in grapes used for wine production.

In addition to its direct impact on crop productivity, stress also affects the resilience of fruit crops to pests and diseases. Stressed plants are often more susceptible to pest infestations and pathogen attacks, as their weakened physiological state reduces their ability to mount effective defense responses. Drought-stressed plants, for example, may be more vulnerable to insect pests that feed on weakened tissues, while waterlogged conditions create favorable environments for soilborne pathogens. Integrated pest and disease management strategies that take stress into account can help minimize pest pressure and reduce the need for chemical inputs. For example, maintaining optimal soil moisture levels can reduce the risk of root rot, while selecting stress-tolerant crop varieties may enhance resistance to certain pests and diseases. By incorporating stress management into pest and disease control practices, growers can reduce their reliance on pesticides, promoting a more sustainable and environmentally friendly approach to fruit production.

Finally, understanding stress and its effects on fruit crops is essential for climate resilience, as climate change is expected to increase the frequency and intensity of abiotic stresses in many regions. Rising temperatures, changing precipitation patterns, and increased occurrence of extreme weather events are likely to exacerbate stress factors such as drought, heat waves, and soil degradation. Developing stress-resilient production systems is therefore a critical component of adapting to climate change and ensuring the long-term sustainability of fruit crops. Advances in plant breeding, genetic engineering, and biotechnology offer promising avenues for improving crop resilience to stress. For example, genetic markers associated with drought or heat tolerance can be used to select crop varieties that are better suited to changing climatic conditions. In addition, emerging technologies like remote sensing and precision agriculture allow for real-time monitoring of crop stress, enabling growers to respond quickly to environmental challenges and optimize resource use.

2. Definition of Plant Stress

General Definition of Stress in Biological Terms

In the broad field of biological sciences, "**stress**" refers to any external or internal condition that challenges the normal functioning of an organism, causing disturbances that affect its overall state of equilibrium, or homeostasis. For plants, stress manifests in unique ways compared to animals or other organisms due to their sessile nature—they cannot relocate to escape the stress and must instead adapt their internal mechanisms to manage adverse conditions. In plants, stress often leads to a series of physiological and biochemical changes aimed at minimizing damage and restoring stability. These adaptations, however, can detract from energy resources needed for growth and reproduction, ultimately impacting plant health, productivity, and survival. Understanding stress in biological terms provides a foundational knowledge base essential for managing plant responses in fields like agriculture and horticulture, where crop resilience and productivity are primary concerns.

1. Biological Basis of Stress: Homeostasis and Disruption

In biological systems, **homeostasis** is the state of internal stability that organisms strive to maintain for optimal functioning. This stability involves regulating factors such as temperature, hydration, pH, and nutrient levels within a range that supports cellular processes and overall health. Stress represents any deviation from these optimal conditions that forces the organism to adapt or react. In plants, stress can arise from both biotic sources (e.g., pests, diseases, competition) and abiotic sources (e.g., water scarcity, temperature extremes, salinity). When a plant is exposed to stress, its homeostatic mechanisms are disrupted, requiring the activation of adaptive responses.

Plants achieve homeostasis through a finely tuned balance of metabolic processes, where enzymes, hormones, and cellular signaling pathways work together to regulate growth, development, and response to environmental changes. Stress disrupts these processes by altering the availability or effectiveness of resources that support homeostasis. For instance, drought stress affects water balance and osmotic pressure, temperature stress disrupts enzyme activity and protein stability, and nutrient deficiencies impede metabolic functions. Understanding stress in terms of homeostasis provides a framework for studying plant responses to external pressures and devising strategies to help plants cope with adverse conditions.

2. Stress as a Stimulus for Adaptation and Survival

From an evolutionary perspective, stress acts as a selective force that shapes the adaptive traits of organisms, including plants. In the wild, plants that can successfully adapt to harsh conditions have a better chance of survival and reproduction. This evolutionary adaptation process has led to the development of various stress-tolerance mechanisms that allow plants to withstand and recover from environmental challenges. For example, plants in arid regions have evolved deep root systems, succulent leaves, and waxy cuticles that minimize water loss and improve drought tolerance. Similarly, plants in saline environments have developed mechanisms to exclude or compartmentalize excess salts, allowing them to survive in high-salinity soils.

In agriculture, stress adaptation is critical because it influences crop resilience and yield stability. Crops that can withstand drought, heat, salinity, and other stresses are essential for food security, especially in regions where extreme environmental conditions are common. Breeding programs and genetic engineering efforts often focus on enhancing stress tolerance by introducing or amplifying adaptive traits found in stress-resistant plant species. This approach highlights the importance of understanding stress as not only a challenge to plant health but also a stimulus for developing resilient traits that ensure long-term survival and productivity.

3. Classification of Stress: Biotic vs. Abiotic Stress

Stress can be broadly classified into two categories: **biotic stress** and **abiotic stress.**

- **Biotic Stress** involves interactions with living organisms, such as pathogens, insects, nematodes, and other plants that compete for resources. Biotic stress can lead to infection, predation, and competition, each of which has unique effects on plant health and productivity.

Abiotic Stress arises from non-living environmental factors, such as temperature extremes, drought, salinity, and heavy metal toxicity, each affecting plants in specific ways. Abiotic stresses are particularly impactful in agriculture as they often cannot be controlled and may vary unpredictably with changing weather patterns and climate conditions.

Abiotic stress is often more difficult to manage in crop systems because it encompasses a wide range of environmental factors that are not always within human control. Droughts, heatwaves, soil degradation, and saline intrusion are examples of abiotic stressors that can cause significant crop loss. For farmers, understanding the different types of stress and their impacts on crops is essential for managing agricultural productivity and sustainability.

4. Characteristics of a Stress Event: Intensity, Duration, and Frequency

The effects of stress on plants depend on the **intensity, duration, and frequency** of the stress event. These characteristics determine how much the plant's physiological processes are affected and whether the plant can recover once the stress is alleviated.

- **Intensity**: The magnitude or strength of the stressor, such as the level of soil salinity or the degree of water scarcity, influences how severely the plant is affected. High-intensity stress often leads to more pronounced physiological disruptions.
- **Duration**: The length of time the plant is exposed to the stressor. Short-term stress may cause only temporary disruption, while prolonged exposure can lead to long-term damage or even death.

» **Frequency**: The recurrence of stress events also influences plant health. Frequent exposure to stress can lead to cumulative damage, weakening the plant's resilience and depleting its energy reserves.

Understanding these characteristics helps scientists predict plant responses to stress and develop management strategies that mitigate the negative impacts of environmental fluctuations.

5. Physiological Responses to Stress: Adjustments in Metabolism and Growth

When plants encounter stress, they undergo several physiological changes to conserve resources, minimize damage, and enhance survival. **Metabolic adjustments** are one of the primary responses to stress, as plants reallocate resources to maintain essential functions.

» **Photosynthesis**: Stress often reduces photosynthetic activity, especially under drought or high temperatures, which cause stomatal closure to conserve water. Reduced photosynthesis limits energy production and impacts growth.

» **Respiration**: Stress conditions often increase respiration rates as plants mobilize energy reserves to support stress responses. However, prolonged high respiration can deplete carbohydrate stores, reducing growth potential.

» **Water and Nutrient Transport**: Water stress directly impacts the plant's ability to uptake and transport nutrients, causing deficiencies that affect growth and development.

Plants that can quickly adapt their physiological processes to mitigate the effects of stress are more likely to survive and maintain productivity. These adjustments, however, come at a cost to growth and reproduction, as resources are diverted from normal functions to stress management.

6. Biochemical Pathways Activated by Stress

Stress in plants triggers a series of biochemical responses, including the activation of pathways involved in the production of **stress-protective compounds**. Some of the key biochemical changes include:

» **Antioxidant Production**: Many stresses, especially those involving high light or drought, increase the production of reactive oxygen species (ROS), which can damage cellular components. Plants produce antioxidants like superoxide dismutase, catalase, and ascorbate to neutralize ROS and protect cells.

» **Osmolytes Accumulation**: Under water-deficit and salt stress, plants accumulate osmolytes like proline, glycine betaine, and sugars to stabilize cellular structures, maintain osmotic balance, and protect enzymes from denaturation.

» **Stress Hormones**: Hormones like abscisic acid (ABA), ethylene, and jasmonic acid play significant roles in modulating stress responses. For instance, ABA is crucial for drought response as it induces stomatal closure to reduce water loss.

These biochemical pathways enable plants to tolerate and survive stress by minimizing cellular damage and maintaining essential functions under adverse conditions.

7. Genetic Basis of Stress Tolerance

» Plants' ability to respond to stress is partly determined by genetic factors. Certain species and cultivars possess **stress-resistant genes** that enhance their resilience to specific stressors. For example, genes involved in drought tolerance may regulate root architecture, osmolyte synthesis, and ABA sensitivity. Advances in molecular biology have enabled the identification of these genes, providing pathways for breeding and genetic engineering to enhance crop resilience.

» **Genetic Variation**: Natural genetic diversity among plants contributes to variation in stress tolerance. Breeding programs capitalize on this diversity by selecting varieties with desirable stress-resistance traits.

» **Transcription Factors**: Proteins like DREB (dehydration-responsive element-binding protein) and HSF (heat shock factor) regulate stress-responsive genes, allowing plants to modulate their stress responses at the transcriptional level.

» **Gene Editing and Biotechnology**: Technologies like CRISPR/Cas9 allow for precise manipulation of stress-resistance genes, offering opportunities to develop crops with enhanced tolerance to specific abiotic stresses.

8. Stress Memory and Acquired Tolerance

Plants can sometimes "remember" exposure to stress, a phenomenon known as **stress memory**. When plants experience mild stress, they can develop a form of acquired tolerance that makes them better equipped to handle future stress of a similar nature.

» **Epigenetic Modifications**: Stress memory often involves epigenetic changes, such as DNA methylation and histone modification, which affect gene expression without altering the DNA sequence.

» **Priming for Future Stress**: Primed plants can respond more rapidly to subsequent stress, as they retain biochemical and molecular "markers" of the initial stress event. This adaptation is especially relevant for crop systems facing recurrent stress events.

How Stress is Specifically Understood in the Context of Plant Physiology and Agriculture?

In plant physiology and agricultural sciences, stress is more than just an adverse condition; it is a multifaceted phenomenon that impacts every aspect of plant function, growth, and productivity. Stress represents any external condition that challenges a plant's metabolic processes, and understanding it involves examining how plants perceive, respond to, and adapt to these adverse conditions. Unlike animals, plants are immobile and must rely entirely on internal physiological and biochemical mechanisms to cope with external stressors. This dependence on adaptive internal responses makes stress a unique and particularly significant concept within plant biology.

The study of plant stress in agriculture is crucial for food production and security, as stresses can directly lead to reductions in crop yield and quality. In an agricultural context, understanding stress goes beyond identifying how it impacts plant growth; it also entails developing and implementing strategies to mitigate these effects through breeding, agronomic practices, and environmental management. The focus on stress in agriculture is a response to the growing demand for high-quality produce in the face of increasingly challenging environmental conditions, including climate change, water scarcity, and soil degradation. By analyzing how plants react to stress at the physiological, molecular, and cellular levels, scientists and farmers can devise methods to enhance crop resilience, ensuring sustainability and stability in food production.

1. Perception of Stress in Plants: The Role of Stress Sensors and Signals

Plants have evolved intricate mechanisms to perceive stress through specialized receptors and signaling pathways. These sensors are primarily located in the cell membrane, where they detect changes in environmental conditions such as temperature, light intensity, soil moisture, and salinity. For instance, certain receptors respond to osmotic stress caused by drought or high salinity, while others are sensitive to temperature fluctuations, signaling the plant to activate heat-shock or cold-stress responses. These stress sensors initiate a cascade of intracellular signals that activate genes involved in stress tolerance, allowing the plant to respond quickly to changes in its environment.

In plant physiology, understanding the process of stress perception is essential because it represents the first step in the plant's response to adverse conditions. The signaling pathways activated by stress sensors lead to the production of secondary messengers, such as calcium ions, reactive oxygen species (ROS), and abscisic acid (ABA). These messengers transmit information to different parts of the plant, inducing responses that help the plant cope with the stressor. For example, under drought stress, ABA levels increase, prompting stomatal closure to reduce water loss. In agriculture, the ability to manipulate these signaling pathways offers potential for enhancing crop tolerance to stress through genetic and biotechnological approaches.

2. Physiological Impacts of Stress on Growth and Development

Stress affects various physiological processes within plants, often leading to reduced growth, delayed development, and compromised reproductive success. Each type of stress—whether it be drought, salinity, heat, or nutrient deficiency—affects specific physiological functions, with broad consequences for plant health. For instance:

» **Water Stress**: Drought stress reduces water availability, leading to decreased cell turgor, stomatal closure, and reduced photosynthesis. This impacts leaf expansion, photosynthetic efficiency, and ultimately, biomass production.

» **Temperature Stress**: High temperatures denature proteins and disrupt enzyme activity, leading to heat damage, while low temperatures can freeze cellular fluids and cause tissue injury.

» **Salinity Stress**: High salt concentrations in the soil create osmotic stress, reducing water uptake and causing ion toxicity that disrupts nutrient balance and cellular integrity.

These physiological impacts collectively result in a reduced capacity for photosynthesis, nutrient uptake, and growth, affecting not only the plant's current health but also its ability to survive future stresses. In agriculture, understanding these physiological impacts allows researchers and farmers to develop strategies to reduce the negative effects of stress, such as irrigation management during droughts or soil amendments in saline environments.

3. Biochemical Responses to Stress: Antioxidants, Osmolytes, and Protective Compounds

When plants encounter stress, they initiate biochemical responses aimed at stabilizing cellular structures, protecting enzymes, and minimizing oxidative damage. Stress often leads to the accumulation of

reactive oxygen species (ROS), which can damage proteins, lipids, and DNA. To mitigate this damage, plants produce antioxidants—such as superoxide dismutase (SOD), catalase, and ascorbate—that neutralize ROS and protect cells.

Plants also produce **osmolytes**—small organic molecules such as proline, glycine betaine, and soluble sugars—that help maintain osmotic balance, stabilize proteins, and protect cellular membranes under conditions of water deficit or high salinity. In addition to these, plants may synthesize **heat shock proteins (HSPs)**, which act as molecular chaperones to protect proteins from denaturation and assist in protein refolding during heat stress.In agriculture, understanding these biochemical responses is valuable for developing crop varieties that can accumulate higher levels of protective compounds under stress, thereby enhancing resilience. For instance, breeding programs have focused on increasing proline accumulation in crops to improve drought tolerance, and similar efforts are being made to enhance antioxidant production to mitigate oxidative stress in various crops.

4. Role of Stress Hormones in Plant Response

Plants produce several hormones that regulate growth, development, and response to stress. Among these, **abscisic acid (ABA)**, **ethylene**, and **jasmonic acid (JA)** are particularly important in stress response.

- **Abscisic Acid (ABA)**: ABA is often referred to as the "stress hormone" because it plays a central role in plant response to drought and other abiotic stresses. Under drought conditions, ABA levels increase, leading to stomatal closure to reduce water loss and activation of genes involved in stress tolerance.
- **Ethylene**: Ethylene mediates responses to various stresses, including flooding, heat, and pathogen attack. It is involved in regulating processes like leaf senescence, root growth, and stress-induced cell death.
- **Jasmonic Acid (JA)**: JA is associated with defense responses, particularly against biotic stressors like insects and pathogens, but also plays a role in modulating responses to abiotic stress.

Understanding how these hormones function and interact is crucial in agriculture, as manipulating hormone pathways can help enhance crop resilience. For example, ABA analogs are sometimes used to induce drought tolerance in crops, and breeding for ethylene-insensitive plants can reduce stress-induced senescence, prolonging crop productivity.

5. Gene Expression and Transcriptional Regulation of Stress Responses

Stress responses in plants are largely governed at the genetic level through the activation of specific genes involved in tolerance and adaptation. **Transcription factors (TFs)**, such as dehydration-responsive element-binding proteins (DREB) and heat shock factors (HSFs), regulate the expression of stress-responsive genes. These TFs bind to promoters of target genes, initiating the production of proteins that help the plant survive stress conditions.

Genetic regulation of stress response is a complex process involving multiple pathways that coordinate the synthesis of protective proteins, antioxidants, and other compounds. For example, under drought stress, DREB proteins activate genes that contribute to osmolyte synthesis, while HSFs upregulate genes encoding heat shock proteins under heat stress. These transcription factors act as master regulators, controlling entire networks of genes that collectively enhance stress tolerance.

In agriculture, the ability to identify and manipulate these stress-related genes is transformative. Through conventional breeding or genetic engineering, researchers can introduce stress-resistance traits into crops, allowing them to tolerate extreme conditions without significant yield loss. This genetic approach to stress management is increasingly important as climate change intensifies the frequency and severity of abiotic stresses in agriculture.

6. Epigenetic Modifications and Stress Memory

Plants are sometimes capable of "remembering" exposure to stress, a phenomenon known as **stress memory**. Epigenetic modifications, such as DNA methylation and histone modifications, play a significant role in this memory, affecting gene expression patterns in response to stress. When a plant experiences mild stress, these modifications can prime the plant to respond more effectively to future stress of a similar nature, enhancing resilience.

Stress memory has important implications for agriculture, as it suggests that exposing crops to mild stress may enhance their tolerance to future stress. Understanding how epigenetic changes contribute to stress resilience could lead to innovative farming practices that leverage natural priming to improve crop health and productivity under adverse conditions.

Key Characteristics That Define a Stress Event, Including Its Intensity and Duration

In plant physiology and agricultural sciences, the characteristics of a stress event, particularly its **intensity, duration, frequency, timing**, and **recovery potential**, are fundamental to understanding the full impact on plants. These factors collectively determine the physiological responses, adaptive mechanisms, and potential survival of a plant under adverse conditions. When plants experience stress, whether due to drought, extreme temperatures, or high salinity, the severity of these stress events plays a crucial role in shaping their response. Furthermore, the timing and recovery potential between stress events add complexity to the plant's ability to manage and overcome adverse conditions. For agronomists, understanding these characteristics is essential to managing crops and ensuring productivity in fluctuating and often challenging environmental conditions.

1. Intensity of Stress:

The intensity of a stress event refers to the strength or severity of the stressor. In plants, stress intensity is directly correlated with the degree of physiological disruption that occurs. For example, a slight drop in water availability during drought may cause mild stress, prompting the plant to close its stomata partially, which reduces water loss but still permits some level of photosynthesis. However, severe drought conditions, where soil moisture levels fall drastically, cause intense stress that forces complete stomatal closure, halting photosynthesis entirely and significantly impacting plant growth and productivity. The higher the intensity, the more immediate and severe the physiological disruptions, often leading to cell death or tissue damage if the plant cannot adapt quickly.

In cases of high-temperature stress, plants may experience rapid protein denaturation and membrane instability as temperatures exceed their physiological thresholds. This disruption affects cellular functions and may impair photosynthesis, respiration, and even DNA stability, all of which can result in significant reductions in growth and productivity. For instance, temperatures above 45°C can quickly lead to a breakdown of essential proteins and enzymes, especially in plants that are not heat-adapted. In response, heat-tolerant plants produce **heat shock proteins (HSPs)** that stabilize other proteins and prevent cellular damage. These protective responses, however, are energy-intensive, diverting resources from growth and reproduction.

Similarly, in cases of salt stress, plants exposed to low salinity levels may only need to adjust osmotic balance by accumulating osmoprotectants like proline or glycine betaine. However, at high salinity levels, sodium and chloride ions accumulate excessively within plant tissues, leading to toxic effects on enzymes and cell membranes. Salt-tolerant plants, such as certain halophytes, can compartmentalize these toxic ions in vacuoles, isolating them from crucial metabolic activities, but even these plants are affected by high-intensity salt stress, which imposes significant energetic costs and may reduce growth.In agricultural contexts, assessing the intensity of stress is essential for implementing appropriate interventions. Farmers in regions prone to extreme weather events, such as prolonged heatwaves or droughts, often rely on high-intensity, stress-tolerant crop varieties. These varieties are selectively bred to withstand adverse conditions, and understanding the impact of stress intensity allows for precise decisions regarding irrigation, shading, and soil management practices to alleviate the effects of high-stress environments.

2. Duration of Stress:

The **duration** of a stress event, or the period for which a plant is exposed to the adverse condition, is another critical characteristic that impacts plant responses. Short-term or transient stress events may cause temporary disruption, but plants can often recover quickly once the stress is relieved. For instance, a brief heat spike or mild drought may only slow growth temporarily, as the plant can resume normal functions with minimal lasting impact once conditions improve. However, when stress is prolonged, such as in cases of extended drought or seasonal temperature extremes, the cumulative effect can cause long-term damage that affects plant structure, growth, and reproductive success.

Prolonged drought, for instance, gradually reduces soil moisture, depleting water reserves and impacting plants more severely over time. As water scarcity continues, cellular dehydration becomes increasingly pronounced, and plants are forced to draw on internal water stores, leading to wilting and cellular damage. Extended drought conditions limit the plant's ability to transport water from roots to shoots, impair nutrient uptake, and disrupt cellular turgor, ultimately resulting in leaf drop, reduced leaf area, and stunted growth. Long-term drought can also lead to premature flowering and reduced fruit size, impacting the quality and yield of crops. In agriculture, understanding the duration of drought stress enables farmers to apply irrigation methods that maintain minimal hydration during critical growth phases, preserving crop viability under long-term water scarcity.

Likewise, **extended exposure to high salinity** levels in soils prevents plants from effectively balancing osmotic pressure and ion toxicity, leading to structural and metabolic damage. Under long-term salt stress, plants may exhaust their capacity to produce osmoprotectants, leading to further ionic imbalances and nutrient deficiencies. Some salt-tolerant plants activate genes that facilitate salt sequestration or ion transport, but these mechanisms are often insufficient when salinity levels remain high over long periods. Salinity-tolerant crops, such as barley or sorghum, can withstand prolonged exposure to moderate salinity, but extreme or extended salinity events often result in compromised growth and yield, emphasizing the need for strategies like controlled irrigation and soil amendments to mitigate long-duration stress.

3. Frequency of Stress Events:

The **frequency** of stress events, or how often a plant is exposed to similar stress conditions, plays an essential role in determining the plant's capacity to recover. Repeated exposure to stress, particularly when recovery periods are insufficient, leads to cumulative damage, depleting the plant's reserves and reducing its resilience. For example, crops exposed to frequent cycles of drought may suffer from stunted growth, nutrient deficiencies, and reduced reproductive capacity as they cannot fully recover between stress episodes. Frequent exposure can also lead to stress-induced senescence, where older leaves and non-essential tissues are sacrificed to conserve resources for new growth, ultimately impacting overall biomass and yield.

In agricultural regions with alternating wet and dry seasons, crops may face multiple cycles of drought followed by rain, creating fluctuating water availability. These repeated stress cycles can induce physiological changes that enhance tolerance to future drought events, a phenomenon known as **stress priming** or **acquired tolerance**. For instance, plants subjected to mild drought conditions may increase root depth, accumulate osmoprotectants, and alter gene expression in ways that better prepare them for subsequent drought. However, this priming effect is effective only up to a certain frequency threshold; beyond that, plants may experience chronic stress effects, such as energy depletion and inhibited growth, as they lack the resources for repeated adaptation.Frequency also impacts **temperature stress** resilience. For example, frequent exposure to mild heat stress can induce the production of heat shock proteins, allowing plants to better withstand subsequent temperature spikes. However, repeated high-intensity heat stress without sufficient recovery may lead to exhaustion of the plant's defense mechanisms, leading to cellular damage, reduced photosynthetic efficiency, and impaired reproductive function. In practical agricultural applications, adjusting planting schedules to avoid periods of high-stress frequency, such as the peak of summer heat, can help reduce the cumulative impact on crops.

4. Timing of Stress about Plant Developmental Stages

The timing of a stress event about a plant's **developmental stage** is a key characteristic that determines its impact on growth and yield. Plants have distinct physiological needs and vulnerabilities at each stage of development, making some stages more sensitive to stress than others. For instance, during **germination**, water availability is crucial for seed imbibition and emergence. Drought or salinity stress at this stage can inhibit germination entirely or lead to weak, poorly established seedlings that are more susceptible to future stresses.

During the **flowering** and **fruiting stages**, stress has a direct impact on reproductive success. For example, drought stress during flowering reduces pollen viability, leading to poor fertilization and, ultimately, low fruit set. Similarly, heat stress during flowering can impair pollen tube growth, reducing fruit quality and yield. In temperature-sensitive crops like wheat and maize, stress during flowering results in fewer seeds per ear or cob, with a significant reduction in yield. To mitigate these effects, farmers often schedule irrigation to ensure adequate water during flowering or use protective coverings to reduce heat exposure during critical reproductive stages.In the **fruit maturation phase,** water stress can reduce fruit size, sugar accumulation, and overall quality. In grape cultivation, for example, mild water stress is sometimes intentionally applied to increase sugar concentration for wine production. However, severe or prolonged stress during this stage can lead to shriveled, underdeveloped fruit that is unsuitable for harvest.

5. Recovery Potential Between Stress Events

Plants have varying capacities for recovery based on the **interval between stress events** and the degree of adaptation they have developed through previous exposure. When plants are given adequate recovery time after a stress event, they can restore their physiological and metabolic functions, regaining resilience. However, if stress events are too frequent or intense, the plant's capacity for recovery diminishes, resulting in cumulative damage that compromises growth and productivity. This recovery potential is closely tied to a phenomenon known as **stress memory**, where plants "remember" previous stress exposure and adjust their response to future stress events accordingly.Stress memory is particularly significant in crops exposed to seasonal stresses. Plants that survive an initial stress event may retain epigenetic marks—modifications to DNA or histones that do not change the genetic code but alter gene expression—that prime them for faster or stronger responses to subsequent stress. For instance, exposure to mild drought can prime plants to respond more effectively to future drought by enhancing root growth or improving water use efficiency. However, if recovery periods are insufficient, the plant may be unable to sustain this adaptive response, leading to exhaustion of energy reserves and impaired growth.In agricultural practices, optimizing the recovery interval between stress events can improve crop resilience. Practices such as **regulated deficit irrigation**—where plants are exposed to mild drought stress followed by re-watering—capitalize on stress memory to enhance drought tolerance in crops like grapes and olives. This approach requires careful timing and intensity management to avoid overstressing the plant, allowing it to benefit from stress memory without compromising productivity.

3. Types of Stress in Horticultural Science

Abiotic vs. Biotic Stress: Differentiation Between Non-Living and Living Stressors

Plants in horticulture and agriculture encounter numerous challenges that can disrupt their growth, productivity, and quality. These challenges, termed "stress," are broadly categorized into **abiotic** and **biotic** stressors. The fundamental difference between abiotic and biotic stress lies in their origins: **abiotic stress** arises from non-living, environmental factors, while **biotic stress** results from interactions with other living organisms. Both types of stress affect plant physiology, biochemistry, and cellular function, and each requires specific adaptive responses to mitigate the adverse effects.These two primary types of stress is essential in horticultural science, as it allows researchers, agronomists, and farmers to develop targeted strategies that enhance crop resilience, minimize productivity losses, and improve quality. Below, we'll delve into abiotic and biotic stress in detail, examining their causes, effects, plant responses, and management practices.

1. Abiotic Stress: Environmental Challenges from Non-Living Factors

Abiotic stress encompasses a variety of environmental factors that are non-living but can still exert profound effects on plant growth and productivity. The primary forms of abiotic stress include **drought, salinity, temperature extremes (both heat and cold), nutrient deficiencies, waterlogging,**

and exposure to radiation. Each of these factors imposes specific physiological and biochemical challenges that require plants to deploy adaptive mechanisms to survive.

Key Abiotic Stressors and Their Effects

» **Drought Stress**: Lack of water is one of the most common and severe abiotic stresses affecting plants. Drought leads to reduced soil moisture and water availability, causing a decrease in leaf turgor, stomatal closure, reduced photosynthesis, and impaired nutrient transport. Under prolonged drought, plants suffer from dehydration and cellular damage, which can lead to stunted growth and, in severe cases, death.

» **Salinity Stress**: High salinity levels in the soil create an osmotic imbalance, preventing plants from absorbing water efficiently. Excess salts, primarily sodium (Na^+) and chloride (Cl^-) ions, accumulate within plant tissues, leading to ion toxicity, nutrient imbalance, and impaired metabolic functions. Salinity stress is particularly prevalent in arid and semi-arid regions, where soil salts accumulate due to high evaporation rates.

» **Temperature Extremes**: Both high and low temperatures impact plants. Heat stress can lead to protein denaturation, membrane instability, and accelerated respiration, while cold stress can cause cell membrane rigidification and ice formation within cells, resulting in tissue damage. Temperature stress affects enzymatic activities and impairs photosynthesis, ultimately reducing growth and yield.

» **Nutrient Deficiencies and Toxicities**: Plants require specific nutrients, such as nitrogen, phosphorus, potassium, and micronutrients like zinc and iron, for optimal growth. Deficiency in any of these nutrients can disrupt metabolic processes, while an excess, often due to over-fertilization, can lead to toxicity and impair root and shoot development.

» **Radiation Stress**: Excessive sunlight, particularly UV-B radiation, can damage DNA, proteins, and cell membranes. Radiation stress causes oxidative damage, leading to a buildup of reactive oxygen species (ROS) that impair cellular functions. Conversely, low light levels limit photosynthesis, affecting energy production and growth.

» **Waterlogging**: Excess water in the root zone reduces oxygen availability, causing hypoxia (oxygen deficiency). Under waterlogged conditions, roots shift to anaerobic respiration, leading to the production of toxic byproducts, such as ethanol, which damage root tissues and inhibit nutrient uptake.

Plants respond to abiotic stress by activating various **adaptive mechanisms** to minimize damage and maintain essential functions. These responses include physiological adjustments, such as stomatal regulation and root growth alterations, biochemical changes like antioxidant production, and molecular responses involving gene expression regulation.In agriculture, managing abiotic stress involves both **proactive and reactive strategies**. Proactive strategies, like selecting stress-tolerant varieties and implementing irrigation systems, prepare crops for adverse conditions, while reactive strategies address stress after its onset. For instance, controlled irrigation can reduce drought stress, while mulching helps conserve soil moisture. Soil amendments, like gypsum or organic matter, improve soil structure and reduce salinity effects. Understanding and managing abiotic stress are crucial for maintaining productivity in regions with harsh climates and limited resources.

2. Biotic Stress: Challenges From Living Organisms

Biotic stress, in contrast to abiotic stress, arises from interactions with other living organisms. These organisms, which include **pathogens (such as bacteria, fungi, viruses), insects, nematodes, and even competing plants**, can cause direct or indirect harm to plants. Biotic stress leads to infection, infestation, competition, and nutrient depletion, affecting plant health, yield, and quality.

Major Types of Biotic Stressors and Their Effects

» **Pathogens (Bacteria, Fungi, and Viruses)**: Pathogens are disease-causing organisms that invade plant tissues, disrupting cellular function and leading to diseases. Fungal pathogens, such as **Fusarium** and **Phytophthora**, often cause root rot, wilting, and leaf blights. Bacterial pathogens, like **Xanthomonas** and **Pseudomonas**, infect leaves, stems, and fruits, causing spots, cankers, and tissue necrosis. Viral pathogens disrupt plant metabolism and often cause stunted growth, mosaic patterns, and reduced fruit quality. Pathogens typically spread through wind, water, soil, or insect vectors, making them challenging to control.

» **Insects and Mites**: Insect pests feed on plant tissues, causing direct damage and often acting as vectors for diseases. Aphids, whiteflies, and thrips, for instance, feed on phloem sap, weakening plants and spreading viral pathogens. Caterpillars and beetles chew leaves and stems, reducing photosynthetic area and injuring plants. Mites, such as spider mites, feed on plant sap and cause stippling, leaf discoloration, and tissue damage. Insect infestations reduce plant vigor, hinder growth, and compromise fruit quality.

» **Nematodes**: Nematodes are microscopic roundworms that primarily affect plant roots, leading to root galls, lesions, and reduced root function. Root-knot nematodes (Meloidogyne spp.) are notorious for causing root deformations that impede nutrient and water uptake, resulting in stunted growth and yield reduction. Nematodes are particularly problematic in crops like tomatoes, peppers, and potatoes, where root health is critical for productivity.

» **Weeds and Parasitic Plants**: Competing plants, particularly invasive weeds, compete for essential resources such as light, water, and nutrients. Weeds, such as Amaranthus and Cynodon, grow aggressively, overshadowing crops and depleting soil nutrients. Parasitic plants, like **Cuscuta** (dodder) and **Orobanche** (broomrape), directly attach to host plants, extracting nutrients and water, which weakens the host and reduces its productivity.

Biotic stress triggers a range of **defensive responses** in plants, from physical barriers like thickened cell walls to chemical responses involving the production of secondary metabolites that deter pests or inhibit pathogen growth. Plants also activate **systemic acquired resistance (SAR)** and **induced systemic resistance (ISR)** pathways, which prepare the plant to respond more effectively to future attacks. SAR is a broad-spectrum defense response often triggered by pathogens, while ISR is usually induced by beneficial microbes like rhizobacteria that enhance plant resistance to insect pests and diseases.

Management of Biotic Stress in Agriculture involves integrated pest management (IPM) strategies that combine cultural, biological, mechanical, and chemical controls. Cultural practices, such as crop rotation, field sanitation, and planting resistant varieties, reduce pest and disease incidence. Biological controls use natural predators, parasitoids, and beneficial microorganisms to suppress pest populations. For example, ladybugs feed on aphids, and Bacillus thuringiensis (Bt) bacteria produce toxins that target specific insect pests without harming beneficial insects. Chemical controls, including pesticides and fungicides, are used selectively to minimize environmental impact and prevent resistance buildup in pests. IPM focuses on sustainable, eco-friendly approaches to manage biotic stress and protect crop health.

3. Key Differences Between Abiotic and Biotic Stress

While both abiotic and biotic stresses adversely affect plant health, productivity, and quality, the **fundamental differences** between these stress types lie in their origins, mechanisms of action, and plant responses. Abiotic stress is environmentally driven, and its effects are often broad and systemic, impacting cellular metabolism, water relations, nutrient balance, and energy production. Biotic stress, by contrast, involves direct interactions with living organisms, which can localize effects in specific tissues, such as infection sites or feeding zones.

- **Nature of Stress**: Abiotic stress is caused by non-living environmental factors, while biotic stress arises from living organisms.
- **Pathways and Responses**: Abiotic stress often triggers cellular and metabolic pathways that adjust physiological functions to manage the environmental change. Biotic stress activates immune and defense responses aimed at recognizing and neutralizing the biological threat.
- **Adaptation and Evolution**: Abiotic stress adaptation often involves genetic and physiological changes that enhance resilience over generations. In contrast, biotic stress adaptation frequently results in co-evolutionary dynamics between plants and their natural enemies, as seen in plant-insect and plant-pathogen interactions.

4. Combined Stress: Interactions Between Abiotic and Biotic Stress

In natural and agricultural settings, plants often encounter **combined stress**, where abiotic and biotic stressors occur simultaneously or sequentially. For instance, drought-stressed plants may become more susceptible to insect pests or fungal pathogens due to weakened defenses. Similarly, plants under heat stress may show increased susceptibility to bacterial infections as high temperatures disrupt cell wall integrity and lower immunity.

Combined stress presents unique challenges because the interaction between abiotic and biotic factors can amplify their negative effects on plant health. For instance, drought stress limits water availability, impacting nutrient uptake and reducing the plant's capacity to produce defense compounds. These weakened defenses make plants more vulnerable to biotic stressors, leading to a compounded impact that is often more severe than either stress alone.Understanding combined stress is crucial in horticultural science, as it reflects the complexity of real-world agricultural conditions. Integrated approaches that consider both abiotic and biotic stress factors—such as breeding for multi-stress tolerance and implementing agroecological practices—are essential for developing resilient cropping systems.

5. Practical Implications of Abiotic and Biotic Stress Management in Horticulture

Effective management of abiotic and biotic stress in horticulture involves **targeted, adaptive strategies** that address the specific stressors affecting each crop. In arid regions, managing abiotic stress may include selecting drought-tolerant varieties and applying mulches to retain soil moisture. In areas prone to pest outbreaks, managing biotic stress may involve IPM strategies, crop rotation, and habitat conservation for natural enemies.Advances in technology, such as **precision agriculture** and **genomics**, have revolutionized stress management, allowing farmers to monitor stress levels in real-time, tailor interventions, and develop crops with improved resistance to multiple stress factors. By understanding the distinctions and interactions between abiotic and biotic stress, horticulturists can make informed decisions that optimize crop health, sustainability, and productivity in increasingly variable climates.

Focus on Abiotic Stress in Horticulture and How It Impacts Fruit Crop Quality and Yield

Abiotic stress, resulting from non-living environmental factors, is one of the most prominent challenges affecting horticultural crops worldwide. For fruit crops in particular, abiotic stressors—including drought, temperature extremes, salinity, waterlogging, and nutrient imbalances—can disrupt growth, reduce yields, and compromise quality attributes that are essential for marketability and consumer appeal. Fruit crops are highly sensitive to environmental conditions during critical growth stages, such as flowering, fruit set, and ripening, and any adverse changes can result in yield losses and suboptimal quality.

1. Major Types of Abiotic Stress Affecting Fruit Crop Production

Abiotic stress in fruit crop production is caused by a variety of environmental factors, each affecting plants in specific ways. The primary types of abiotic stress include:

» **Drought Stress**: Drought is one of the most severe forms of abiotic stress and occurs when water availability is below the plant's requirements for optimal growth. In fruit crops, drought stress reduces leaf turgor, triggers stomatal closure, and reduces photosynthesis. This restricts the plant's ability to produce energy and allocate resources to fruit development, leading to smaller, less flavorful fruits with low sugar content. Prolonged drought can result in flower and fruit drop, directly impacting yield.

» **Heat Stress**: Excessive heat disrupts cellular functions and increases water loss through transpiration. High temperatures often cause protein denaturation and destabilize membrane structures, affecting photosynthesis and respiration. In fruit crops, heat stress leads to reduced fruit size, sunburn, and accelerated ripening, which shortens the shelf life and decreases the market value of fruits.

» **Cold and Frost Stress**: Cold stress affects plants in various ways, depending on the severity and timing of exposure. Chilling injury occurs in non-freezing but low temperatures, disrupting enzymatic activities and photosynthesis. Frost, or temperatures below freezing, can cause ice crystal formation in plant tissues, leading to cellular dehydration, damage, and tissue death. In sensitive fruit crops like citrus, bananas, and tomatoes, cold stress reduces fruit quality, texture, and flavor, often rendering them less marketable.

» **Salinity Stress**: Saline soils, often found in coastal areas or regions with high evaporation rates, create osmotic stress and ion toxicity in plants. High salt levels prevent roots from absorbing water efficiently and cause toxic ions, such as sodium and chloride, to accumulate in plant tissues. Salinity stress reduces leaf area, inhibits growth, and stunts fruit development. In fruit crops, salinity negatively impacts fruit size, flavor, and nutrient composition.

» **Nutrient Deficiency and Toxicity**: Nutrient imbalances, such as deficiencies or toxicities of essential elements like nitrogen, phosphorus, potassium, and micronutrients, impact metabolic functions and plant health. For instance, nitrogen deficiency reduces leaf chlorophyll content, resulting in poor growth and diminished yield. Conversely, excess nitrogen can lead to excessive vegetative growth at the expense of fruiting. Micronutrient deficiencies, such as iron or zinc, cause chlorosis and stunted growth, affecting fruit color, size, and overall quality.

Each of these abiotic stressors has unique and specific impacts on plant physiology, which influence growth patterns, flowering, and fruit sets. The extent of the damage often depends on the timing, intensity, and duration of the stress event, as well as the resilience of the crop variety.

2. Physiological Impacts of Abiotic Stress on Fruit Crop Growth and Development

Abiotic stress affects several physiological processes essential for fruit crop productivity, including photosynthesis, transpiration, water relations, and nutrient uptake. These disruptions impact plant growth, leading to compromised fruit quality and yield:

» **Photosynthesis and Energy Production**: Photosynthesis is highly sensitive to abiotic stress. Drought, salinity, and high temperatures often lead to stomatal closure, reducing the intake of carbon dioxide (CO_2) needed for photosynthesis. This limits the plant's energy production and slows down growth, reducing the size and sugar content of the fruits. For instance, heat stress causes thermal deactivation of photosynthetic enzymes and damages chlorophyll, impairing the photosynthetic apparatus. Reduced photosynthesis leads to smaller fruits with lower sugar content and compromised flavor.

» **Transpiration and Water Relations**: Water stress due to drought or salinity disrupts the plant's ability to maintain water balance, leading to dehydration and reduced cellular turgor pressure. As a result, cells become flaccid, and leaf wilting occurs. In fruit crops, water stress during fruit formation and ripening leads to smaller fruits with reduced juice content. Excess water loss can also lead to a condition known as **blossom-end rot** in tomatoes and peppers, where calcium uptake is impaired, resulting in blackened, sunken lesions on the fruit.

» **Nutrient Uptake and Transport**: Abiotic stress affects the plant's root function, limiting nutrient absorption and transport to the aerial parts of the plant. Under saline conditions, high concentrations of sodium and chloride ions interfere with the uptake of essential nutrients like potassium, magnesium, and calcium, which are critical for plant growth and development. These nutrient imbalances result in chlorosis (yellowing of leaves), poor fruit set, and weak plants that are less resilient to further stress. For instance, potassium deficiency under drought stress impacts carbohydrate metabolism and reduces sugar levels in fruits, compromising their sweetness and market value.

» **Respiration and Energy Expenditure**: Under stress conditions, plants often experience a shift in energy allocation, where resources are directed toward survival mechanisms instead of growth. Increased respiration rates, especially under heat stress, lead to carbohydrate depletion, reducing resources available for fruit development and ripening. Prolonged abiotic stress conditions exhaust the plant's energy reserves, reducing productivity and compromising fruit yield.

The cumulative effects of these physiological disruptions can significantly reduce fruit crop yield, as well as affect key quality attributes such as flavor, texture, and color. Understanding how abiotic stress impacts these physiological processes enables farmers and horticulturists to make strategic decisions that support plant health and productivity.

3. Biochemical Responses to Abiotic Stress in Fruit Crops

When exposed to abiotic stress, fruit crops activate a range of biochemical pathways to help mitigate damage and survive adverse conditions. These biochemical responses include the production of **antioxidants, osmolytes, stress hormones**, and **protective proteins**.

» **Antioxidant Production**: Abiotic stress, particularly drought, salinity, and high temperatures, leads to the accumulation of reactive oxygen species (ROS), such as superoxide radicals, hydrogen peroxide, and hydroxyl ions, which cause oxidative damage to lipids, proteins, and nucleic acids. To counter ROS, plants produce antioxidant enzymes like superoxide dismutase (SOD), catalase, and peroxidase, which neutralize these harmful molecules. Additionally, non-enzymatic antioxidants such as ascorbate (vitamin C), glutathione, and carotenoids help protect cellular structures from oxidative damage. Enhanced antioxidant production helps maintain fruit quality, protecting color, flavor, and nutritional value from oxidative degradation.

» **Osmolyte Accumulation**: Under drought and salinity stress, plants accumulate small organic molecules called osmolytes, such as proline, glycine betaine, and soluble sugars. Osmolytes stabilize proteins, maintain cellular osmotic balance, and protect membranes from damage due to dehydration. This osmotic adjustment allows plants to survive with minimal water, reducing water loss and ensuring that fruits retain some turgidity. Proline accumulation is particularly crucial in drought-stressed crops, as it helps retain water, allowing fruits to develop adequately even under water-limited conditions.

» **Hormonal Regulation**: Plant hormones such as abscisic acid (ABA), ethylene, and jasmonic acid (JA) regulate responses to abiotic stress. ABA levels increase in response to drought and salinity, prompting stomatal closure to conserve water and activating stress-responsive genes. Ethylene production also increases under stress, which accelerates ripening but can lead to premature fruit drop and reduced shelf life. Managing these hormonal responses through genetic selection and breeding can help reduce the adverse impacts of stress on fruit crops, allowing for longer ripening times and better post-harvest quality.

» **Heat Shock and Cold-Regulated Proteins**: Under extreme temperatures, plants produce specific proteins that protect cellular structures. Heat shock proteins (HSPs) stabilize proteins and repair damaged ones, while cold-regulated (COR) proteins improve membrane fluidity and reduce freezing damage. These proteins play an essential role in maintaining cellular function and protecting fruit quality under extreme temperatures.

These biochemical responses to stress help fruit crops maintain cellular stability and minimize the impact on fruit quality and yield. However, the energy required to activate these defenses often reduces growth and reproductive output, highlighting the need for management practices that minimize stress without excessive energy expenditure.

4. Impact of Abiotic Stress on Fruit Quality Attributes and Yield

Abiotic stress has a profound impact on the quality and marketability of fruit crops, influencing attributes like **size, color, texture, flavor, aroma**, and **nutritional composition**:

- **Size and Shape**: Water and nutrient stress during critical phases of fruit development often result in smaller, misshapen fruits that are less appealing to consumers and command a lower market price. In crops like apples, tomatoes, and citrus, reduced water availability leads to poor cell expansion, resulting in compact, undersized fruits. Additionally, nutrient imbalances due to drought or salinity stress can cause structural abnormalities, such as blossom-end rot in tomatoes and peppers, where a lack of calcium uptake leads to dark, sunken lesions on the fruit. These size and shape deformities not only impact visual appeal but also reduce crop weight, ultimately decreasing market yield.
- **Color and Pigmentation**: The color of fruit is an important quality marker, as it influences consumer perception and marketability. Abiotic stresses, particularly temperature extremes, affect the synthesis of pigments such as chlorophyll, anthocyanins, and carotenoids. For instance, excessive heat stress can cause sunburn on fruit surfaces, resulting in brown or pale patches that detract from the uniformity of color. In fruits like apples, grapes, and berries, cool nights promote anthocyanin synthesis, leading to vibrant colors; however, unseasonal high temperatures can reduce pigment production, yielding duller colors. Conversely, chilling stress can inhibit chlorophyll breakdown in certain fruits, resulting in incomplete color development and impacting appearance at harvest.
- **Texture and Firmness**: Texture and firmness are key determinants of fruit quality, as they influence the sensory experience and storage life of fruits. Abiotic stress affects cellular turgor and cell wall composition, both of which play a crucial role in determining texture. Drought and salinity stress reduce turgor pressure, resulting in fruits that are less juicy and appear shriveled. In peaches and plums, for instance, water stress leads to lower juice content, impacting consumer satisfaction. Heat stress can also soften fruits by accelerating the breakdown of cell wall polysaccharides, shortening shelf life and making fruits more susceptible to mechanical damage during transport and handling.
- **Flavor and Aroma**: The flavor and aroma of fruits are complex attributes derived from sugars, organic acids, and volatile compounds. Abiotic stress, particularly drought and nutrient deficiencies, impacts these components by altering carbohydrate metabolism and secondary metabolite synthesis. Drought stress can reduce sugar levels by limiting photosynthesis, resulting in less sweet fruits. In crops like grapes and apples, moderate water stress is sometimes applied to enhance flavor by increasing sugar concentration; however, excessive drought reduces sugar accumulation, diminishing the richness of flavor. Furthermore, salinity stress can increase the concentration of certain organic acids, making fruits taste more acidic or bitter. Heat stress can accelerate the breakdown of aromatic compounds, reducing the characteristic aroma of fruits such as strawberries, mangoes, and citrus.

» **Nutritional Composition**: Nutritional quality, including the content of vitamins, minerals, and antioxidants, is significantly influenced by abiotic stress. In many fruits, drought and mild salinity stress can increase the concentration of phenolic compounds and antioxidants as part of the plant's defense mechanism. This can enhance the health benefits of fruits like blueberries, pomegranates, and tomatoes. However, severe stress may have the opposite effect, reducing nutrient content by inhibiting nutrient uptake and biosynthesis. For example, vitamin C levels decrease under prolonged heat stress, reducing the nutritional value of crops like oranges, bell peppers, and guava. Nutrient stress, whether due to deficiency or toxicity, also affects the mineral content of fruits, impacting consumer health and nutritional intake.

» **Post-Harvest Shelf Life**: Abiotic stress can accelerate ripening and senescence, reducing the post-harvest quality and shelf life of fruits. Ethylene, a hormone associated with ripening, is often produced in greater quantities under stress conditions, hastening fruit maturation and making fruits more prone to spoilage. In crops like bananas, avocados, and tomatoes, stress-induced ripening shortens storage time and increases susceptibility to decay, impacting marketability. Managing post-harvest quality is essential for stress-affected fruits, as even slight reductions in firmness or color uniformity can influence consumer choice and economic returns.

The negative impacts of abiotic stress on fruit quality emphasize the importance of pre-harvest and post-harvest strategies to optimize quality and ensure that fruits retain their value from field to market. In horticulture, abiotic stress management strategies focus on minimizing the severity of stress during critical growth stages, which helps preserve these quality attributes.

5. Strategies for Managing Abiotic Stress in Fruit Crop Production

To address the challenges posed by abiotic stress, horticulturists and farmers employ various strategies that include **crop selection, cultural practices, soil amendments, and advanced technology**. Effective management of abiotic stress requires an integrative approach that considers both environmental conditions and the specific requirements of each fruit crop.

» **Selection of Stress-Tolerant Varieties**: Developing and cultivating stress-tolerant crop varieties is one of the most effective ways to manage abiotic stress in horticulture. Breeding programs focus on traits such as drought tolerance, heat resistance, and salt tolerance. For instance, salt-tolerant tomato varieties are bred to thrive in saline soils, while drought-tolerant grape cultivars are better suited to arid regions. Additionally, advances in genetic engineering allow researchers to introduce genes that enhance stress resilience, such as those encoding for antioxidant production or osmolyte accumulation. By selecting stress-tolerant varieties, farmers can ensure better crop performance under challenging environmental conditions.

» **Irrigation Management and Water Conservation**: Efficient irrigation practices are essential for managing drought stress in fruit crops. Techniques like **drip irrigation** and **subsurface irrigation** deliver water directly to the root zone, minimizing evaporation and optimizing water use. **Regulated deficit irrigation** (RDI) is sometimes applied in crops like grapes and olives, where mild drought stress can enhance flavor by concentrating sugars and aromatic compounds. Mulching and soil moisture monitoring also help conserve water, reduce evaporation, and maintain consistent soil moisture levels, reducing the risk of drought stress during critical growth phases.

» **Soil Amendments and Nutrient Management**: Soil health plays a significant role in stress resilience. Adding organic matter, such as compost or biochar, improves soil structure, increases water-holding capacity, and enhances nutrient availability, helping plants tolerate drought and salinity. In saline soils, gypsum applications can help leach excess sodium, reducing soil salinity and improving nutrient uptake. Additionally, precise nutrient management through fertilization ensures that plants have adequate access to essential nutrients, minimizing the risk of nutrient deficiency under stress conditions.

» **Use of Protective Structures**: Protective structures, such as **shade nets, windbreaks**, and **greenhouses**, help control environmental conditions and reduce the impact of heat, wind, and excessive sunlight. Shade nets protect fruits from sunburn and reduce temperature fluctuations, improving fruit quality in hot climates. Windbreaks minimize wind stress, which can damage delicate fruit crops and exacerbate moisture loss. In regions with frequent temperature extremes, greenhouse cultivation provides a controlled environment that shields crops from abiotic stressors and allows for optimal growth conditions.

» **Precision Agriculture and Stress Monitoring**: Advances in precision agriculture have enabled the use of technologies such as soil moisture sensors, temperature monitors, and remote sensing to monitor stress conditions in real time. Data collected through these tools helps farmers make informed decisions about irrigation, nutrient application, and harvest timing. For instance, soil moisture sensors can trigger automated irrigation systems to prevent drought stress, while thermal imaging from drones can detect heat stress in large orchards. By closely monitoring environmental conditions, farmers can reduce the frequency and intensity of abiotic stress, ensuring that fruit crops receive consistent care tailored to their needs.

» **Biostimulants and Plant Growth Regulators**: Biostimulants, such as seaweed extracts, amino acids, and microbial inoculants, are increasingly used to improve plant resilience to abiotic stress. These compounds stimulate root growth, enhance nutrient uptake, and boost antioxidant levels, helping plants better withstand environmental challenges. Plant growth regulators, such as abscisic acid (ABA) and cytokinins, are also applied to manage stress responses, promoting adaptive mechanisms like stomatal closure during drought or delayed ripening under heat stress.

Classification of Abiotic Stress Based on Environmental Source and Plant Response

Abiotic stress encompasses a range of environmental factors that impact plant health, productivity, and quality. In horticultural science, abiotic stressors are often classified based on their environmental sources—such as water availability, temperature, soil composition, and atmospheric conditions—and on the unique physiological and biochemical responses plants exhibit to survive these challenging conditions.

1. Classification Based on Environmental Sources

Each type of abiotic stress arises from specific environmental conditions, with each impacting plants in distinct ways. Here are some primary sources of abiotic stress, each creating unique challenges for plant growth and development:

» **Water-Related Stress (Drought and Waterlogging)**: Water availability is crucial for cellular hydration, nutrient transport, and photosynthesis. Water stress can arise from both scarcity (drought) and excess (waterlogging).

- **Drought Stress**: Drought occurs when water supply is insufficient to meet the plant's needs, leading to reduced leaf turgor, stomatal closure, and reduced photosynthesis. Prolonged drought stress causes cellular dehydration, leading to structural damage, reduced growth, and decreased yield. Drought is particularly common in arid and semi-arid regions and can result in economic losses in crops sensitive to water deficits, such as grapevines, tomatoes, and citrus.
- **Waterlogging Stress**: Waterlogging occurs when soil is saturated with water for extended periods, depriving roots of oxygen. Without adequate oxygen, roots switch to anaerobic respiration, which produces toxic byproducts like ethanol, leading to root damage, reduced nutrient uptake, and stunted growth. Crops like peaches and avocados are especially susceptible to waterlogging, which leads to root rot, reduced growth, and poor fruit quality.

» **Temperature-Related Stress (Heat and Cold Stress)**: Temperature extremes disrupt cellular and enzymatic functions, affecting plant metabolism and growth.

- **Heat Stress**: High temperatures, especially those above a plant's physiological threshold, cause protein denaturation, enzyme inactivation, and membrane instability. Heat stress leads to increased respiration and water loss through transpiration, reducing energy available for growth and causing dehydration. In fruit crops like strawberries, apples, and grapes, heat stress can cause sunburn on fruits, accelerate ripening, and reduce post-harvest quality.
- **Cold and Frost Stress**: Cold stress includes both chilling (non-freezing, low temperatures) and freezing temperatures. Chilling stress affects cell membranes, reducing fluidity and enzymatic function, while freezing stress causes ice crystal formation in cells, leading to dehydration, tissue damage, and cell death. Cold-sensitive crops, like bananas, citrus, and tomatoes, experience chilling injuries, which cause discoloration, softening, and reduced flavor.

» **Soil Composition-Related Stress (Salinity and Soil pH Imbalances)**: Soil properties, including salt content and pH, significantly impact plant growth by influencing water absorption, ion transport, and nutrient availability.

- **Salinity Stress**: Saline soils contain high concentrations of dissolved salts, primarily sodium (Na^+) and chloride (Cl^-) ions, which disrupt water uptake by roots and cause ion toxicity. Excess salts cause osmotic stress, preventing roots from accessing water and affecting nutrient balance. Salt-sensitive crops, such as strawberries, beans, and citrus, show signs of leaf burn, stunted growth, and reduced fruit size under saline conditions.
- **Soil pH Imbalance**: Soil pH affects the solubility and availability of nutrients. Alkaline soils

can lead to micronutrient deficiencies, such as iron, zinc, and manganese, while acidic soils may cause aluminum toxicity. Nutrient imbalances in the soil impact chlorophyll production, photosynthesis, and root health, affecting the overall vigor of the plant and reducing yield.

» **Atmospheric Stress (Light Intensity, Radiation, and Air Pollution)**: Atmospheric conditions, including light levels, radiation exposure, and pollutants, also create stress.

- **Excessive Light and UV Radiation**: High light intensity and ultraviolet (UV) radiation cause photo-oxidative stress, leading to chlorophyll degradation, DNA damage, and production of reactive oxygen species (ROS). Sensitive crops like lettuce, peppers, and strawberries may experience leaf burn, reduced growth, and poor quality under intense sunlight and UV exposure.
- **Air Pollution**: Pollutants such as ozone (O_3), sulfur dioxide (SO_2), and nitrogen oxides (NO_x) damage plant tissues, impair photosynthesis, and reduce growth. Ozone exposure, for example, causes leaf chlorosis, premature senescence, and oxidative stress, which reduces productivity and fruit quality in crops such as soybeans, grapes, and tomatoes.

Each environmental source creates specific challenges for plants, and different species or crop varieties have varying levels of tolerance to these stressors. Addressing these sources of abiotic stress requires both understanding plant responses and adopting strategies to enhance resilience.

2. Plant Responses to Abiotic Stress

Plants have developed a range of adaptive responses to cope with abiotic stress, including **physiological, biochemical, and molecular mechanisms** that help maintain essential functions and prevent damage. These responses vary depending on the type of stressor and its severity.

» **Physiological Responses**: At the physiological level, plants adjust their growth and development patterns to conserve resources, reduce water loss, and manage nutrient intake. Some common physiological responses include:

- **Stomatal Regulation**: In response to drought and high temperatures, plants close their stomata (tiny pores on leaf surfaces) to reduce transpiration and water loss. This response helps conserve water but limits CO_2 intake, reducing photosynthesis.
- **Root System Modification**: Under water scarcity, plants may develop deeper and more extensive root systems to access water from deeper soil layers. For instance, drought-resistant crops often have thicker, deeper roots that increase water uptake capacity, enhancing survival during dry periods.
- **Leaf Rolling and Wilting**: To minimize surface area and water loss, plants may roll or wilt leaves during extreme heat or water deficit conditions. This response helps reduce sunlight exposure and transpiration, allowing plants to retain more water in their tissues.

» **Biochemical Responses**: Biochemical adjustments help plants stabilize cellular structures and manage stress-induced oxidative damage. Common biochemical responses to abiotic stress include:

- **Osmolyte Accumulation**: Under drought, salinity, and cold stress, plants accumulate osmolytes—small organic compounds like proline, glycine betaine, and soluble sugars. These compounds stabilize cell membranes, proteins, and other cellular structures, helping maintain osmotic balance and protecting against dehydration.

- **Antioxidant Production**: Many abiotic stressors increase the production of reactive oxygen species (ROS), which damage cellular components. Plants respond by producing antioxidants, such as superoxide dismutase (SOD), catalase, and ascorbate, that neutralize ROS and prevent oxidative damage to cells. The production of non-enzymatic antioxidants, including glutathione and carotenoids, further protects plant cells and tissues from oxidative stress.
- **Hormonal Adjustments**: Plant hormones play crucial roles in regulating stress responses. Abscisic acid (ABA), ethylene, and jasmonic acid (JA) are particularly important in abiotic stress regulation. ABA levels increase in response to drought and salinity, promoting stomatal closure, activating stress-response genes, and inducing osmolyte accumulation. Ethylene, which regulates growth and senescence, often increases under stress, accelerating ripening and senescence processes. Jasmonic acid helps modulate defense responses, particularly under oxidative stress conditions.

» **Molecular Responses**: At the molecular level, stress conditions activate specific genes that help plants adapt to adverse environments. These molecular responses include:

- **Stress-Induced Gene Expression**: Stress activates genes involved in the synthesis of protective compounds, including heat shock proteins (HSPs) under high temperatures and cold-regulated proteins (COR) under low temperatures. These proteins stabilize cellular components, reduce protein denaturation, and improve membrane stability, enhancing tolerance to temperature extremes.
- **Signaling Pathways and Transcription Factors**: Abiotic stress triggers cellular signaling pathways that involve transcription factors like DREB (dehydration-responsive element-binding protein), which regulate genes associated with drought and salinity tolerance. These transcription factors play a key role in stress adaptation, activating genes that produce osmolytes, antioxidants, and protective proteins.
- **Epigenetic Modifications**: Some plants retain a "memory" of previous stress exposure through epigenetic changes, such as DNA methylation and histone modifications, that modify gene expression without altering the DNA sequence. This stress memory helps plants respond more effectively to future stress events, enhancing resilience over successive generations.

3. Interaction Between Environmental Sources and Plant Response

The relationship between the environmental source of stress and plant response is dynamic and complex, as plants often face multiple abiotic stresses simultaneously. For example, high temperatures can coincide with drought, intensifying water stress and increasing the need for adaptive mechanisms like stomatal regulation and antioxidant production. Similarly, salinity often occurs with nutrient imbalance, requiring plants to adjust nutrient transport mechanisms in addition to managing osmotic stress.

4. Mechanisms of Stress Tolerance and Adaptation in Plants

How Plants Perceive and Respond to Stress

Plants, as immobile organisms, are constantly subjected to various environmental stresses they cannot escape. These stresses include drought, salinity, temperature extremes, nutrient imbalances, and even excess or deficit light conditions. To survive and thrive in their environments, plants have developed sophisticated systems to perceive stress, signal its presence, and initiate responses that protect against damage. These systems involve the perception of external stimuli, signal transduction pathways that communicate these signals and adaptive physiological, biochemical, and molecular responses that help plants tolerate and survive stress.

1. Perception of Stress

The perception of stress begins when a plant detects changes in its external environment. This perception process relies on specialized receptors and sensors located on the cell membrane, cytoplasm, and even organelles. These sensors are sensitive to specific environmental stimuli, such as water availability, ion concentration, temperature fluctuations, and light intensity.

» **Osmosensors**: Osmosensors detect changes in osmotic pressure, particularly under drought and salinity stress, where water availability is limited or salt concentrations are high. These sensors are essential in initiating responses to water deficits by perceiving the reduction in turgor pressure and triggering adjustments in osmotic balance. For instance, during drought conditions, osmosensors activate signals that lead to stomatal closure, a crucial response to reduce water loss and conserve moisture.

» **Thermosensors**: Temperature fluctuations, both high and low, impose significant stress on plants by affecting cellular structure and enzymatic activity. Plants possess thermosensors that detect shifts in temperature, triggering responses to help cope with heat or cold stress. During heat stress, thermosensors stimulate the production of heat shock proteins (HSPs) that protect cells from protein denaturation and membrane instability. In cold conditions, these sensors activate cold-regulated genes that produce protective proteins, aiding in maintaining membrane fluidity and preventing ice crystal formation within cells.

» **Photoreceptors**: Photoreceptors respond to light intensity and quality, helping plants adjust to excess light or shading. Photoreceptors such as phytochromes, cryptochromes, and phototropins are sensitive to specific wavelengths and regulate processes like chloroplast movement, leaf orientation, and the synthesis of antioxidants under excessive light or ultraviolet (UV) radiation. Photoreceptors are crucial for managing light stress, as excess light can lead to photodamage and reactive oxygen species (ROS) generation in the photosynthetic apparatus.

» **Ion Sensors**: Under salinity stress, the concentration of specific ions, particularly sodium (Na^+) and chloride (Cl^-), increases in the soil. Ion sensors detect high concentrations of these ions and initiate responses that regulate ion transport. In salt-sensitive plants, ion sensors activate mechanisms to compartmentalize sodium and chloride ions within vacuoles, isolating them from sensitive cellular processes to avoid toxicity. These sensors play an essential role in minimizing the negative effects of salt stress on cellular metabolism.

The sensors responsible for perceiving stress are the plant's primary detection system. When activated, these sensors initiate the signaling pathways that transmit information about the stressor throughout the plant, enabling timely and coordinated responses to ensure survival and reduce potential damage.

2. Signal Transduction Pathways

After the initial perception of stress, plants rely on complex **signal transduction pathways** to convey information from the sensors to various cellular components. This process involves an array of secondary messengers, protein kinases, transcription factors, and hormones that work together to amplify the initial signal, coordinating the plant's response at multiple levels.

Key Components of Signal Transduction Pathways

» **Secondary Messengers**: These molecules play an intermediary role in signaling by amplifying the initial stress signal and enabling rapid communication within the cell. Key secondary messengers in plant stress responses include:

- **Calcium Ions (Ca^{2+})**: Calcium plays a central role in signal transduction, acting as a versatile messenger that responds to a wide range of stress stimuli, including drought, salinity, and cold. Upon sensing stress, calcium ions are released into the cytosol, where they bind to calcium-binding proteins, which activate specific protein kinases. This chain reaction enables calcium to regulate various aspects of cellular metabolism and stress-related gene expression.
- **Reactive Oxygen Species (ROS)**: Although ROS can be damaging in excess, they serve as vital signaling molecules in moderation. Under stress conditions, ROS levels rise, and these molecules act as secondary messengers that initiate antioxidant production, regulate stress-related genes, and promote cell survival mechanisms. For example, in response to drought or high light intensity, ROS signaling helps activate genes that protect chloroplasts from photodamage.
- **Nitric Oxide (NO)**: Nitric oxide is a gaseous signaling molecule involved in both abiotic and biotic stress responses. It interacts with ROS and other messengers to regulate antioxidant enzyme activity, maintain redox balance, and enhance stress tolerance, particularly under drought and salinity stress conditions.

» **Protein Kinases**: Protein kinases are enzymes that phosphorylate other proteins, thereby activating or modifying their function. Among them, **Mitogen-Activated Protein Kinases (MAPKs)** are essential for stress response pathways, amplifying signals and coordinating downstream responses. MAPKs play a crucial role in activating transcription factors and enzymes involved in adaptive mechanisms, enabling plants to respond to stress by altering gene expression.

» **Transcription Factors**: Transcription factors bind to specific DNA sequences and regulate the expression of genes associated with stress tolerance. Some of the key transcription factors involved in abiotic stress responses are:

- **DREB (Dehydration-Responsive Element-Binding Protein)**: DREB is essential in responses to drought and salinity. It activates genes that produce osmoprotectants, antioxidants, and protective proteins to help plants cope with water deficits and salt stress.

- **NAC (NAM, ATAF1/2, and CUC2)**: NAC transcription factors are involved in cellular repair, programmed cell death, and stress adaptation, providing protection against various abiotic stresses by activating genes that help cells recover from damage.
- **bZIP (Basic Leucine Zipper)**: bZIP transcription factors play a role in abscisic acid (ABA)-mediated responses, particularly in drought stress tolerance, by regulating genes that help plants conserve water and maintain cellular homeostasis.

» **Plant Hormones**: Plant hormones, or phytohormones, are signaling compounds that regulate stress responses at the whole-plant level. Each hormone plays a specific role in coordinating the plant's adaptive responses to abiotic stress:

- **Abscisic Acid (ABA)**: Often termed the "stress hormone," ABA levels rise under drought, salinity, and cold stress. ABA induces stomatal closure to conserve water, activates genes that improve osmotic adjustment, and enhances the production of protective compounds.
- **Ethylene**: Ethylene regulates responses to flooding, salinity, and temperature stress. While it promotes leaf senescence under prolonged stress, it also enhances tolerance to oxidative stress by regulating antioxidant defenses.
- **Jasmonic Acid (JA) and Salicylic Acid (SA)**: These hormones are involved in regulating responses to both biotic and abiotic stress. JA and SA promote antioxidant production and modulate responses to drought and salinity, helping to prevent oxidative damage.

Through these signal transduction components, plants coordinate responses that range from local cellular adjustments to whole-plant changes, balancing survival with growth. This process ensures that the plant responds in a targeted, efficient manner, minimizing resource expenditure while addressing the immediate threat.

3. Adaptive Responses: Physiological, Biochemical, and Molecular Mechanisms

Once stress signals are transmitted, plants activate a series of **adaptive responses** that help protect cellular structures, maintain homeostasis, and ensure survival. These responses include physiological changes, biochemical defenses, and molecular adaptations.

Physiological Adjustments

Plants make structural and morphological changes to conserve resources, optimize water usage, and minimize the effects of stress. Common physiological responses include:

» **Stomatal Closure and Water Conservation**: Under drought and salinity stress, plants reduce transpiration by closing their stomata, thereby conserving water and maintaining turgor pressure. Stomatal closure helps limit water loss, although it reduces CO_2 intake, impacting photosynthesis. Plants balance the need for water conservation with energy production, often shifting resources to root growth to enhance water uptake from deeper soil layers.

» **Root System Modification**: Under limited water conditions, plants may develop deeper and more extensive root systems, increasing their capacity to absorb water from lower soil layers. This morphological adjustment is common in drought-resistant crops, allowing them to access deeper moisture reserves. Root architecture adjustments also help avoid areas with high salt concentrations in saline soils, mitigating salt stress.

» **Leaf Area Reduction and Structural Changes**: Some plants reduce leaf area through leaf rolling, wilting, or shedding to minimize exposure to sunlight and reduce water loss. This response limits transpiration and helps plants conserve moisture. Leaf modifications, such as thicker cuticles or smaller leaf sizes, are common in arid environments, where water conservation is essential for survival.

Biochemical Defenses

In addition to physiological adjustments, plants activate a range of biochemical defenses to mitigate cellular damage caused by stress. These responses include the accumulation of protective compounds, antioxidant production, and osmotic adjustments.

» **Osmoprotectant Accumulation**: Osmoprotectants are small organic molecules, such as proline, glycine betaine, and soluble sugars, that accumulate within plant cells in response to osmotic stress, particularly under drought and salinity conditions. These compounds play a crucial role in stabilizing cell membranes and enzymes, protecting cells from dehydration and maintaining cellular function. By balancing the osmotic pressure, osmoprotectants prevent water loss and allow the plant to retain its structure and integrity even in low-water conditions. This osmotic adjustment is vital for plants in arid and semi-arid environments, where prolonged drought is a common challenge.

» **Antioxidant Defense System**: Many types of abiotic stress, including high light, drought, salinity, and temperature extremes, increase the production of reactive oxygen species (ROS) in plants. While ROS at low levels function as signaling molecules, excessive ROS can cause oxidative damage to lipids, proteins, and DNA. To counteract this oxidative stress, plants produce a range of antioxidant enzymes, such as superoxide dismutase (SOD), catalase, and peroxidase, which neutralize ROS and protect cellular structures. Non-enzymatic antioxidants, including ascorbate (vitamin C), glutathione, and carotenoids, also play crucial roles in reducing ROS levels and safeguarding plant tissues from oxidative stress. Enhanced antioxidant production helps plants maintain cellular homeostasis and reduce damage under adverse conditions, improving resilience and survival rates.

» **Compatible Solute Production**: Compatible solutes, such as trehalose and polyamines, are molecules that stabilize proteins and cell membranes, allowing plants to maintain cellular function during stress conditions like dehydration and high salinity. Trehalose, a disaccharide, protects cell membranes by preserving their structure, while polyamines contribute to membrane stability and regulate ion transport, helping cells adapt to osmotic changes. By accumulating compatible solutes, plants protect their cells from the osmotic and ionic imbalances that arise under water-limiting or high-salinity conditions.

» **Stress-Induced Metabolites**: Many plants produce specific metabolites under stress conditions that help them tolerate adverse environments. For example, phenolic compounds and flavonoids accumulate in response to UV radiation and serve as UV protectants by absorbing excess light and reducing photodamage. These compounds also have antioxidant properties, adding an extra layer of protection against oxidative damage induced by high light or drought stress. In addition to protecting cellular structures, these stress-induced metabolites may enhance the nutritional and health benefits of fruits and vegetables, providing a secondary benefit for human consumption.

Molecular Adaptations

At the molecular level, plants activate specific genes and proteins that support cellular integrity and enhance survival. These molecular adaptations include the synthesis of protective proteins, regulation of stress-responsive genes, and, in some cases, epigenetic modifications that enhance stress tolerance.

- **Heat Shock Proteins (HSPs):** Under high temperatures, plants produce heat shock proteins (HSPs) that function as molecular chaperones, stabilizing other proteins and preventing denaturation. HSPs assist in the refolding of damaged proteins and prevent the aggregation of misfolded proteins, thereby maintaining cellular function even under thermal stress. HSPs are critical for plants in hot climates, where high temperatures are common, as they enable plants to cope with heat by protecting the structural integrity of cellular components.
- **Cold-Regulated Proteins (COR)**: In response to cold and frost stress, plants synthesize cold-regulated proteins (COR) that help maintain membrane fluidity and protect cells from freezing damage. COR proteins prevent ice crystal formation within cells, reducing the risk of cellular dehydration and rupture. Plants adapted to cold environments, such as alpine and boreal species, rely heavily on COR proteins to survive freezing temperatures. By maintaining cell membrane stability, these proteins allow plants to continue metabolic processes despite low temperatures.
- **Ion Transporters and Compartmentalization**: In saline environments, the high concentration of sodium and chloride ions can be toxic to plants. To manage these ions, plants activate ion transporters that compartmentalize sodium and chloride within vacuoles, isolating them from sensitive cellular processes and reducing their toxic effects. Transporters such as HKT (high-affinity potassium transporter) and NHX (sodium/hydrogen exchanger) play crucial roles in regulating ion balance and maintaining cellular homeostasis under salt stress. This compartmentalization strategy is particularly important for halophytes (salt-tolerant plants), allowing them to thrive in saline soils.
- **Epigenetic Modifications and Stress Memory**: Some plants develop a "stress memory" through epigenetic changes, such as DNA methylation and histone modifications, which alter gene expression without changing the DNA sequence. This memory enables plants to respond more effectively to future stress by "priming" stress-responsive genes. For example, a plant exposed to mild drought may retain epigenetic modifications that make it more resilient to future droughts. Stress memory allows plants to acclimate to recurring stresses and improves their ability to survive in challenging environments. This phenomenon is especially relevant in perennial crops and trees that face repeated stress events over their long lifespans.
- **Transcriptional Regulation of Stress-Responsive Genes**: Key transcription factors, such as DREB (dehydration-responsive element-binding protein), NAC, and bZIP, regulate the expression of genes involved in stress tolerance. These transcription factors activate genes that

produce osmoprotectants, antioxidants, and proteins involved in cellular repair and survival. For instance, DREB transcription factors are essential in drought and salinity responses, promoting the synthesis of compounds that enhance water retention and reduce ion toxicity. Transcriptional regulation enables plants to tailor their response to specific types of stress, optimizing resource allocation and energy use for survival.

Practical Applications of Understanding Stress Mechanisms

The knowledge of how plants perceive and respond to stress has practical applications in agriculture and horticulture, where stress-tolerant crops are essential for stable and high-yield production in variable environments. Breeding and biotechnology focus on enhancing these stress-tolerance mechanisms in crops to improve resilience, yield stability, and crop quality under adverse conditions.

1. **Breeding for Stress Tolerance**: Traditional breeding programs focus on selecting varieties with naturally high levels of drought, salinity, or temperature tolerance. These programs identify plants with effective stress-perception mechanisms, efficient signaling pathways, and strong adaptive responses. For example, crops bred for deep root systems are more effective in accessing water under drought conditions, while those with higher antioxidant levels show improved resilience to oxidative stress from high temperatures.
2. **Genetic Engineering and Biotechnology**: Advances in genetic engineering allow scientists to introduce genes responsible for stress tolerance directly into crop genomes. For instance, genes that enhance osmoprotectant accumulation or activate antioxidant pathways can be inserted into susceptible crops, improving their performance under stress. The use of CRISPR/Cas9 and other gene-editing technologies allows for precise modifications, enabling the development of crops that are better equipped to withstand abiotic stress.
3. **Biostimulants and Plant Growth Regulators**: Biostimulants, such as seaweed extracts, amino acids, and microbial inoculants, can enhance plant tolerance to abiotic stress by stimulating root growth, improving nutrient uptake, and activating antioxidant pathways. Plant growth regulators, such as abscisic acid (ABA) and jasmonic acid (JA), are sometimes applied to modulate stress responses, helping plants conserve water during drought or delay senescence under high temperatures. These treatments offer practical solutions for enhancing stress resilience in crops without the need for genetic modifications.
4. **Precision Agriculture and Stress Monitoring**: Sensor technologies and remote sensing enable real-time monitoring of environmental conditions, allowing farmers to respond to stress as it occurs. Soil moisture sensors, for example, can trigger automated irrigation systems to prevent drought stress, while drones equipped with thermal cameras detect heat stress in large fields. By continuously monitoring stress indicators, precision agriculture enables targeted interventions, minimizing resource use and enhancing crop resilience.
5. **Soil and Water Management**: Soil amendments, mulching, and controlled irrigation techniques help manage soil moisture levels, nutrient availability, and salinity, reducing the impact of abiotic stress on crops. Mulching conserves soil moisture, while drip irrigation minimizes water loss, helping plants avoid drought stress. In saline soils, amendments like gypsum can reduce salt buildup, allowing plants to absorb nutrients more effectively.

Tolerance Mechanisms at Molecular, Cellular, and Whole-Plant Levels

Plants are constantly exposed to various abiotic stresses, including drought, high salinity, extreme temperatures, and nutrient deficiencies. These environmental challenges can disrupt growth and productivity, but plants have evolved complex tolerance mechanisms that operate at molecular, cellular, and whole-plant levels to mitigate stress impacts and enhance survival.

1. Molecular-Level Tolerance Mechanisms

At the molecular level, plants activate specific genes, signaling molecules, and proteins that enable them to detect and respond to stress. Molecular tolerance mechanisms are crucial for initiating and regulating downstream responses at the cellular and whole-plant levels. These mechanisms include the production of protective proteins, activation of signaling pathways, and transcriptional regulation of stress-responsive genes.

Key Molecular Responses

- **Gene Expression and Transcription Factors**: The activation of stress-responsive genes is a fundamental molecular mechanism that allows plants to produce proteins and compounds essential for stress tolerance. Various transcription factors, such as DREB (Dehydration-Responsive Element-Binding), NAC, and bZIP (Basic Leucine Zipper), play key roles in regulating these genes. For example, DREB transcription factors are involved in drought and salinity tolerance, activating genes that produce osmoprotectants and antioxidants. These transcription factors allow plants to fine-tune their response based on the type and severity of stress.
- **Stress Signaling Pathways**: Stress signaling begins when plants perceive changes in their environment through specialized receptors. Secondary messengers, including calcium ions (Ca^{2+}), reactive oxygen species (ROS), and nitric oxide (NO), transmit these signals to activate specific protein kinases, such as MAPKs (Mitogen-Activated Protein Kinases). This cascade amplifies the stress signal, allowing for a rapid and coordinated response. For instance, calcium ions play a key role in drought response by activating pathways that promote stomatal closure and water conservation.
- **Protective Proteins**: Plants produce a range of proteins in response to stress, including heat shock proteins (HSPs) and cold-regulated proteins (CORs). HSPs function as molecular chaperones, helping other proteins maintain their structure and preventing denaturation under high temperatures. COR proteins help stabilize cellular structures and maintain membrane fluidity under cold conditions, reducing the risk of freezing damage. These protective proteins are crucial for maintaining cellular integrity and metabolic function in extreme environments.
- **Hormonal Regulation**: Plant hormones, such as abscisic acid (ABA), ethylene, jasmonic acid (JA), and salicylic acid (SA), regulate stress responses at the molecular level. ABA, known as the "stress hormone," plays a central role in drought and salinity tolerance by promoting stomatal closure, conserving water, and activating stress-related genes. Ethylene and JA contribute to salinity and oxidative stress responses, while SA plays a role in antioxidant

production. Hormonal signaling enables plants to coordinate stress responses across tissues, optimizing resource allocation for survival.

» **Epigenetic Modifications**: Plants can "remember" past stress events through epigenetic changes, such as DNA methylation and histone modification. These modifications allow plants to adjust gene expression in response to recurring stress, enhancing tolerance. For instance, a plant exposed to mild drought may retain epigenetic changes that improve resilience in future drought events. This stress memory provides a lasting adaptation mechanism, especially beneficial for perennial plants that face seasonal stresses.

Molecular mechanisms form the first line of defense, enabling plants to detect stress and initiate immediate responses. These mechanisms are foundational for the cellular-level changes that follow, as they provide the instructions and signals necessary for cellular adaptation and resilience.

2. Cellular-Level Tolerance Mechanisms

The cellular level is where molecular signals translate into biochemical and structural changes that protect plant cells from the damaging effects of stress. Cellular tolerance mechanisms help stabilize cellular structures, maintain osmotic balance, and reduce oxidative damage, ensuring that essential functions continue under stress.

Key Cellular Responses

» **Osmotic Adjustment and Osmoprotectants**: In response to drought and salinity stress, plants adjust their internal osmotic balance to prevent water loss and maintain cellular hydration. Cells accumulate osmoprotectants, such as proline, glycine betaine, and soluble sugars, which help retain water by balancing osmotic pressure. These compounds stabilize proteins and cell membranes, protecting them from damage caused by dehydration or high salt concentrations.

» **Antioxidant Defense System**: Stress conditions, particularly high light, drought, and temperature extremes, increase the production of reactive oxygen species (ROS) in cells, leading to oxidative stress. To counteract this, plants activate antioxidant defense mechanisms that include enzymes like superoxide dismutase (SOD), catalase, and peroxidase. These enzymes neutralize ROS, preventing damage to lipids, proteins, and DNA. Non-enzymatic antioxidants, such as ascorbate (vitamin C), glutathione, and carotenoids, also contribute to cellular protection by scavenging ROS.

» **Membrane Stability and Fluidity**: Cell membranes are highly sensitive to temperature and ionic stress. In cold conditions, membranes can become rigid, while heat and salinity stress increase membrane permeability. Plants produce compounds like unsaturated fatty acids and compatible solutes to maintain membrane fluidity and stability. This adjustment prevents cellular contents from leaking out, ensuring that the cell retains essential ions and metabolites. Additionally, certain proteins, such as HSPs and COR proteins, help stabilize membranes, reducing the risk of cellular damage.

» **Ion Compartmentalization**: High salinity stress increases the concentration of sodium and chloride ions, which are toxic to plant cells at high levels. To prevent toxicity, plants use vacuolar compartmentalization to sequester excess ions away from critical cellular functions. Transport proteins, such as Na^+/H^+ antiporters, actively pump sodium ions into vacuoles, isolating them from the cytoplasm and reducing their harmful effects. This ion compartmentalization is particularly important for salt-tolerant species, or halophytes, which thrive in saline environments.

» **Cell Wall Reinforcement**: Under mechanical stress, such as wind or water deficiency, plants reinforce their cell walls to improve structural support. The deposition of lignin, cellulose, and pectin strengthens the cell wall, making it more resistant to mechanical damage and water loss. For instance, in drought conditions, reinforced cell walls help maintain cell shape and prevent wilting, contributing to overall plant stability.

The cellular mechanisms activated under stress help protect the structural and functional integrity of plant cells, ensuring that metabolism, photosynthesis, and nutrient transport continue despite adverse conditions. These cellular responses support the larger-scale changes observed at the whole-plant level.

3. Whole-Plant Level Tolerance Mechanisms

At the whole-plant level, tolerance mechanisms include physiological and morphological adjustments that help the plant as a whole cope with stress. These changes involve shifts in resource allocation, growth patterns, and structural modifications, enabling plants to optimize survival under stress conditions.

Key Whole-Plant Responses

» **Stomatal Regulation and Water Conservation**: Stomata, the tiny pores on leaf surfaces, control water loss and gas exchange. Under drought and salinity stress, plants close their stomata to reduce transpiration and conserve water. Although this limits CO_2 intake and reduces photosynthesis, stomatal closure helps prevent dehydration, preserving water within plant tissues. Stomatal regulation is particularly critical for drought-resistant species that must conserve water over extended periods.

» **Root System Adjustment**: Plants adapt their root architecture to improve water and nutrient uptake under stress. In response to drought, plants may develop deeper roots to access water from lower soil layers, while under nutrient deficiency, lateral root growth increases to explore a larger soil volume. This plasticity in root structure helps plants optimize resource acquisition, making them more resilient in low-water and low-nutrient environments.

» **Leaf Area Reduction and Structural Changes**: To minimize water loss and reduce exposure to extreme temperatures, plants often reduce their leaf area by shedding older leaves, rolling leaves, or producing smaller leaves with thicker cuticles. These changes reduce the surface area exposed to sunlight, lowering transpiration rates and protecting against heat and water loss. In arid environments, these structural adaptations are essential for long-term survival.

» **Phenological Shifts**: Under certain stress conditions, plants may alter their life cycle timing, a phenomenon known as phenological shift. For instance, drought stress may cause some plants to accelerate flowering and fruiting to complete their reproductive cycle before water availability declines further. This shift allows plants to produce seeds even under stress, ensuring species survival. Phenological adjustments are particularly important in annual crops that rely on a single growing season for reproduction.

» **Resource Allocation and Growth Trade-offs**: In response to prolonged stress, plants may redirect resources from growth to survival. For example, plants experiencing nutrient deficiency may allocate fewer resources to leaf growth and more to root expansion to enhance nutrient uptake. Similarly, under drought stress, plants may reduce shoot growth to conserve water and allocate resources to root development. These trade-offs allow plants to balance growth with survival, ensuring that essential functions continue even under resource-limited conditions.

» **Symbiotic Relationships**: Some plants enhance their stress tolerance by forming symbiotic relationships with soil microbes, such as mycorrhizal fungi and nitrogen-fixing bacteria. These microbes improve nutrient and water absorption, enhance root growth, and boost the plant's overall resilience to abiotic stress. For example, mycorrhizal fungi increase the surface area for nutrient uptake, helping plants access phosphorus and other minerals that are often limited in drought or saline conditions.

4. Integrating Tolerance Mechanisms Across Levels

The mechanisms of stress tolerance in plants are not isolated but rather interconnected across molecular, cellular, and whole-plant levels. For instance, molecular signaling pathways initiate cellular responses that, in turn, trigger physiological and structural changes at the whole-plant level. This integration enables plants to mount a coordinated and adaptive response to stress, balancing growth, resource use, and survival.

» **Example 1: Drought Stress Response**

Under drought conditions, osmosensors at the molecular level detect reduced water availability, triggering ABA production and stomatal closure at the cellular level. This response minimizes water loss, while the accumulation of osmoprotectants protects cellular structures. At the whole-plant level, reduced leaf area and enhanced root growth further optimize water use, enabling the plant to survive prolonged drought.

» **Example 2: Salinity Stress Response**

In saline environments, ion sensors at the molecular level detect high sodium levels, activating transporters that compartmentalize sodium ions within vacuoles. This reduces ion toxicity at the cellular level, while antioxidant production protects cells from ROS. At the whole-plant level, modified root architecture and symbiotic relationships with soil microbes improve nutrient and water uptake, enhancing salt tolerance.

Implications for Agriculture and Crop Improvement

Understanding plant tolerance mechanisms at molecular, cellular, and whole-plant levels has practical implications for agriculture. Breeding programs, genetic engineering, and precision agriculture techniques aim to enhance these mechanisms to improve crop resilience, stability, and yield under environmental stress.

» **Breeding for Resilience**: Breeding programs target specific tolerance traits, such as deep root systems, antioxidant production, and efficient ion transport. Selecting and breeding varieties that excel in these traits allows for the development of crops that perform better under stress conditions.

» **Genetic Engineering**: Genetic modification techniques, such as CRISPR, enable the introduction of stress-tolerance genes, including those encoding transcription factors, transporters, and osmoprotectants, into crops that are otherwise vulnerable to stress. This approach accelerates the development of stress-resilient crops.

» **Precision Agriculture**: Sensors and data analytics enable real-time monitoring of plant health, environmental conditions, and stress indicators. By providing timely information, precision agriculture allows for targeted interventions that enhance plant resilience, such as optimized irrigation, nutrient management, and pest control.

5. Abiotic Stress: Categories and Definitions

a. Water Stress: Classification into Drought and Flooding, and Physiological Effects on Fruit Crops

Water stress is a major environmental factor affecting the growth, productivity, and quality of fruit crops. It occurs in two contrasting forms: drought (water deficit) and flooding (water excess). Both types of water stress can lead to significant physiological disruptions, affecting critical processes such as photosynthesis, nutrient uptake, and fruit development. Understanding how drought and flooding impact fruit crops is essential for developing management strategies to maintain productivity, especially as climate variability increases the frequency and severity of extreme weather events.

Water Stress: Classification into Drought and Flooding

Water stress, a critical factor impacting plant growth and productivity, is broadly classified into two main types based on the availability of water in the root zone: **drought** (water deficit) and **flooding** (water excess). Both conditions create significant physiological challenges, disrupting key processes such as photosynthesis, nutrient transport, and cellular function. The effects of these distinct types of water stress and how plants respond to each can help develop effective strategies for managing water-related challenges in crop cultivation, especially in fruit crops that are sensitive to fluctuations in water availability.

Classification of Water Stress: Drought vs. Flooding

Water stress occurs when there is an imbalance between the water needs of the plant and the availability of water in the environment. This imbalance leads to two main types of water stress:

1. **Drought (Water Deficit)**: Occurs when the water supply in the soil is insufficient to meet the plant's requirements. Drought stress is characterized by reduced water potential in plant cells, leading to dehydration and impacting nearly all physiological processes. Drought is prevalent in regions with low or erratic rainfall, high temperatures, or inadequate irrigation.

Flooding (Water Excess): Occurs when the soil is saturated with water for extended periods, leading to oxygen deprivation in the root zone. Flooding stress can result from heavy rainfall, poor drainage, or over-irrigation. It disrupts the plant's ability to respire and absorb nutrients, often leading to a cascade of physiological challenges.

Both drought and flooding stress affect plant health and productivity but in contrasting ways. Below, we'll explore each type of water stress in detail, examining their causes, specific physiological impacts, and the adaptations plants employ to cope.

1. Drought (Water Deficit)

Drought, or water deficit, is a condition in which the water available in the soil is below the level needed to sustain optimal plant growth. This type of water stress is especially harmful during critical growth phases, such as flowering and fruit development, and is exacerbated by environmental factors like high temperatures, wind, and low humidity.

» **Characteristics of Drought**: Drought occurs when water loss due to transpiration exceeds water uptake by the roots. This leads to reduced water potential in plant tissues, resulting in dehydration at the cellular level. Depending on the intensity and duration, drought stress may be temporary (acute drought) or prolonged (chronic drought), with varying impacts on plant growth and productivity. In fruit crops, drought is particularly detrimental, as it can directly affect yield and quality.

» **Causes of Drought**:

- **Climate and Weather Patterns**: In regions with seasonal or unpredictable rainfall, drought is common, particularly during the dry season. Climate change has also increased the frequency and severity of drought events, further impacting agricultural productivity.
- **Soil Composition and Water-Holding Capacity**: Soil type plays a significant role in water retention. Sandy soils, for instance, have low water-holding capacity, making plants more susceptible to drought stress. Conversely, clayey soils hold more water but are often challenging for root penetration.
- **Water Management Practices**: Inefficient irrigation systems or over-reliance on rainfed agriculture can exacerbate drought stress. Inadequate water management limits water availability, especially during critical growth stages, impacting plant resilience.

» **Physiological Effects of Drought on Plants**:

- **Reduced Photosynthesis**: Drought stress forces plants to close their stomata to reduce water loss. However, this also limits CO_2 uptake, which is essential for photosynthesis. Reduced CO_2 intake decreases the rate of photosynthesis, leading to a shortage of carbohydrates that stunts growth and reduces fruit size and yield. In severe drought, leaf senescence (aging) and abscission (shedding) can further reduce photosynthetic capacity.
- **Decreased Cellular Turgor and Cell Expansion**: Cellular turgor, or water pressure within cells, is essential for cell expansion and plant rigidity. Under drought conditions, reduced water availability decreases cellular turgor, causing wilting and reduced growth. In fruit crops, this translates to smaller, less juicy fruits with tougher textures, which impacts both yield and marketability.
- **Altered Hormone Levels**: Drought stress triggers the production of abscisic acid (ABA), a hormone that regulates water conservation by closing stomata. Increased ABA levels also inhibit growth-related hormones such as cytokinins and gibberellins,

slowing down cell division and expansion. The hormonal shifts induced by drought lead to stunted growth and premature ripening in fruits, which may cause reduced fruit size and altered flavor profiles.

- **Disruption in Nutrient Uptake**: Water is the medium through which essential nutrients travel from soil to plant roots. Drought restricts water flow, limiting nutrient mobility and absorption. Nitrogen, calcium, and potassium deficiencies are common under drought stress, as these nutrients require adequate moisture for uptake. In fruit crops, calcium deficiency can lead to physiological disorders like blossom-end rot in tomatoes and peppers, compromising fruit quality.
- **Osmotic Adjustment and Accumulation of Osmoprotectants**: To combat dehydration, plants accumulate compatible solutes like proline, glycine betaine, and soluble sugars. These compounds help retain cellular water and stabilize proteins and membranes. While osmotic adjustment helps plants endure drought, it requires substantial energy, which diverts resources from growth and reproduction.
- **Increased Production of Reactive Oxygen Species (ROS)**: Drought stress often leads to the accumulation of ROS, such as superoxide radicals and hydrogen peroxide. Excessive ROS can cause oxidative damage to lipids, proteins, and DNA. Plants activate antioxidant defense systems, producing enzymes like superoxide dismutase (SOD) and catalase to neutralize ROS. However, prolonged drought can overwhelm these defenses, leading to cellular damage.
- **Effects on Fruit Quality**: Drought stress affects fruit size, texture, and flavor. Reduced water availability during fruit development leads to smaller, less juicy fruits, as well as a tougher texture. In some cases, mild drought stress may increase sugar concentration in fruits, enhancing sweetness. However, severe drought can alter flavor profiles negatively, resulting in bitterness in fruits like apples and oranges.

Drought stress requires plants to employ a range of adaptive mechanisms to retain water, protect cellular structures, and mitigate dehydration effects. While these mechanisms improve drought tolerance, they come at an energy cost, reducing the plant's ability to allocate resources to growth and yield.

2. Flooding (Water Excess)

Flooding, or waterlogging, occurs when the root zone is saturated with water for prolonged periods, leading to oxygen deficiency. Roots rely on oxygen for aerobic respiration, which is necessary for energy production. Excess water in the root zone limits oxygen availability, forcing roots to switch to less efficient anaerobic respiration and leading to a range of physiological challenges.

» **Characteristics of Flooding**: Flooding can be short-term (lasting hours or days) or prolonged (lasting weeks or months). Short-term flooding may cause temporary stress, which plants can recover from. However, prolonged flooding leads to hypoxia (oxygen deficiency) or anoxia (complete lack of oxygen) in the root zone, damaging root tissues and limiting nutrient uptake. Flooding stress is particularly harmful to crops with shallow root systems, as these roots are more susceptible to submersion and oxygen deprivation.

» **Causes of Flooding**:

- **Excessive Rainfall and Extreme Weather**: Intense rainfall, particularly during monsoon or rainy seasons, can cause waterlogged conditions. Extreme weather events, such as hurricanes, contribute to flooding in agricultural regions.
- **Poor Soil Drainage**: Soils with high clay content have low drainage capacity, retaining water for extended periods. Compacted soils, often resulting from heavy machinery use, further restrict water movement, making flooding more severe.
- **Irrigation Practices**: Over-irrigation, particularly in fields without adequate drainage systems, can lead to localized flooding. Flooding is common in areas with frequent irrigation and insufficient monitoring of soil moisture.

» **Physiological Effects of Flooding on Plants**:

- **Oxygen Deficiency and Anaerobic Respiration**: Flooded soils are low in oxygen due to limited gas exchange with the atmosphere. Roots require oxygen for aerobic respiration, which produces energy (ATP). Under anaerobic conditions, roots switch to less efficient anaerobic respiration, which produces toxic byproducts like ethanol and lactate. These byproducts damage root tissues, limiting water and nutrient uptake.
- **Impaired Nutrient Uptake and Transport**: Flooding disrupts root function, reducing the plant's ability to absorb essential nutrients, particularly nitrogen, phosphorus, and potassium. Nutrient deficiencies manifest as chlorosis (leaf yellowing) and reduced growth. For instance, nitrogen deficiency causes poor vegetative growth and low chlorophyll content, reducing photosynthetic efficiency.
- **Accumulation of Ethylene**: Flooding stress increases ethylene production, which accelerates processes like leaf senescence, fruit ripening, and abscission (shedding of leaves and fruit). Elevated ethylene levels can cause premature fruit drop, affecting yield and quality. Excessive ethylene also inhibits root growth, compounding the negative effects of flooding on nutrient and water uptake.
- **Altered Hormone Levels**: Flooding-induced hypoxia raises abscisic acid (ABA) levels, promoting stomatal closure to prevent water uptake. Increased ethylene and ABA levels inhibit growth-promoting hormones like cytokinins and gibberellins, further slowing growth and delaying recovery.

» **Damage to Root and Shoot Systems**: Prolonged flooding leads to hypoxia-induced root damage, where cells in the root tissue begin to deteriorate due to a lack of oxygen and the buildup of toxic byproducts like ethanol. Damaged roots are more susceptible to pathogenic infections, particularly root rot diseases caused by fungi and bacteria, which can further degrade root health. As root systems decline, the plant's overall stability and ability to absorb water and nutrients are compromised. In extreme cases, flooding-induced root damage can result in shoot dieback, wilting, and even plant death due to the loss of essential nutrient transport pathways.

» **Disruption of Photosynthesis and Gas Exchange**: Flooding stress can also affect above-ground tissues, as reduced oxygen availability and increased ethylene levels cause stomatal closure in leaves. This limits gas exchange, particularly CO_2 intake, which is essential for photosynthesis. Reduced photosynthetic activity limits the production of carbohydrates, which are critical for growth, development, and fruit formation. The decline in photosynthesis also impacts energy availability, leading to stunted growth, poor flowering, and reduced fruit set, all of which negatively affect overall crop productivity.

» **Imbalance of Plant Metabolites**: Under flooded conditions, plants experience significant shifts in metabolic processes due to the lack of oxygen and energy deficits caused by anaerobic respiration. As roots struggle to respire aerobically, metabolic waste products, such as ethanol and lactate, accumulate. These compounds disrupt normal metabolic pathways and, in high concentrations, are toxic to plant cells. Furthermore, an imbalance in plant hormones, especially with increased levels of ethylene and ABA, interferes with the synthesis of essential metabolites, leading to poor growth and an overall decline in plant health.

» **Impact on Fruit Quality and Post-Harvest Shelf Life**: Flooding stress adversely affects the quality and shelf life of fruits, particularly in crops such as tomatoes, strawberries, and peppers. Excessive water uptake during fruit development can lead to a waterlogged or "watery" texture, reducing fruit firmness and making them more susceptible to bruising and decay. Flooding-induced ethylene production can lead to premature ripening, resulting in fruits with a lower sugar concentration and blander flavor. Additionally, flooding increases the risk of post-harvest diseases due to the weakened structural integrity of fruit tissues, which shortens shelf life and reduces marketability.

Comparative Summary of Drought and Flooding Effects on Fruit Crops

Both drought and flooding create unique physiological challenges for plants, impacting their growth, development, and yield. Despite their contrasting nature, both types of water stress necessitate that plants activate adaptive responses that affect cellular, tissue, and whole-plant functions.

1. **Water Conservation vs. Oxygen Deficiency**:
   - **Drought** induces a water conservation response, where plants close stomata to reduce water loss. This conserves internal water but also limits photosynthesis due to reduced CO_2 intake.
   - **Flooding** creates oxygen deficiency in the root zone, leading to anaerobic respiration, which is less efficient and produces toxic byproducts that harm root cells.
2. **Impact on Nutrient Uptake:**
   - **Drought** limits nutrient uptake due to restricted water flow, causing deficiencies in essential nutrients, such as nitrogen, potassium, and calcium.
   - **Flooding** disrupts nutrient transport due to root damage and lack of oxygen, leading to deficiencies in nitrogen, phosphorus, and potassium. This can cause visible symptoms like chlorosis and stunted growth.
3. **Hormonal Imbalance:**
   - **Drought** stress raises abscisic acid (ABA) levels, promoting stomatal closure and inhibiting growth-related hormones, which leads to reduced growth and accelerated ripening.

- **Flooding** increases both ABA and ethylene levels, causing root growth inhibition, premature fruit drop, and leaf senescence, which reduces the plant's resilience and recovery ability.

4. **Effects on Photosynthesis and Energy Production:**
   - **Drought** reduces photosynthesis due to stomatal closure, which limits CO_2 availability and carbohydrate production.
   - **Flooding** affects photosynthesis indirectly by damaging root systems, reducing nutrient uptake, and causing stomatal closure in leaves due to high ethylene levels.

5. **Impact on Fruit Quality:**
   - **Drought** can reduce fruit size and juiciness but may increase sugar concentration, leading to sweeter fruits if the stress is mild.
   - **Flooding** causes fruits to have a watery texture and reduced sweetness, resulting in a bland flavor. The compromised structural integrity of fruits under flooding stress also shortens shelf life and increases susceptibility to decay.

Adaptive Mechanisms to Mitigate Water Stress

To survive drought and flooding, plants have evolved adaptive mechanisms that help them cope with these stresses, ensuring survival and, in some cases, continued productivity. Understanding these adaptive responses is crucial for developing resilient fruit crops that can better tolerate water stress in variable climates.

Adaptive Mechanisms in Response to Drought

» **Osmotic Adjustment**: Plants accumulate osmoprotectants, such as proline and glycine betaine, to adjust their internal osmotic balance and retain cellular water. These compounds protect cellular structures from dehydration and allow metabolic processes to continue.

» **Root System Modifications**: Under drought conditions, plants may develop deeper and more extensive root systems to access water from lower soil layers. This adaptation is particularly beneficial for drought-resistant fruit crops that grow in arid regions.

» **Leaf Morphology Adjustments**: Plants under drought stress may reduce leaf area, increase leaf thickness, or produce waxy cuticles to minimize water loss through transpiration. Some plants exhibit leaf rolling or wilting, reducing surface exposure to sunlight and conserving moisture.

Adaptive Mechanisms in Response to Flooding

» **Aerenchyma Formation**: Certain plants develop specialized air spaces, known as aerenchyma, in root tissues, allowing oxygen to reach submerged roots. This adaptation is particularly useful in plants that frequently encounter waterlogged conditions, such as rice.

» **Adventitious Root Formation**: In response to flooding, some plants produce adventitious roots near the soil surface, where oxygen availability is higher. These roots improve gas exchange and help the plant survive prolonged waterlogging.

- » **Hormonal Regulation and Ethylene Management**: Flooded plants may modulate hormone levels to delay senescence and manage ethylene levels. For instance, maintaining lower ethylene levels can prevent excessive leaf drop and premature fruit ripening under flooded conditions.

Practical Implications for Agriculture

The classification and physiological effects of water stress informs agricultural practices aimed at mitigating drought and flooding impacts. Effective water management, crop selection, and the use of soil amendments can enhance resilience to water stress in fruit crops.

1. **Drought Management Strategies**:
   - **Irrigation Techniques**: Drip irrigation and regulated deficit irrigation (RDI) optimize water use by delivering precise amounts of water directly to the root zone, reducing water wastage and preventing water stress.
   - **Mulching and Soil Amendments**: Applying organic mulches helps retain soil moisture and reduce evaporation. Soil amendments, like organic compost, improve soil structure and increase water-holding capacity, enhancing drought resilience.
   - **Drought-Resistant Varieties**: Breeding programs focus on developing fruit crops with drought-resistant traits, such as deep root systems, high water-use efficiency, and osmoprotectant accumulation.
2. **Flooding Management Strategies**:
   - **Improving Drainage**: Installing drainage systems in fields prone to flooding prevents waterlogging by directing excess water away from the root zone. Raised beds or mounds can also help elevate crops above the water-saturated soil.
   - **Soil Aeration and Compaction Prevention**: Regular soil aeration and minimizing soil compaction with reduced machinery use improve soil structure and drainage capacity, reducing the risk of flooding.
   - **Flood-Tolerant Varieties**: Selecting or breeding flood-tolerant rootstocks that exhibit traits like aerenchyma formation and adventitious root growth can help crops survive in waterlogged conditions.
3. **Precision Agriculture**: The use of sensors, drones, and data analytics enables real-time monitoring of soil moisture levels, plant stress indicators, and water use. This data allows farmers to make informed decisions about irrigation scheduling, drainage adjustments, and stress mitigation practices, ensuring optimal water management.

Physiological Effects of Drought on Fruit Crops

Drought stress imposes severe physiological limitations on fruit crops, affecting their growth, yield, and fruit quality. Some of the primary physiological effects of drought on fruit crops include:

1. **Reduction in Photosynthesis**: Drought stress leads to reduced stomatal conductance as plants close their stomata to minimize water loss through transpiration. However, this also limits the intake of CO_2, which is essential for photosynthesis. Reduced CO_2 availability decreases the rate of photosynthesis, which in turn limits the production of carbohydrates.

This reduction in photosynthesis affects energy availability, leading to stunted growth and smaller fruit sizes. Prolonged drought stress can even lead to leaf senescence and drop, further reducing the photosynthetic capacity of the plant.

2. **Inhibition of Cell Expansion and Fruit Growth**: Water plays a crucial role in maintaining cell turgor, which is necessary for cell expansion. Under drought conditions, reduced cellular turgor pressure limits cell enlargement, leading to stunted shoot and root growth. In fruit crops, reduced cell expansion directly affects fruit size and weight. For example, in grapes, tomatoes, and apples, drought stress results in smaller fruits with reduced juice content, affecting both yield and marketability.
3. **Alteration of Hormone Levels**: Drought triggers the synthesis of abscisic acid (ABA), a stress hormone that regulates stomatal closure and root growth. Increased ABA levels help plants conserve water, but they also inhibit growth-related hormones like gibberellins and cytokinins. This hormonal shift reduces vegetative growth and diverts resources toward survival rather than fruit production. Additionally, the elevated ABA levels accelerate the ripening process, which may lead to premature fruit drop and reduced fruit quality.
4. **Impact on Nutrient Uptake and Transport**: Drought reduces the availability of water in the soil, which limits nutrient solubility and uptake by roots. As a result, nutrient deficiencies, especially of nitrogen, potassium, and calcium, are common under drought stress. Nutrient deficiencies lead to issues like chlorosis (yellowing of leaves) and reduced growth. In fruit crops, insufficient calcium uptake can cause physiological disorders, such as blossom-end rot in tomatoes and peppers, impacting fruit quality and reducing market value.
5. **Accumulation of Osmoprotectants and Compatible Solutes**: To cope with dehydration, plants accumulate osmoprotectants such as proline, glycine betaine, and soluble sugars. These compounds help maintain cellular osmotic balance and protect cellular structures from damage caused by water deficit. While this osmotic adjustment allows cells to retain water and maintain metabolic activity, it also requires energy, diverting resources away from growth and fruit production.
6. **Increased Production of Reactive Oxygen Species (ROS)**: Drought stress often leads to the overproduction of reactive oxygen species (ROS), such as hydrogen peroxide and superoxide radicals. High levels of ROS cause oxidative stress, damaging cellular components like lipids, proteins, and DNA. To counteract ROS, plants activate antioxidant defense systems, producing enzymes like superoxide dismutase (SOD) and catalase. Although these antioxidant responses help protect cells, they consume energy that could otherwise support growth and fruit development.
7. **Effect on Fruit Quality**: Drought stress affects the quality attributes of fruits, including size, texture, flavor, and nutrient content. For instance, reduced water availability during fruit development can lead to smaller, less juicy fruits with a tougher texture. In some cases, drought stress increases sugar concentration, leading to a sweeter flavor, as seen in grapes for winemaking. However, severe drought can reduce flavor complexity and cause bitterness in fruits like apples and oranges. Additionally, the nutritional quality of fruits, such as vitamin C and antioxidant levels, may be compromised under prolonged drought.

Physiological Effects of Flooding on Fruit Crops

Flooding stress imposes a different set of physiological challenges on fruit crops, as the excess water in the root zone reduces oxygen availability and disrupts normal metabolic processes. Key physiological effects of flooding on fruit crops include:

1. **Oxygen Deficiency and Anaerobic Respiration**: Flooding reduces oxygen availability in the root zone, leading to hypoxia or anoxia. Roots rely on oxygen for aerobic respiration to produce energy (ATP). Under flooded conditions, plants switch to anaerobic respiration, which is much less efficient and produces toxic byproducts like ethanol and lactate. The accumulation of these compounds can cause root tissue damage, leading to root rot and impaired nutrient uptake. The energy deficit caused by anaerobic respiration limits growth and reduces yield in fruit crops.

2. **Reduced Nutrient Absorption and Transport**: Flooding disrupts root function, making it difficult for plants to absorb essential nutrients from the soil. The lack of oxygen affects root membrane permeability, reducing nutrient uptake and leading to deficiencies of nitrogen, phosphorus, and potassium. This nutrient deficiency manifests as chlorosis, stunted growth, and reduced fruit size. For example, nitrogen deficiency under flooding can lead to poor vegetative growth and low chlorophyll content, reducing photosynthetic efficiency.

3. **Accumulation of Ethylene**: Flooding increases ethylene production in plants, which accelerates leaf senescence, fruit ripening, and abscission (shedding of leaves, flowers, or fruits). In fruit crops, elevated ethylene levels can lead to premature fruit ripening and drop, affecting fruit quality and yield. High ethylene levels also inhibit root growth, compounding the negative effects of flooding on nutrient and water uptake.

4. **Altered Hormone Balance**: Flooding stress induces hormonal changes that affect growth and stress response. Besides increasing ethylene, flooding raises abscisic acid (ABA) levels, which helps close stomata to conserve water. However, high ABA and ethylene levels inhibit growth-promoting hormones like cytokinins and gibberellins, leading to reduced shoot and root growth. This hormonal imbalance limits the plant's ability to recover from flooding once water levels subside.

5. **Root and Shoot Damage**: Prolonged flooding causes root tissue damage due to oxygen deprivation and toxic byproduct accumulation. Damaged roots become susceptible to pathogens, leading to root rot and overall decline in plant health. Reduced root function limits the plant's ability to take up water and nutrients, resulting in wilting, even in flooded conditions. Additionally, shoot tissues may develop symptoms of chlorosis and necrosis due to nutrient deficiencies and hormonal imbalance.

6. **Effect on Photosynthesis and Leaf Gas Exchange**: Under flooded conditions, stomatal conductance often decreases to reduce water uptake and prevent further saturation. However, this also restricts CO_2 intake, leading to reduced photosynthesis and energy production. In fruit crops, reduced photosynthesis limits carbohydrate production, affecting fruit growth and yield. Flooding stress can cause leaf chlorosis and senescence, further reducing photosynthetic capacity and diminishing crop productivity.

7. **Impact on Fruit Quality**: Flooding stress affects fruit quality attributes such as flavor, texture, and post-harvest shelf life. Excessive water during fruit development can lead to water-soaked tissues, resulting in soft, low-sugar fruits with reduced flavor. For example, in tomatoes and strawberries, flooding often results in fruits with a watery texture and bland taste. Flooding also reduces post-harvest quality, making fruits more susceptible to rot and decay during storage and transport.

Table 1: Effect of Flooding on Fruit Trees

S. N.	Name of Fruit Crops/Cultivars	Effect of Flooding
1.	Apple	Resulting in severe injury to the root system. Tree roots in waterlogged soils stop growing, minerals are not absorbed, leaves turn yellow and remain small, and finally roots begin to die as infections of phytopathogen microorganisms occurs in "wet-feet" conditions. Often, apple trees in poorly drained areas are infected by Phytophthora root and crown rot, slowly declining over one or more years.
2.	Cherry	Fruit cracking
3.	Pineapple	Excess moisture may cause the yellowing of the leaves followed by red and it reduces the length
4.	Papaya	Papaya is susceptible to fungal root diseases. Killing of waterlogged plants.
5.	Mango	Reduce root dry weight, resulting in an increased shoot to root ratio, reduced net carbon dioxide assimilation rates.

Comparative Summary of Drought and Flooding Effects on Fruit Crops

Water stress, whether from drought or flooding, poses substantial challenges for fruit crops, impacting their growth, development, and quality in distinct ways. While drought induces water conservation mechanisms to counteract dehydration, flooding creates oxygen deficiency in the root zone, leading to metabolic disruptions. Both conditions affect nutrient uptake, ethylene production, and, ultimately, fruit quality and yield, albeit through different physiological pathways. Understanding these differences helps inform strategies to mitigate water stress and improve resilience in fruit crop production.

1. Water Conservation vs. Oxygen Deficiency

Drought and flooding have contrasting effects on the availability of resources essential for plant metabolism, namely water and oxygen.

» **Drought (Water Conservation Mechanisms)**: Under drought stress, plants activate a series of water conservation mechanisms to prevent dehydration and maintain cellular function. One of the primary responses is **stomatal closure**, where plants reduce water loss by closing stomata (small pores on leaf surfaces). However, stomatal closure also limits carbon dioxide (CO_2) intake, reducing photosynthesis and carbohydrate production. Plants under drought conditions also accumulate **osmoprotectants**—small organic compounds like proline, glycine betaine, and sugars—that help retain water within cells and maintain osmotic balance. While these mechanisms help plants survive low water conditions, they come at a cost to growth, as resources are diverted from reproductive processes like fruit formation.

» **Flooding (Oxygen Deficiency and Anaerobic Respiration)**: In contrast to drought, flooding stress creates oxygen deficiency in the root zone, leading to **hypoxia** (low oxygen levels) or **anoxia** (absence of oxygen). Roots require oxygen for aerobic respiration, which produces the energy (ATP) necessary for nutrient absorption and root function. Under flooded conditions, roots switch to **anaerobic respiration**, a less efficient process that produces toxic byproducts like ethanol and lactate. These toxic compounds accumulate in root tissues, damaging cell membranes, impairing root function, and potentially leading to root rot. The energy deficit and root damage caused by oxygen deprivation limit the plant's ability to absorb water and nutrients, impacting overall growth and fruit development.

The primary distinction between drought and flooding lies in how plants attempt to balance essential resources: **drought forces plants to conserve water**, while **flooding deprives roots of oxygen**, limiting their ability to perform essential functions.

2. Nutrient Deficiencies

Both drought and flooding affect nutrient uptake, but for different reasons. Water availability directly influences nutrient solubility and transport in soil, while oxygen availability affects root health and nutrient absorption efficiency.

» **Drought (Reduced Nutrient Solubility and Transport)**: In drought conditions, limited water availability reduces nutrient solubility in the soil, restricting the movement of essential ions like nitrogen (N), calcium (Ca), and potassium (K) to the roots. Drought-induced stomatal closure further reduces the rate of transpiration, which slows the upward movement of nutrients through the xylem (water-conducting tissue). As a result, nutrient deficiencies are common in drought-stressed plants. Calcium deficiency is particularly problematic, as calcium is immobile and depends heavily on transpiration for transport. In fruit crops, calcium deficiency can lead to physiological disorders such as **blossom-end rot** in tomatoes and peppers, which directly impacts fruit quality.

» **Flooding (Root Damage and Reduced Absorption)**: Flooding affects nutrient uptake by damaging roots and reducing their ability to absorb nutrients. The lack of oxygen disrupts root membrane permeability and inhibits the activity of nutrient transport proteins. Consequently, plants in flooded conditions often suffer from nitrogen, phosphorus (P), and potassium deficiencies, as roots are unable to actively absorb these nutrients under hypoxic conditions. Nitrogen deficiency is especially detrimental to fruit crops, as nitrogen is essential for chlorophyll production and photosynthesis. Flooding-induced nitrogen deficiency leads to chlorosis (leaf yellowing) and stunted growth, reducing the plant's ability to produce energy and support fruit development.

In both drought and flooding, nutrient deficiencies lead to weakened plants, lower fruit quality, and reduced yields. However, while drought stress primarily limits nutrient mobility, flooding stress directly impairs root function, compounding nutrient uptake challenges.

3. Ethylene Accumulation

Ethylene is a plant hormone that regulates processes such as fruit ripening, leaf senescence, and response to stress. Both drought and flooding increase ethylene production, but the outcomes vary depending on the type of water stress.

» **Drought (Accelerated Ripening and Senescence)**: Under drought conditions, elevated ethylene production often leads to **accelerated ripening** and **leaf senescence** (aging). Ethylene promotes fruit maturation, which can be advantageous in mild drought conditions, as it may concentrate sugars and enhance flavor in fruits like grapes. However, in severe drought, excessive ethylene levels cause premature ripening, leading to smaller, less developed fruits with reduced shelf life. High ethylene levels in drought-stressed plants can also accelerate leaf senescence, further limiting photosynthesis and carbohydrate production, which are crucial for fruit development and overall productivity.

» **Flooding (Leaf Drop and Root Growth Inhibition)**: Flooding stress also triggers ethylene production, but with different effects. In flooded conditions, ethylene promotes **leaf abscission** (leaf drop), **premature fruit ripening**, and **inhibition of root growth**. Leaf drop reduces transpiration and helps prevent excess water uptake, but it also limits photosynthetic capacity, affecting the plant's ability to produce energy. In fruit crops, elevated ethylene levels can cause premature fruit ripening, resulting in lower fruit quality and reduced storage life. Flooding-induced ethylene production also inhibits root elongation, further compromising the plant's ability to recover from stress once water levels recede.

While both drought and flooding induce ethylene production, the outcomes reflect each type of stress: **drought-triggered ethylene accelerates fruit ripening and leaf aging**, while **flooding-triggered ethylene causes leaf and fruit drop and inhibits root growth**, both of which negatively affect fruit quality and yield.

4. Fruit Quality and Yield

Drought and flooding have contrasting effects on fruit quality attributes such as size, texture, flavor, and storage life, ultimately influencing yield and marketability.

» **Drought (Size Reduction and Sugar Concentration)**: Drought stress typically reduces fruit size due to limited cell expansion, as dehydration restricts cell turgor and inhibits growth. However, mild drought stress may increase **sugar concentration** in fruits, enhancing sweetness and flavor in crops like grapes, apples, and tomatoes. This concentration effect occurs as reduced water content concentrates sugars and other soluble solids, which can improve fruit quality in controlled drought conditions. However, severe drought often results in bitter flavors, tough texture, and low fruit weight, which negatively impact both yield and market value.

» **Flooding (Soft Texture and Low Sugar Content)**: Flooding stress produces fruits with a **soft, watery texture** and low sugar content due to excess water uptake and reduced photosynthesis. For example, flooded tomatoes and strawberries often have a bland, watery taste and lack the rich flavor consumers expect. Flooded fruits also have reduced firmness, making them more susceptible to bruising and mechanical damage during transport. Additionally, the compromised structural integrity of flooded fruits increases susceptibility to post-harvest diseases, reducing storage life and marketability.

Drought and flooding create significant challenges for fruit crops by impacting water and oxygen availability, nutrient uptake, ethylene production, and fruit quality. Although they affect different physiological processes, both types of water stress necessitate energy-intensive adaptive responses, reducing the plant's overall growth potential and productivity.

» **Drought** triggers water conservation mechanisms that limit photosynthesis and nutrient transport, leading to smaller fruits that may have concentrated sugars in mild cases. Severe drought, however, reduces fruit size, causes bitterness, and leads to nutrient deficiencies that compromise yield.

» **Flooding** causes oxygen deprivation, forcing plants to rely on inefficient anaerobic respiration, leading to root damage, reduced nutrient uptake, and leaf and fruit drop due to elevated ethylene levels. Flooding also reduces fruit firmness and sweetness, creating watery fruits with limited storage potential.

Agricultural Implications

Effective management of water stress involves tailored strategies to address the unique impacts of drought and flooding. For example:

» **Drought Management**: Employing drip irrigation, mulching, and soil amendments to retain moisture, selecting drought-resistant varieties, and managing irrigation timing are effective strategies to mitigate drought effects.

» **Flooding Management**: Improving soil drainage, implementing raised beds or mounds, and selecting flood-tolerant rootstocks are practices that can reduce the impact of flooding on crops. Additionally, the use of real-time monitoring and data-driven decisions through precision agriculture technologies can help anticipate and manage water stress before it impacts crop yield.

b. Temperature Stress: High and Low-Temperature Effects on Cellular Processes

Temperature stress, resulting from either excessively high or low temperatures, disrupts cellular functions and metabolic pathways in plants. This stress affects essential processes like photosynthesis, respiration, enzyme activity, and membrane stability, all of which are critical to plant growth and development.

High-Temperature Stress: Effects on Cellular Processes

High temperatures, particularly those exceeding a plant's optimal growth range, subject plants to heat stress, which affects cellular and metabolic processes essential for growth and development. This stress disrupts biochemical reactions, compromises cellular integrity, and ultimately impacts yield and quality, especially in temperature-sensitive fruit crops. One of the first and most critical responses of plants to heat stress is the denaturation of enzymes, which leads to a reduction in metabolic efficiency and disrupts energy production, growth, and reproduction.

1. Enzyme Denaturation and Reduced Metabolic Efficiency

Enzymes play a central role in almost all biochemical processes within plant cells. They act as catalysts, speeding up reactions necessary for growth, energy production, and nutrient assimilation. Each enzyme has an optimal temperature range in which it functions most effectively. When temperatures rise beyond this range, enzymes can denature, leading to a loss of structure and, consequently, a decline in activity. This enzyme denaturation has far-reaching consequences for the plant's physiology and productivity.

» **Structural Sensitivity of Enzymes to Temperature**:

- Enzymes are complex proteins whose function depends on their three-dimensional structure. This structure is stabilized by weak bonds, such as hydrogen bonds, hydrophobic interactions, and ionic bonds, which hold the protein in its functional shape.
- High temperatures increase the kinetic energy of molecules, which can disrupt these stabilizing bonds. As these bonds break, enzymes lose their specific structure—a process known as **denaturation**. Denatured enzymes cannot catalyze reactions efficiently, leading to a slowdown in essential metabolic processes.
- In many plants, critical enzymes begin to lose activity at temperatures above 30–35°C, with severe denaturation occurring at higher temperatures (above 40°C), depending on the species.

» **Impact on Photosynthetic Enzymes, Particularly Rubisco**:

- Photosynthesis is one of the most temperature-sensitive processes in plants. The enzyme **Rubisco** (Ribulose-1,5-bisphosphate carboxylase/oxygenase) is responsible for fixing carbon dioxide (CO_2) during the Calvin cycle, a key step in photosynthesis.
- Rubisco is highly sensitive to heat, and as temperatures rise, its activity declines. High temperatures also promote photorespiration (a process where Rubisco binds to oxygen instead of CO_2), which consumes energy without producing sugars. As a result, the efficiency of photosynthesis decreases under heat stress.
- Reduced Rubisco activity limits the production of carbohydrates, the primary energy source for growth and fruit formation. This carbohydrate deficit affects all energy-dependent processes, including cell expansion, root growth, and fruit development. In fruit crops, lower carbohydrate availability can lead to smaller fruits, reduced sweetness, and compromised quality.

» **Respiratory Enzymes and Metabolic Imbalance**:

- Respiration is another essential process affected by heat stress. During respiration, enzymes in the mitochondria break down carbohydrates to produce ATP (adenosine triphosphate), the energy currency of the cell. High temperatures disrupt these respiratory enzymes, leading to increased respiration rates.
- Increased respiration consumes more sugars, which are already limited under heat stress due to reduced photosynthesis. This creates a **metabolic imbalance** where energy production is insufficient to meet the demands of growth, maintenance, and reproduction.

- Additionally, as respiratory enzymes denature, the efficiency of ATP production declines, leading to energy deficits that can compromise root growth, leaf expansion, and flower development. In fruit crops, this results in stunted growth, poor fruit set, and reduced yield.

» **Effects on Key Growth and Developmental Processes**:

- **Cell Expansion**: High temperatures affect the enzymes responsible for cell wall loosening, which is necessary for cell expansion. Enzyme denaturation under heat stress limits the ability of cells to expand, leading to reduced leaf and fruit size.
- **Hormone Synthesis and Balance**: Enzymes involved in hormone biosynthesis, such as those producing auxins, gibberellins, and cytokinins, are also sensitive to high temperatures. Reduced activity of these enzymes leads to lower hormone levels, impacting processes like cell division, fruit development, and seed formation.
- **Protein Synthesis:** High temperatures inhibit the activity of ribosomes, the cellular machinery responsible for protein synthesis. This affects the production of new proteins required for growth and repair, compromising the plant's ability to respond to stress and recover.

» **Cascading Effects on Plant Health and Productivity**:

- **Delayed Flowering and Fruit Set**: Enzyme denaturation affects the synthesis of floral hormones, delaying flowering and reducing fruit set in many fruit crops. For instance, high temperatures during the flowering stage of tomatoes and peppers can lead to poor pollination, resulting in lower yields.
- **Reduced Sugar Accumulation in Fruits**: With limited photosynthesis and increased respiration, the accumulation of sugars in fruits decreases, impacting taste and nutritional value. In crops like grapes and apples, high temperatures can reduce sweetness and alter flavor profiles, affecting marketability.
- **Premature Senescence and Reduced Lifespan**: Enzyme denaturation accelerates senescence, as plants are unable to maintain cellular function and repair damage. This leads to early leaf yellowing and drop, reducing photosynthetic capacity and further limiting energy production.

Cellular Mechanisms to Counteract Enzyme Denaturation

Plants activate several protective mechanisms to mitigate the effects of enzyme denaturation under heat stress. These mechanisms include the production of **heat shock proteins (HSPs)**, the activation of **antioxidant defenses**, and **osmoprotectant accumulation**.

» **Heat Shock Proteins (HSPs):**

- HSPs are specialized proteins produced in response to high temperatures. They function as molecular chaperones, helping to stabilize and refold denatured proteins, preventing them from aggregating and causing cellular damage.
- HSPs assist in maintaining enzyme function by refolding partially denatured enzymes and protecting critical cellular components. Their production, however, requires significant energy and resources, which can strain the plant's metabolic reserves.

- While HSPs provide temporary relief from heat stress, prolonged exposure to high temperatures exhausts cellular resources, limiting the plant's ability to produce new enzymes and proteins needed for growth.

» **Antioxidant Defense Activation**:

- High temperatures lead to the accumulation of reactive oxygen species (ROS), which exacerbate enzyme denaturation and cellular damage. Plants respond by activating antioxidant enzymes such as superoxide dismutase (SOD), catalase, and peroxidase to neutralize ROS.
- Antioxidants help protect enzymes and cellular structures from oxidative damage, preserving metabolic function under heat stress. However, maintaining high antioxidant levels is energy-intensive, further taxing the plant's already limited resources under stress.

» **Accumulation of Osmoprotectants**:

- Plants produce osmoprotectants like proline, glycine betaine, and trehalose in response to heat stress. These molecules stabilize proteins and cell membranes, protecting them from denaturation and dehydration.
- Osmoprotectants also help maintain cellular osmotic balance, reducing water loss from cells. This adaptation is particularly beneficial in arid climates where high temperatures are often accompanied by low humidity.

2. Membrane Fluidity and Leakage

High temperatures directly affect the physical and functional stability of cellular membranes in plants, impacting processes essential for maintaining cellular homeostasis. Cell membranes are primarily composed of a lipid bilayer, interspersed with proteins, that serves as a selective barrier controlling the movement of ions, water, and other molecules in and out of cells. Under heat stress, changes in membrane fluidity disrupt this selective permeability, leading to issues like ion leakage, osmotic imbalance, and nutrient transport inefficiencies, which can collectively impair cellular function and affect plant growth and productivity.

1. **Temperature-Induced Changes in Membrane Fluidity**

Cell membranes are typically fluid structures that allow for a degree of flexibility, enabling membrane proteins and other components to move laterally within the bilayer. This fluidity is crucial for cellular functions like signaling, transport, and interaction with the extracellular environment. The degree of membrane fluidity is influenced by environmental factors, particularly temperature.

- **High-Temperature Impact on Lipid Bilayer**: The lipid bilayer of cell membranes is composed of phospholipids with hydrophobic fatty acid tails. As temperatures rise, the kinetic energy of these lipids increases, causing the membrane to become more fluid. When this fluidity exceeds optimal levels, the membrane structure becomes compromised, losing its integrity. This excessive fluidity leads to increased permeability, as lipids begin to spread further apart, creating gaps that ions and molecules can freely pass through.
- **Effect on Saturated vs. Unsaturated Fatty Acids**: The fluidity of the membrane is

also influenced by the composition of fatty acids in phospholipids. Membranes with a higher proportion of unsaturated fatty acids (with double bonds in their structure) are more fluid and flexible, which is advantageous in cooler conditions. However, under high temperatures, unsaturated fatty acids increase the likelihood of excessive fluidity. Plants in hot environments often adapt by increasing saturated fatty acids in their membrane composition to counteract fluidity, but rapid temperature changes can still cause significant stress as membrane composition adjustments take time.

2. **Compromised Membrane Integrity and Ion Leakage**

As membrane fluidity increases, the physical structure of the membrane becomes unstable. The once tightly packed lipids and proteins start to spread apart, compromising the selective permeability of the membrane. This increased permeability causes the uncontrolled leakage of essential ions and molecules, which is detrimental to cellular stability.

- **Ion Leakage**: Ions such as potassium (K^+) and calcium (Ca^{2+}) play critical roles in cellular signaling, osmotic balance, and metabolic reactions. Potassium is essential for maintaining cell turgor and enzyme activation, while calcium is vital for cell wall stability, signal transduction, and enzyme regulation. Under heat stress, these ions leak out of the cell, disrupting osmotic balance and leading to cellular dehydration. This ion loss triggers a series of stress responses that divert resources from growth and reproduction to repair.
- **Impact on Osmotic Balance**: The leakage of ions and small molecules through a destabilized membrane impacts the cell's osmotic potential, leading to a state of water stress even when external water availability is sufficient. This internal water stress can cause plasmolysis, where the plasma membrane pulls away from the cell wall due to water loss. Plasmolysis leads to wilting, reduced cell expansion, and impaired photosynthesis, particularly affecting developing fruits, which are highly sensitive to changes in cellular hydration.

3. **Effects on Membrane Proteins and Transport Mechanisms**

Cell membranes contain various proteins, including transporters, channels, and receptors, embedded in the lipid bilayer. These proteins are responsible for the selective movement of ions, nutrients, and signaling molecules across the membrane, regulating cellular homeostasis. High temperatures impact both the structural stability and functional capacity of these proteins, compromising nutrient and water transport.

- **Destabilization of Membrane Proteins**: Just as high temperatures increase lipid fluidity, they can also affect the structure of membrane proteins, many of which are temperature-sensitive. Heat stress can cause these proteins to lose their functional shape (denaturation), impairing their ability to transport ions and other molecules effectively. For example, aquaporins, which facilitate water movement across the membrane, may become less efficient or even dysfunctional under heat stress, further contributing to water stress within the cell.
- **Disruption of Ion Transport Mechanisms**: Ion transport proteins, such as potassium channels and calcium pumps, are essential for maintaining ionic gradients across the membrane. Heat-induced membrane destabilization impairs these proteins, reducing their

ability to control ion flow. Potassium and calcium gradients are particularly crucial for cell signaling, growth regulation, and stress responses. When these transport mechanisms fail, cells lose the ability to regulate internal ion concentrations, resulting in unregulated ion influx and efflux that destabilizes cellular metabolism.

4. **Increased Susceptibility to Oxidative Damage**

High-temperature stress increases reactive oxygen species (ROS) production, which exacerbates membrane damage. The lipid bilayer is highly susceptible to oxidative damage, as ROS readily react with unsaturated fatty acids, causing **lipid peroxidation**—a process that leads to the breakdown of lipid molecules within the membrane. This oxidative damage further compromises membrane integrity, creating a vicious cycle of ROS accumulation and membrane deterioration.

- **Effects of Lipid Peroxidation on Membrane Stability**: Lipid peroxidation produces toxic byproducts, such as malondialdehyde (MDA), which disrupts cellular structures and inhibits enzymatic function. These byproducts accumulate within the cell, intensifying oxidative stress and damaging other cellular components, including nucleic acids and proteins.
- **Antioxidant Responses to Protect Membrane Integrity**: Plants activate antioxidant defense mechanisms under heat stress to counteract ROS and protect membrane integrity. Antioxidant enzymes like superoxide dismutase (SOD), catalase, and peroxidase play critical roles in neutralizing ROS, reducing the impact of oxidative damage on membranes. However, prolonged heat exposure overwhelms these defenses, leading to extensive oxidative stress and cell damage.

5. **Nutrient Imbalance and Reduced Cellular Stability**

Disrupted membrane function and ion leakage under heat stress affect nutrient and water transport across the membrane, leading to nutrient imbalances within the cell. The loss of essential ions like potassium and calcium, along with inhibited nutrient uptake, creates metabolic imbalances that impact cellular stability and plant growth.

- **Impaired Nutrient Uptake**: The destabilization of membrane transport proteins under heat stress reduces the plant's ability to absorb nutrients from the soil, particularly potassium, calcium, and magnesium. Potassium is critical for enzyme activation, protein synthesis, and photosynthesis, while calcium is essential for cell wall stability and signal transduction. A deficiency in these nutrients leads to poor cellular stability and growth retardation.
- **Effects on Growth and Reproduction**: Nutrient imbalance due to membrane instability directly affects cell division, expansion, and differentiation, which are essential for tissue growth and fruit formation. In fruit crops, this translates to smaller fruits, irregular ripening, and compromised flavor and texture, reducing overall yield and quality. Additionally, as nutrient uptake decreases, metabolic processes become inefficient, leading to energy deficits that further hinder growth and development.

6. **Impact on Hormonal Signaling and Stress Responses**

Membranes play an important role in mediating hormonal signaling, particularly for stress-response hormones like abscisic acid (ABA) and ethylene. Heat stress-induced membrane instability affects the localization and function of receptors embedded in the membrane, which are crucial for perceiving and responding to hormonal signals.

- **ABA-Mediated Stomatal Closure**: ABA, a key hormone in stress responses, promotes stomatal closure under heat stress to reduce water loss. However, membrane instability can impair the function of ABA receptors, reducing the efficiency of this response. As a result, plants may be unable to conserve water effectively, intensifying the effects of heat stress.
- **Ethylene Production and Premature Senescence**: High temperatures increase ethylene production, which accelerates senescence (aging) and ripening processes. Ethylene receptors embedded in membranes become destabilized, leading to erratic hormone signaling that can cause premature fruit ripening, leaf drop, and a reduction in photosynthetic capacity. These effects compromise both the longevity and productivity of fruit crops.

7. **Adaptive Responses to Stabilize Membrane Fluidity**

Plants employ several adaptive mechanisms to stabilize membranes under high temperatures, aiming to restore fluidity and protect cellular function.

- **Modification of Membrane Lipid Composition**: Plants can adjust the ratio of saturated to unsaturated fatty acids in membrane lipids to control fluidity. Increasing the proportion of saturated fatty acids under high temperatures helps reduce excessive fluidity, stabilizing membrane structure.
- **Production of Compatible Solutes and Osmoprotectants**: Plants produce compatible solutes like proline, glycine betaine, and trehalose, which stabilize membrane proteins and lipids. These compounds maintain membrane integrity, protect cellular structures, and help the plant retain water.
- **Activation of Heat Shock Proteins (HSPs)**: Heat shock proteins help prevent membrane protein denaturation, preserving transport and signaling functions. HSPs act as molecular chaperones, refolding damaged proteins and ensuring that essential membrane functions continue under stress.

3. Increased Production of Reactive Oxygen Species (ROS)

Reactive oxygen species (ROS) are highly reactive molecules derived from oxygen, including superoxide radicals (O_2•–), hydrogen peroxide (H_2O_2), and hydroxyl ions (•OH). While ROS are naturally produced in cellular processes, high temperatures accelerate their production, causing an overabundance that can be harmful to cellular components. Elevated ROS levels under heat stress lead to oxidative damage to essential cellular structures, impairing plant growth, metabolism, and overall productivity. Plants rely on complex antioxidant defense systems to counter ROS, but these defenses are often insufficient under prolonged or extreme heat, resulting in oxidative stress.

1. **High Temperatures and Accelerated ROS Production**

Under normal conditions, ROS are byproducts of metabolic processes such as photosynthesis and respiration, where they play essential roles in signaling and cellular regulation. However, high temperatures elevate metabolic rates, increasing ROS production to levels that surpass the plant's ability to manage them safely.

- **Photosynthetic Electron Transport Chain (ETC)**: In chloroplasts, ROS are primarily generated in the electron transport chain (ETC) during photosynthesis. High temperatures disrupt the balance in the ETC, causing "leakage" of electrons that react with oxygen, forming superoxide radicals. This is particularly pronounced in the photosystem II (PSII) complex, where heat-induced stress impairs the efficient flow of electrons, leading to an excess of superoxide radicals and hydrogen peroxide.
- **Mitochondrial Respiration**: ROS production also increases in the mitochondria during respiration. High temperatures accelerate respiratory activity, increasing the electron flux through the mitochondrial ETC. This heightened activity elevates ROS production, as more electrons escape the ETC and react with oxygen, forming superoxide radicals and hydrogen peroxide. Excessive ROS in mitochondria can damage the organelle's membrane, compromising ATP production and reducing the energy available for cellular processes.
- **Peroxisomes and Photorespiration**: Peroxisomes are another source of ROS, particularly during photorespiration, which increases under heat stress. Photorespiration occurs when the enzyme Rubisco binds to oxygen instead of CO_2, producing toxic byproducts that are processed in the peroxisomes, where hydrogen peroxide is produced as a byproduct. This ROS production adds to the oxidative stress experienced by cells under high-temperature conditions.

2. **Oxidative Damage to Cellular Structures**

ROS are highly reactive and can oxidize essential cellular components, including lipids, proteins, and DNA. This oxidative damage leads to a cascade of cellular disruptions that impair membrane integrity, protein function, and genomic stability, ultimately impacting plant health and productivity.

- **Lipid Peroxidation and Membrane Damage**: ROS readily oxidize unsaturated fatty acids in the lipid bilayer, causing lipid peroxidation. This process generates toxic byproducts, such as malondialdehyde (MDA), which disrupt membrane structure and function. Lipid peroxidation increases membrane permeability, leading to ion leakage, osmotic imbalance, and cellular dehydration. Damage to cell membranes is particularly detrimental in fruit crops, where membrane stability is essential for maintaining fruit texture, flavor, and nutrient content.
- **Protein Oxidation and Dysfunction**: ROS can modify proteins by oxidizing amino acid side chains, altering their structure and function. Oxidized proteins may lose their enzymatic activity or become misfolded, resulting in aggregation and cellular toxicity. For instance, oxidative damage to photosynthetic enzymes, such as Rubisco, reduces the efficiency of carbon fixation, compromising energy production and carbohydrate synthesis. Additionally, oxidation of respiratory enzymes disrupts ATP production, leading to an energy deficit that affects growth and development.
- **DNA Damage and Genomic Instability**: ROS can induce DNA strand breaks and mutations by reacting with nucleotides. Oxidative DNA damage impairs replication and transcription, potentially leading to genomic instability. DNA lesions, such as

8-oxoguanine, interfere with gene expression and cellular regulation, causing mutations that may affect cellular function and survival. In fruit crops, such genomic instability can disrupt normal growth patterns, reduce yield, and impact fruit quality.

3. Antioxidant Defense Systems to Counter ROS

Plants have evolved sophisticated antioxidant defense systems to neutralize ROS and protect cellular structures from oxidative damage. These systems include both enzymatic and non-enzymatic antioxidants that work together to scavenge ROS and maintain redox balance under stressful conditions.

» **Enzymatic Antioxidants**: Key antioxidant enzymes include superoxide dismutase (SOD), catalase, and peroxidase. These enzymes catalyze reactions that convert ROS into less reactive molecules, preventing oxidative damage.

- **Superoxide Dismutase (SOD)**: SOD catalyzes the conversion of superoxide radicals into hydrogen peroxide and oxygen. This reaction is crucial for controlling superoxide levels in chloroplasts and mitochondria, where ROS production is high.
- **Catalase (CAT)**: Catalase breaks down hydrogen peroxide into water and oxygen, preventing the accumulation of hydrogen peroxide in peroxisomes, chloroplasts, and mitochondria. Catalase is particularly important in detoxifying hydrogen peroxide produced during photorespiration.
- **Peroxidases (POD)**: Peroxidases reduce hydrogen peroxide by using various electron donors, such as ascorbate or glutathione. This process further detoxifies hydrogen peroxide, protecting cells from oxidative damage. Ascorbate peroxidase (APX) is especially active in chloroplasts, where it neutralizes hydrogen peroxide generated during photosynthesis.

» **Non-Enzymatic Antioxidants**: Non-enzymatic antioxidants, such as ascorbate (vitamin C), glutathione, carotenoids, and flavonoids, provide additional ROS-scavenging capacity. These molecules directly neutralize ROS and regenerate enzymatic antioxidants, helping maintain redox balance under heat stress.

- **Ascorbate and Glutathione**: Ascorbate and glutathione form the ascorbate-glutathione cycle, a key pathway for detoxifying hydrogen peroxide. This cycle operates in chloroplasts, mitochondria, and cytosol, providing continuous ROS-scavenging support across cellular compartments.
- **Carotenoids**: Carotenoids in chloroplasts protect the photosynthetic apparatus by quenching singlet oxygen and other ROS. These pigments absorb excess light energy, preventing ROS formation in the first place and reducing the oxidative load on the cell.
- **Flavonoids**: Flavonoids are secondary metabolites with antioxidant properties. They scavenge ROS and modulate stress responses, enhancing the plant's overall resistance to heat-induced oxidative stress.

4. Limitations of Antioxidant Defenses under Prolonged Heat Stress

While antioxidant defenses are effective in neutralizing ROS under moderate stress, prolonged or extreme heat can overwhelm these systems, leading to oxidative stress and cellular damage that cannot be fully counteracted.

- **Enzyme Inactivation and Resource Depletion**: Under prolonged heat stress, the efficiency of antioxidant enzymes declines due to enzyme denaturation and oxidative inactivation. This reduces the plant's ability to scavenge ROS, allowing oxidative damage to accumulate.
- **Energy and Resource Constraints**: Antioxidant defense systems require ATP and metabolic resources to function. Under high temperatures, the increased demand for ATP to repair oxidative damage strains the plant's energy reserves, which are already compromised due to inhibited photosynthesis and respiration. This energy deficit limits the production of antioxidants, exacerbating oxidative stress.
- **Accumulation of Oxidative Damage**: As oxidative damage accumulates, cellular structures become increasingly compromised, reducing membrane integrity, enzyme activity, and DNA stability. This damage impairs growth, development, and yield in fruit crops, particularly during sensitive growth stages like flowering and fruit set.

5. **Implications for Fruit Crop Health and Productivity**

In fruit crops, the effects of ROS overproduction and oxidative damage under high temperatures can significantly impact quality, yield, and marketability.

- **Reduced Photosynthetic Efficiency and Growth**: ROS-induced damage to chloroplasts and photosynthetic enzymes reduces the plant's ability to generate carbohydrates, leading to stunted growth and smaller fruits. In crops like tomatoes, grapes, and berries, reduced carbohydrate production affects flavor, sweetness, and overall quality.
- **Poor Fruit Texture and Shelf Life**: Oxidative damage to cell membranes reduces fruit firmness and increases susceptibility to bruising, impacting post-harvest quality. Fruits affected by oxidative stress have a shorter shelf life due to increased susceptibility to decay and nutrient loss.
- **Altered Nutritional and Flavor Profiles**: High ROS levels can disrupt metabolic pathways involved in nutrient and flavor compound synthesis. For instance, oxidative damage can reduce the production of antioxidants like vitamin C, carotenoids, and phenolic compounds in fruits, affecting their nutritional value and flavor complexity.

4. Photosynthetic Impairment

Photosynthesis, the process by which plants convert light energy into chemical energy, is highly sensitive to temperature changes. Heat stress disrupts the efficiency of photosynthesis, affecting the structure and function of chloroplasts—the organelles where photosynthesis occurs. High temperatures impair photosynthetic machinery, leading to decreased carbohydrate production, energy shortages, and, ultimately, reduced fruit growth and quality. Given that fruit crops heavily depend on carbohydrates for flavor, size, and nutritional value, high-temperature-induced photosynthetic impairment can significantly impact agricultural productivity and crop marketability.

1. **Impact of High Temperatures on Chloroplast Structure and Function**

Chloroplasts are specialized organelles that house the molecular machinery required for photosynthesis, including pigments, enzymes, and electron transport chains. These structures are highly sensitive to temperature fluctuations, and high temperatures can disrupt chloroplast integrity and function.

- **Disruption of Thylakoid Membranes**: Within chloroplasts, photosynthetic processes occur on thylakoid membranes, which contain chlorophyll pigments and protein complexes essential for capturing and converting light energy. Heat stress disrupts the fluidity and stability of thylakoid membranes, causing structural alterations that reduce the efficiency of light absorption and electron transport. This damage to the thylakoid structure limits the chloroplast's ability to produce ATP and NADPH—molecules essential for the energy-intensive Calvin cycle where carbohydrates are synthesized.
- **Destabilization of Photosystem Complexes**: Photosystems I and II are protein complexes in the thylakoid membrane responsible for absorbing light and driving electron flow during photosynthesis. High temperatures destabilize these photosystem complexes, particularly photosystem II (PSII), which is highly temperature-sensitive. PSII damage disrupts electron transport, leading to reduced production of ATP and NADPH and hindering the Calvin cycle's ability to produce carbohydrates. As PSII function declines, photosynthesis becomes less efficient, directly impacting the plant's energy supply.

2. Photoinhibition and Reduced Light Capture

Excessive heat can lead to **photoinhibition**, a condition where the light-harvesting complex in chloroplasts becomes damaged. Photoinhibition occurs when the rate of light absorption exceeds the plant's capacity to process and use this energy, causing photooxidative damage to the photosynthetic apparatus.

- **Damage to the D1 Protein in PSII**: PSII relies on a specific protein, D1, which plays a crucial role in capturing and transferring electrons. High temperatures make the D1 protein particularly susceptible to degradation. As D1 degrades, the efficiency of PSII declines, and the chloroplast's ability to capture and utilize light energy is reduced. The plant may attempt to replace damaged D1 proteins, but prolonged or intense heat stress overwhelms the plant's repair mechanisms, resulting in chronic photoinhibition.
- **Generation of Excessive Reactive Oxygen Species (ROS)**: During photoinhibition, the disrupted flow of electrons within PSII leads to the generation of reactive oxygen species (ROS), such as singlet oxygen. ROS are highly reactive and cause oxidative damage to the thylakoid membrane, further impairing the chloroplast's capacity to perform photosynthesis. ROS accumulation leads to a self-reinforcing cycle of damage, as oxidative stress exacerbates photoinhibition, limiting the plant's ability to recover from high-temperature stress.
- **Reduced Light Absorption and Energy Conversion**: Photoinhibition reduces the chloroplast's ability to absorb and convert light energy, leading to a decline in photosynthetic efficiency. As less light energy is converted into chemical energy, the production of ATP and NADPH declines, directly impacting carbohydrate synthesis and energy availability for the plant's growth, reproductive development, and stress responses.

3. Impact on Carbohydrate Production and Energy Availability

Photosynthetic impairment under heat stress reduces the production of carbohydrates—vital compounds that provide energy for growth, development, and fruit production. Carbohydrates are synthesized during the Calvin cycle, which requires the ATP and NADPH produced by the light reactions. When high temperatures disrupt these light-dependent processes, the Calvin cycle cannot operate efficiently, leading to a significant reduction in carbohydrate output.

- **Reduction in Sugar Accumulation**: Lower carbohydrate production under heat stress limits the accumulation of sugars in developing fruits. In fruit crops like tomatoes, grapes, and apples, sugar content is a major factor in determining flavor, sweetness, and market appeal. High temperatures during the fruit development stage reduce sugar accumulation, leading to fruits with less sweetness and inferior taste. This reduction in sugar content affects not only fruit quality but also the overall nutritional value of the produce.
- **Energy Deficits for Growth and Reproductive Development**: Carbohydrates serve as the primary source of energy for cellular processes such as cell division, cell expansion, and the synthesis of essential biomolecules. Reduced carbohydrate production leads to energy deficits that slow down growth, resulting in smaller leaves, shorter stems, and reduced root expansion. In fruit crops, these energy shortages can impact flower development and fruit set, resulting in lower yields.
- **Impact on Fruit Size and Quality**: High-temperature stress, by limiting carbohydrate availability, reduces the amount of energy allocated to fruit development. Fruits require a substantial supply of carbohydrates for cell expansion, which determines fruit size. Reduced carbohydrate supply leads to smaller fruits with less water content and firmer textures. In fruits like berries and peaches, the lack of sufficient carbohydrates can also affect juiciness, making the fruits less appealing to consumers.

4. **Secondary Effects on Plant Metabolism and Health**

Photosynthetic impairment under heat stress affects more than just carbohydrate production; it has secondary effects that influence the plant's overall metabolic health and ability to withstand stress.

- **Reduction in Defensive Compounds**: Carbohydrates are also precursors for secondary metabolites, such as phenolic compounds and flavonoids, which help plants resist stress and defend against pathogens. With limited carbohydrate production, the synthesis of these defensive compounds declines, leaving the plant more vulnerable to secondary stress factors like disease or pest attacks. In fruit crops, reduced production of phenolic compounds affects flavor complexity and antioxidant content, reducing the nutritional and health value of the fruits.
- **Hormonal Imbalances**: Photosynthesis and carbohydrate levels influence the balance of plant hormones, which regulate growth and stress responses. For example, a reduction in photosynthetic activity can increase the levels of abscisic acid (ABA), a stress hormone that promotes stomatal closure and inhibits growth. Elevated ABA levels, combined with reduced carbohydrate availability, slow down vegetative growth and delay fruit development, which can impact yield and crop cycles in commercial fruit production.
- **Reduced Tolerance to Future Stress**: Carbohydrates serve as an energy reserve that plants can mobilize during periods of environmental stress. Photosynthetic impairment depletes these reserves, leaving the plant less equipped to cope with subsequent stresses, such as drought or nutrient deficiencies. This cumulative effect of reduced resilience

impacts the plant's ability to thrive and maintain productivity in fluctuating conditions.

5. **Adaptive Mechanisms to Mitigate Photosynthetic Impairment**

Plants employ various strategies to adapt to high temperatures and protect photosynthetic efficiency. These adaptations are critical for maintaining productivity in hot climates, where high temperatures are frequent.

- **Heat Shock Proteins (HSPs)**: Plants produce heat shock proteins (HSPs) in response to high temperatures. These proteins act as molecular chaperones, helping to stabilize photosynthetic enzymes and proteins in the thylakoid membrane. HSPs assist in preventing the denaturation of photosynthetic components, preserving function under moderate heat stress.
- **Adjustment of Chlorophyll and Carotenoid Content**: Under heat stress, plants may adjust the levels of chlorophyll and carotenoids, pigments that play essential roles in light absorption and ROS quenching. Carotenoids, such as xanthophylls, protect chlorophyll from photooxidation by dissipating excess light energy as heat, preventing photoinhibition. By modifying pigment composition, plants can better manage light capture and reduce ROS formation, enhancing photosynthetic resilience.
- **Increased Production of Antioxidants**: Antioxidants, including ascorbate, glutathione, and specific antioxidant enzymes, help detoxify ROS produced during photosynthesis. These antioxidants protect chloroplast structures and enzymes from oxidative damage, mitigating the effects of photoinhibition and oxidative stress under heat conditions.
- **Modification of Leaf Orientation and Structure**: Some plants reduce heat absorption by modifying leaf orientation or structure. Leaf rolling or positioning leaves vertically can minimize direct exposure to sunlight, reducing the thermal load on chloroplasts and decreasing the risk of photoinhibition.

5. Protein Aggregation and Heat Shock Response

Proteins are essential molecules that facilitate nearly every cellular function, from catalyzing metabolic reactions to stabilizing cellular structures. However, proteins are highly sensitive to environmental conditions, particularly temperature. High temperatures cause proteins to lose their native structure, leading to misfolding and aggregation—processes that are toxic to cells. Plants respond to this stress by producing specialized proteins known as heat shock proteins (HSPs), which act as molecular chaperones to protect, refold, and stabilize proteins under heat stress. Although HSPs provide critical protection, their production requires significant cellular resources, and prolonged heat stress can exhaust these reserves, ultimately impacting plant growth, development, and productivity.

1. **High Temperatures and Protein Misfolding**

Proteins are complex structures with specific three-dimensional shapes that are essential for their function. This structure is held together by weak bonds, such as hydrogen bonds and van der Waals interactions, which can be disrupted by temperature fluctuations. High temperatures increase the kinetic energy of molecules, which destabilizes these bonds, causing proteins to lose their functional shape—a process known as **denaturation**.

- **Effects of Denaturation on Enzyme Activity**: Many enzymes, which are proteins that catalyze biochemical reactions, become inactive when denatured. For instance, Rubisco, the enzyme responsible for carbon fixation in photosynthesis, is highly sensitive to temperature and loses efficiency as temperatures rise. As key metabolic enzymes denature, cellular processes slow down or halt, leading to metabolic imbalance and reduced energy production.
- **Protein Aggregation and Cellular Toxicity**: Misfolded proteins tend to aggregate, forming clusters that disrupt cellular structures and functions. Protein aggregates interfere with normal cellular processes, blocking pathways and impeding transport. Aggregates can also clog cellular compartments, such as the endoplasmic reticulum, which further stresses the cell. If aggregates accumulate in chloroplasts, for example, they impair photosynthesis, decreasing the plant's ability to produce energy and grow.

2. **Heat Shock Proteins (HSPs) as Molecular Chaperones**

To counteract the effects of heat-induced protein misfolding and aggregation, plants produce heat shock proteins (HSPs). HSPs are a group of highly conserved proteins that act as **molecular chaperones**. They assist in maintaining protein structure, preventing aggregation, and facilitating the refolding of damaged proteins.

- **HSP Families and Their Functions**: There are several families of HSPs, classified by their molecular weight, each with specific roles:
    - ◊ **HSP70**: This family is one of the most studied and functions in recognizing and binding to misfolded proteins. HSP70 helps prevent aggregation by stabilizing unfolded proteins and guiding them toward refolding pathways.
    - ◊ **HSP90**: These chaperones play a role in stabilizing partially folded proteins, particularly those involved in signal transduction. HSP90 interacts with proteins that regulate growth and stress responses, helping maintain cell function under heat stress.
    - ◊ **Small HSPs (sHSPs)**: Small heat shock proteins aggregate around denatured proteins, forming a protective shield to prevent further damage. sHSPs are particularly abundant in plants and provide a first line of defense under sudden temperature increases.
- **Refolding and Degradation of Damaged Proteins**: HSPs not only stabilize misfolded proteins but also facilitate their proper refolding. If refolding is not possible, HSPs guide damaged proteins to the cell's proteolytic machinery, such as the proteasome, where they are degraded and recycled. This selective degradation prevents the accumulation of dysfunctional proteins, allowing the cell to maintain homeostasis under heat stress.

3. **Mechanisms of HSP Activation Under Heat Stress**

HSP production is rapidly upregulated in response to high temperatures through a process known as the **heat shock response (HSR)**. The HSR is triggered by the activation of heat shock transcription factors (HSFs), which bind to heat shock elements (HSEs) in the promoter regions of HSP genes.

- **Heat Shock Transcription Factors (HSFs)**: Under normal conditions, HSFs are inactive, bound to HSPs that prevent them from triggering HSP gene expression. When heat stress causes proteins to unfold, HSFs are released and become activated. Once activated, HSFs migrate to the nucleus, bind to HSEs, and initiate the transcription of HSP genes. This rapid upregulation ensures that HSPs are available to protect cellular proteins soon after heat stress begins.
- **HSF Regulation and Feedback Mechanisms**: As HSPs refold and stabilize proteins, they also bind to HSFs, forming a feedback loop that reduces HSP expression once the cell has adapted to the stress. This regulatory system prevents unnecessary HSP production, conserving energy and resources when stress levels decrease.

4. Energy Costs and Resource Allocation in HSP Production

While HSPs are essential for managing protein stability under high temperatures, their production and function are energy-intensive, requiring ATP for both the synthesis and activity of HSPs.

- **ATP Dependence in HSP Activity**: Many HSPs, including HSP70 and HSP90, require ATP to bind and release misfolded proteins. This ATP is generated through cellular respiration and photosynthesis, processes that are themselves affected by heat stress. As ATP availability decreases under prolonged stress, HSP activity becomes limited, leading to a buildup of damaged proteins.
- **Resource Trade-offs and Growth Limitations**: The synthesis of HSPs consumes amino acids, ATP, and other cellular resources that would otherwise support growth and reproduction. In conditions of prolonged heat stress, the allocation of resources toward HSP production limits the availability of these resources for other cellular functions, such as cell division and nutrient assimilation. This trade-off often results in stunted growth and reduced productivity, as plants prioritize survival over growth.

5. Limitations of Heat Shock Response Under Prolonged or Severe Heat Stress

While HSPs provide critical support for protein stability under moderate heat stress, prolonged or extreme high temperatures can overwhelm this system, leading to cellular damage and reduced resilience.

- **Accumulation of Unstable Proteins and Aggregates**: Under extended heat stress, the cellular machinery becomes less efficient at refolding proteins and preventing aggregation. Over time, misfolded proteins accumulate faster than HSPs can manage, leading to a buildup of protein aggregates that disrupt cellular processes.
- **Exhaustion of Energy Reserves**: Continuous production of HSPs depletes cellular ATP reserves, reducing the cell's ability to maintain other vital functions. As ATP levels decline, HSPs themselves lose effectiveness, exacerbating the accumulation of misfolded proteins and aggregates. This energy depletion impacts all aspects of cellular metabolism, from nutrient uptake to growth regulation.
- **Diminished Growth and Productivity**: The energy and resource drain associated with sustained HSP production leads to diminished growth, delayed flowering, and reduced fruit set. In fruit crops, this translates to smaller, lower-quality fruits, as well as decreased

yields. Prolonged stress can even lead to premature senescence, where plants initiate aging processes to conserve energy, further reducing crop productivity.

6. **Adaptive Mechanisms to Enhance Heat Tolerance and Minimize Protein Aggregation**

Some plants have evolved additional mechanisms to improve protein stability under high temperatures, enhancing their resilience in hot climates.

- **Genetic Modifications and Heat-Tolerant Varieties**: Breeding programs and genetic engineering have focused on developing crops with enhanced HSP production or modified protein structures that are more heat-resistant. By selecting for naturally heat-tolerant varieties or introducing heat-resistant genes, researchers aim to improve crop resilience under high temperatures.
- **Modification of Protein Structure**: Plants may alter amino acid composition to improve protein stability under heat stress. Proteins with higher proportions of hydrophobic amino acids, for example, are more resistant to heat-induced denaturation. By increasing the stability of essential enzymes, such as those involved in photosynthesis and respiration, plants can better maintain metabolic functions under high temperatures.
- **Enhanced Antioxidant Systems**: High-temperature stress increases the production of reactive oxygen species (ROS), which exacerbate protein aggregation. Plants with robust antioxidant systems, such as high levels of superoxide dismutase (SOD) and catalase, are better equipped to mitigate ROS-related protein damage, reducing the burden on HSPs and enhancing cellular resilience.

6. Hormonal Imbalance

Plant hormones, or phytohormones, play a critical role in regulating growth, development, and stress responses. Under optimal conditions, a balance among various hormones enables plants to grow, reproduce, and adapt to minor environmental fluctuations. However, high temperatures disrupt this balance, triggering a cascade of hormonal changes that affect cellular homeostasis. Heat stress causes increased production of stress-related hormones like abscisic acid (ABA) and ethylene, each of which has specific roles in managing stress responses. While these hormonal changes help plants cope with immediate stress, they often come at the cost of growth and productivity, particularly in fruit crops.

1. **Heat-Induced Increase in Abscisic Acid (ABA)**

Abscisic acid, often referred to as the "stress hormone," is crucial in helping plants respond to adverse conditions such as drought and high temperatures. Under heat stress, ABA levels rise significantly, triggering a series of physiological and molecular responses aimed at conserving water and maintaining cellular stability.

- **Stomatal Closure and Water Conservation**: One of ABA's primary roles under stress is to induce stomatal closure. Stomata, small pores on the leaf surface, facilitate gas exchange by allowing CO_2 to enter and oxygen and water vapor to exit. During heat stress, elevated ABA levels signal the guard cells surrounding each stoma to close, thereby reducing transpiration (water loss) and conserving internal water reserves. While stomatal closure helps prevent dehydration, it also limits CO_2 intake, which reduces photosynthesis and energy production. Over time, the restricted photosynthetic capacity can impact carbohydrate synthesis, leading to smaller, lower-quality fruits and stunted growth.

- **Impact on Nutrient Uptake and Transport**: Stomatal closure also disrupts the flow of water and nutrients through the xylem. Under reduced transpiration, nutrient transport slows, leading to deficiencies of essential nutrients such as nitrogen, potassium, and calcium. These deficiencies impair vital processes like protein synthesis and cell division, further limiting growth and fruit quality.
- **Inhibition of Growth-Promoting Hormones**: High levels of ABA under heat stress inhibit the synthesis and activity of growth-promoting hormones such as auxins, gibberellins, and cytokinins. Auxins regulate cell elongation, gibberellins promote stem and leaf growth, and cytokinins encourage cell division and delay senescence. By suppressing these growth hormones, elevated ABA levels shift the plant's resources away from growth and reproduction toward stress survival. In fruit crops, this hormonal shift can lead to reduced fruit size, delayed ripening, and diminished overall yield.

2. Ethylene Production and Its Effects on Plant Growth and Development

Ethylene is a gaseous plant hormone associated with ripening, senescence, and stress responses. While ethylene plays a positive role in normal fruit ripening, its production under heat stress can lead to premature ripening and other adverse effects that reduce fruit quality and crop productivity.

- **Promotion of Leaf Senescence and Premature Aging**: High temperatures stimulate ethylene synthesis, which accelerates leaf senescence (aging). This process involves chlorophyll degradation and the breakdown of cellular structures, causing leaves to yellow and drop prematurely. Senescence reduces the plant's photosynthetic capacity, limiting energy production and carbohydrate availability for fruit development. In fruit crops, premature leaf senescence can lead to smaller fruits and reduced overall yield, as the plant lacks sufficient photosynthetic output to support both vegetative and reproductive growth.
- **Premature Fruit Ripening and Fruit Drop**: Ethylene is central to the ripening process in climacteric fruits (those that continue to ripen after harvest), such as tomatoes, bananas, and apples. Under heat stress, ethylene levels can rise excessively, leading to accelerated ripening and early fruit drop. Premature ripening often results in smaller, underripe fruits with lower sugar content, affecting flavor, texture, and marketability. Excess ethylene can also weaken the attachment of fruits to the plant, increasing the likelihood of fruit drop before maturity, further impacting yield.
- **Reduced Post-Harvest Quality and Shelf Life**: High ethylene levels during heat stress contribute to rapid fruit softening, which reduces the firmness and texture of fruits. This softening makes fruits more susceptible to bruising and decay during transport and storage, reducing shelf life and post-harvest quality. The accumulation of ethylene-related compounds in fruits can also impact flavor, leading to less desirable taste profiles. In commercial fruit production, these changes translate to lower-quality fruits with limited storage potential, reducing both consumer appeal and economic value.

3. Cumulative Effects of Hormonal Imbalance on Plant Health and Productivity

The hormonal responses to heat stress are complex and interconnected, often leading to a chain reaction of physiological changes that collectively impact plant growth, reproduction, and yield.

- **Disruption of Cellular Homeostasis**: The increase in ABA and ethylene under high temperatures disrupts cellular homeostasis, or the stable internal environment necessary for optimal cellular function. By inducing stomatal closure, ABA limits the availability of CO_2 for photosynthesis, causing energy shortages that impair growth and development. Ethylene-induced senescence and fruit ripening further accelerate the depletion of cellular resources, forcing the plant into a survival mode that prioritizes immediate stress responses over long-term productivity.
- **Impact on Reproductive Success and Fruit Set**: Hormonal imbalances due to heat stress can reduce flower development, pollination success, and fruit set in fruit crops. High ABA levels inhibit floral hormone synthesis, leading to fewer flowers and poor fruit set. Additionally, heat-induced ethylene production can cause flower drop, reducing the number of potential fruits. In crops like tomatoes, peppers, and grapes, reduced fruit set and increased flower drop under heat stress can significantly lower yields and limit profitability.
- **Altered Flavor Profiles and Nutritional Quality**: Hormonal shifts under heat stress can affect the synthesis of secondary metabolites, such as flavonoids, carotenoids, and other antioxidant compounds. Ethylene stimulates the production of carotenoids, which influence fruit color and antioxidant content, but excess ethylene can alter flavor and nutritional profiles, often resulting in fruits with less complex flavors and reduced health benefits. In crops like strawberries and blueberries, high temperatures can lead to a loss of key antioxidants, impacting both taste and nutritional quality.

4. Adaptive Mechanisms and Breeding for Hormonal Resilience

Some plants exhibit adaptive responses that help them mitigate the effects of hormonal imbalance under heat stress, enhancing their ability to maintain productivity.

- **Hormone Modulation and Cross-Talk**: Plants regulate hormone levels through a network of signaling pathways and feedback loops. For example, plants may increase cytokinin levels to counteract ABA-induced growth inhibition, promoting cell division and delaying senescence. Hormone cross-talk, or interaction between hormones, enables plants to balance growth and stress responses, maintaining growth when conditions improve.
- **Breeding and Genetic Modification for Heat-Resilient Hormone Regulation**: Plant breeders have developed heat-tolerant crop varieties with modified hormonal responses to high temperatures. For instance, some tomato and grape varieties are bred to produce lower levels of ethylene under stress, reducing the risk of premature ripening and fruit drop. Genetic modification techniques also allow scientists to enhance hormone regulation pathways, reducing sensitivity to ABA and ethylene, which helps plants maintain growth and productivity under heat stress.
- **Use of Plant Growth Regulators (PGRs)**: In commercial agriculture, plant growth

regulators (PGRs) can be applied to modify hormone levels and improve crop resilience under stress. For instance, applying cytokinins to counteract ABA's inhibitory effects can promote cell division and delay senescence, enhancing growth even under high temperatures. Similarly, anti-ethylene compounds like silver thiosulfate (STS) can reduce ethylene effects, delaying ripening and extending shelf life in fruit crops.

Table 2: Effect of High Temperature on Various Growth Stages of Plant

Growth Stage	Impact of High Temperature
Vegetative growth	High temperature exposure (43oC) in Almond bud failure like symptoms appear in which there is complete separation of distorted and compressed cells from surrounding tissues takes place by a periderm like layer Hellali and Kester, 1979).
Flowering	Abnormal pattern of bud break and development in temperate fruit trees. In apricot, the warm temperature (15.9oC) 3-5oC higher than the normal results in underdeveloped pistils. Rodrigo and Herrero (2002).
Fruit set and Yield	» The higher temperature during fruit let stage also affect the fruit retention. » On trifoliate orange rootstocks, when exposed to 30-35°C for 48h in a controlled environment increased the fruit drop.

Table 3: Impact of High Temperature on Fruit Crops

S.No	Name of Crop	Impact of High Temperature
1.	Apple	Sunburn, bitter pit
2.	Apricot	Reduction in the no. of buds, flowers and fruits
3.	Avacado	Reduced fruit set
4.	Cherry	Accelerated pollen tube growth
5.	Citrus	Poor colour development
6.	Mango	Spongy tissues and fruit cracking
7.	Pineapple	Sun Scald

Low-Temperature Stress: Effects on Cellular Processes

Low temperatures can cause cold stress (non-freezing) or freezing stress (below 0°C), each affecting plants in specific ways. While cold stress slows down metabolic processes, freezing temperatures create physical damage within cells, severely impacting plant structure and function.

1.Reduced Enzyme Activity and Slowed Metabolism

Enzymes are proteins that catalyze biochemical reactions, facilitating nearly every cellular process required for growth, energy production, and development. Each enzyme has an optimal temperature range within which it functions most efficiently. When temperatures drop below this range, enzyme activity declines, slowing down critical metabolic processes. Low temperatures affect various aspects of plant metabolism, from photosynthesis to respiration, leading to an overall reduction in energy production and growth. This metabolic slowdown is particularly challenging for fruit crops, which require substantial energy for cell expansion, fruit set, and ripening.

1. Temperature Sensitivity of Enzymes and Metabolic Reactions

Enzymes rely on specific three-dimensional structures and dynamic interactions to perform their functions. Low temperatures reduce the kinetic energy of molecules, causing enzymes to become less flexible and slowing down the rates of enzyme-substrate interactions.

- **Reduced Reaction Rates**: Enzyme-mediated reactions slow significantly at low temperatures because enzyme-substrate binding becomes less efficient. Enzymes that catalyze metabolic pathways in photosynthesis, respiration, and nutrient assimilation require optimal temperatures to maintain adequate reaction rates. When temperatures drop, reaction rates decrease, leading to slower metabolic processes throughout the plant.
- **Impact on Temperature-Sensitive Enzymes**: Certain enzymes are more sensitive to low temperatures than others. For instance, enzymes involved in cell wall modification, such as those needed for cell expansion and growth, are particularly susceptible to cold. This sensitivity limits cell growth and elongation, leading to smaller plant structures and delayed development.

2. Decline in Photosynthetic Rates Due to Reduced Enzyme Activity

Photosynthesis is one of the most temperature-sensitive processes in plants. Low temperatures slow down the activity of photosynthetic enzymes, particularly those involved in the Calvin cycle—the cycle that converts CO_2 into carbohydrates. This reduction in enzyme activity directly affects the plant's ability to produce energy and grow.

- **Impact on Rubisco and Carbon Fixation**: Rubisco (Ribulose-1,5-bisphosphate carboxylase/oxygenase) is a critical enzyme in the Calvin cycle responsible for capturing CO_2 and converting it into organic compounds. Rubisco operates optimally within a specific temperature range, typically between 20–30°C for most plants. At low temperatures, Rubisco's activity declines, reducing the rate of carbon fixation and carbohydrate production. This decrease in photosynthetic efficiency limits the availability of sugars that fuel growth, reproduction, and cellular maintenance.
- **Chloroplast Structure and Function**: Low temperatures can also disrupt the structure and function of chloroplasts, where photosynthesis occurs. Chloroplast membranes become more rigid in cold conditions, affecting the electron transport chain and the production of ATP and NADPH—molecules essential for the Calvin cycle. This disruption further impairs carbon fixation and carbohydrate synthesis, compounding the effects of reduced enzyme activity.
- **Photoinhibition Under Low Temperatures**: In addition to slowing down the Calvin cycle, low temperatures can cause **photoinhibition**—a condition where the light-harvesting complexes in chloroplasts become damaged. When enzymes in the Calvin cycle are slowed, plants struggle to process the absorbed light energy efficiently, leading to an accumulation of excess energy that can damage photosystem II (PSII) in the

chloroplasts. Photoinhibition reduces the plant's ability to capture and convert light energy, further limiting carbohydrate production.

3. Impact on Cellular Respiration and ATP Production

Cellular respiration is the process by which plants break down carbohydrates to produce ATP, the primary energy currency of the cell. ATP is required for nearly all cellular functions, from growth and repair to nutrient transport and stress responses. Low temperatures slow down respiration rates, reducing ATP availability and limiting the plant's capacity to sustain growth and metabolic activities.

- **Glycolysis and Mitochondrial Activity**: Cellular respiration begins with glycolysis, a process that breaks down glucose to produce pyruvate, which then enters the mitochondria for further processing in the citric acid cycle (Krebs cycle). Low temperatures reduce enzyme activity in both glycolysis and the mitochondrial pathways, slowing down ATP production. Inadequate ATP availability affects cellular maintenance, as energy-intensive processes like protein synthesis, nutrient uptake, and cell division are hindered.
- **Effects on Growth and Development**: ATP is essential for cell division, growth, and other biosynthetic processes. Reduced ATP production due to slowed respiration limits these processes, resulting in stunted growth, delayed flowering, and poor fruit set in crops. For instance, in crops like tomatoes and peppers, low temperatures during the flowering stage can reduce pollination success, leading to fewer fruits and lower yields.
- **Alternative Respiratory Pathways**: In response to low temperatures, some plants activate alternative respiratory pathways to produce ATP, though these pathways are less efficient. For instance, plants may rely on the alternative oxidase (AOX) pathway in mitochondria, which generates ATP with reduced efficiency. This pathway helps maintain a baseline level of ATP under cold conditions but does not fully meet the energy demands for growth, making it a temporary solution rather than a sustainable one.

4. Cascading Effects on Growth, Reproduction, and Productivity

The combined effects of reduced enzyme activity, decreased photosynthesis, and slowed respiration have far-reaching implications for plant growth, reproductive success, and yield, particularly in fruit crops.

- **Delayed Growth and Development**: Low temperatures slow down cell division and expansion, leading to shorter plants with smaller leaves and stems. In fruit crops, delayed vegetative growth impacts overall plant size and strength, which are essential for supporting fruit-bearing structures. Reduced leaf area also limits photosynthesis, creating a feedback loop where slower growth further reduces energy availability.
- **Impact on Flowering and Fruit Set**: Energy shortages due to reduced ATP production affect flowering and fruit set, both of which are energy-intensive processes. In crops like strawberries, apples, and cherries, low temperatures during the reproductive phase can delay flowering, reduce pollination success, and result in smaller, lower-quality fruits. The combined effects of reduced photosynthesis and respiration under cold stress translate into fewer flowers and a decrease in fruit production.
- **Decreased Fruit Size and Quality**: Carbohydrate availability is critical for fruit expansion,

sugar accumulation, and flavor development. Low temperatures limit carbohydrate production, leading to smaller fruits with reduced sugar content and less desirable flavor profiles. In berries, citrus, and stone fruits, this can impact both taste and nutritional quality, reducing marketability and consumer appeal. Additionally, reduced cellular respiration affects the production of secondary metabolites, such as flavonoids and anthocyanins, which influence color and antioxidant content in fruits.

5. Adaptive Mechanisms to Enhance Cold Tolerance and Minimize Metabolic Slowdown

To cope with low temperatures and maintain metabolic function, some plants have evolved adaptations that help sustain enzyme activity and energy production under cold conditions.

- **Production of Cold-Responsive Proteins**: Cold-responsive proteins, including enzymes that function more efficiently at low temperatures, help maintain metabolic activity. For example, cold-tolerant plants produce versions of Rubisco and other Calvin cycle enzymes that have higher activity at low temperatures, allowing them to sustain some level of photosynthesis even under cold stress.
- **Cryoprotectant Accumulation**: Plants synthesize cryoprotectants, such as proline, glycine betaine, and soluble sugars, which help stabilize proteins and cellular membranes. These compounds reduce the risk of enzyme denaturation and maintain cellular structure, enabling essential metabolic reactions to continue. Cryoprotectants also act as osmolytes, helping to retain cellular water and prevent dehydration, which can occur under cold conditions due to osmotic imbalance.
- **Membrane Modification**: Cold-tolerant plants alter the lipid composition of their membranes to maintain fluidity at low temperatures. By increasing the proportion of unsaturated fatty acids in the lipid bilayer, plants prevent membranes from becoming too rigid, which helps enzymes and transport proteins embedded in the membrane continue functioning. This adaptation is particularly important for maintaining photosynthetic efficiency and nutrient uptake under cold stress.

2. Membrane Rigidity and Loss of Fluidity

Cellular membranes play a critical role in maintaining the integrity and functionality of plant cells. Composed primarily of lipid bilayers, these membranes serve as barriers that regulate the movement of water, ions, nutrients, and signaling molecules into and out of cells. The fluidity of these membranes is vital for their function, as it allows embedded proteins, such as transporters and receptors, to move and interact efficiently. However, cold temperatures reduce membrane fluidity, causing membranes to become rigid and less permeable. This change in membrane dynamics can significantly impair plant growth, nutrient uptake, and stress responses, especially in fruit crops that are sensitive to temperature fluctuations.

1. Cold-Induced Membrane Rigidity and Its Effects on Fluidity

Membranes are composed of phospholipids with hydrophilic (water-attracting) heads and hydrophobic (water-repelling) tails. These phospholipids are arranged in a bilayer, with the hydrophobic tails facing inward and the hydrophilic heads facing the aqueous environment outside and inside the cell. The fluidity of this bilayer is influenced by temperature, with colder conditions causing a reduction in the kinetic energy of lipid molecules.

- **Phase Transition and Reduced Fluidity**: As temperatures drop, the lipid bilayer undergoes a phase transition from a fluid, flexible state to a more solid, gel-like state. This phase transition reduces membrane fluidity, making the lipid molecules less dynamic and more tightly packed. The rigidity that results from this transition hinders the lateral movement of proteins and lipids within the membrane, disrupting the dynamic interactions required for cellular processes like nutrient transport and signaling.
- **Effect on Membrane Functionality**: Reduced fluidity affects the function of integral membrane proteins, such as channels, transporters, and receptors, which rely on the flexibility of the lipid bilayer for their function. For example, nutrient transporters that facilitate the movement of ions like potassium and calcium across the membrane become less efficient, limiting the plant's ability to absorb essential nutrients from the soil. The loss of membrane fluidity also impacts processes like exocytosis and endocytosis, where vesicles fuse with the plasma membrane to transport materials in and out of the cell.

2. Formation of Solid Patches and Phase Separation in Membrane Lipids

Cold stress can cause **phase separation** in the lipid bilayer, leading to the formation of solid patches or domains within the otherwise fluid membrane. These domains consist of tightly packed lipids, which create areas of rigidity that disrupt the uniformity of the membrane.

- **Phase Separation and Lipid Packing**: Phase separation occurs when certain lipid molecules pack more tightly than others under cold conditions, forming solidified regions within the membrane. These regions are often rich in saturated fatty acids, which have straight chains that pack closely together, increasing rigidity. In contrast, unsaturated fatty acids, which contain one or more double bonds, have kinked structures that prevent tight packing and promote fluidity. As temperatures drop, the proportion of solid-like regions increases, leading to an overall decrease in membrane flexibility.
- **Membrane Cracking and Leakage**: The formation of rigid, solid patches within the membrane can cause **cracking**, where small fractures develop as the membrane becomes brittle. These cracks compromise the membrane's integrity, leading to the leakage of ions, water, and small metabolites. The loss of essential ions like potassium (K^+) and calcium (Ca^{2+}) can disrupt osmotic balance, leading to cellular dehydration and reduced turgor pressure. For instance, potassium loss affects enzyme activation and cell turgor, while calcium is crucial for maintaining cell wall stability and signaling pathways. The leakage of metabolites, such as amino acids and sugars, further depletes the energy reserves needed for stress responses and recovery.
- **Consequences for Cellular Compartmentalization**: The plasma membrane is not the only membrane affected by cold stress; internal membranes, such as those surrounding organelles like mitochondria, chloroplasts, and the endoplasmic reticulum, also experience reduced fluidity. This affects the compartmentalization of cellular processes, leading to impaired function of organelles involved in energy production (mitochondria),

photosynthesis (chloroplasts), and protein folding (endoplasmic reticulum). The disruption of organelle function contributes to the overall decline in metabolic efficiency under cold conditions.

3. **Impact of Reduced Membrane Fluidity on Receptor Function and Cell Signaling**

Membranes house various receptors that detect environmental signals and initiate responses to stress conditions, such as changes in temperature, light, and water availability. The functionality of these receptors depends on their ability to move within the membrane and interact with other signaling molecules. Cold-induced membrane rigidity interferes with these processes, impairing the plant's ability to perceive and respond to stress.

- **Inhibition of Receptor Mobility**: In a more rigid membrane, receptors embedded in the lipid bilayer become less mobile, reducing their ability to cluster and form signaling complexes. This limited mobility affects the ability of receptors to interact with ligands (signaling molecules), leading to delayed or weakened signal transduction. For example, receptors involved in detecting temperature changes or osmotic stress may fail to activate downstream signaling pathways effectively, delaying the plant's adaptive responses to cold.
- **Disruption of Stress Signaling Pathways**: Cold temperatures also affect the function of **calcium channels**, which play a pivotal role in signaling during stress responses. Calcium ions act as secondary messengers in many signaling pathways, and their controlled release into the cytosol is crucial for activating stress-response genes. Membrane rigidity reduces the function of calcium channels, limiting the influx of calcium ions and thereby impairing the signaling cascade required for activating cold-responsive genes. This disruption hinders the plant's ability to synthesize protective proteins, such as cold-responsive (COR) proteins, that help stabilize cellular structures and maintain metabolic function under stress.
- **Impaired Hormonal Signaling**: Membrane rigidity under cold stress can also impact the perception and transport of plant hormones, such as abscisic acid (ABA) and ethylene, which regulate responses to cold and other environmental stresses. ABA receptors in the plasma membrane become less efficient at sensing stress-induced ABA levels, delaying stomatal closure and water conservation mechanisms. This impairment can leave plants vulnerable to additional stresses, such as drought or freezing, further reducing their resilience and productivity.

4. **Adaptive Mechanisms to Restore Membrane Fluidity**

Plants have developed several strategies to counteract the effects of cold-induced membrane rigidity, allowing them to maintain membrane function and fluidity during low-temperature exposure.

- **Modification of Lipid Composition**: To maintain membrane fluidity under cold conditions, plants alter the composition of membrane lipids by increasing the proportion of unsaturated fatty acids. Unsaturated fatty acids contain double bonds that create kinks in their structure, preventing tight packing of lipids and maintaining a more fluid membrane state. This lipid modification helps maintain the flexibility needed for normal membrane function, allowing transport proteins and receptors to continue operating.

- **Synthesis of Cryoprotectants**: Plants produce cryoprotectants, such as proline, glycine betaine, and sugars, which help stabilize membranes and prevent phase separation. These compounds accumulate in the cytoplasm and interact with membrane lipids, reducing the formation of solid patches and minimizing the risk of cracking. Cryoprotectants also help retain cellular water, reducing the risk of dehydration caused by ion leakage.
- **Cold Acclimation Processes**: Cold acclimation is a process by which plants gradually adapt to lower temperatures by altering their physiology and biochemistry. During acclimation, plants adjust their membrane composition, increase the synthesis of protective proteins, and upregulate genes involved in cold tolerance. This allows them to survive sudden drops in temperature by maintaining cellular function and avoiding the damage typically caused by membrane rigidity.

5. **Implications for Fruit Crop Quality and Yield**

Membrane rigidity under cold stress has significant implications for the health and productivity of fruit crops. Reduced membrane fluidity affects not only growth and nutrient uptake but also the quality and storage potential of fruits.

- **Delayed Growth and Reduced Fruit Set**: Cold stress slows nutrient uptake due to reduced membrane permeability, leading to nutrient deficiencies that impair vegetative growth and flowering. In fruit crops like strawberries, citrus, and grapes, delayed growth and reduced flower development translate to fewer fruits, affecting yield.
- **Impact on Fruit Texture and Storage**: Membrane rigidity can affect fruit texture by influencing water content and cellular integrity. Fruits that develop under cold stress may have a firmer texture due to reduced cell expansion but are also more prone to physiological disorders such as chilling injury during storage. Chilling injury occurs when fruits are exposed to temperatures that cause membrane damage, leading to pitting, browning, and water-soaked areas, which reduce marketability.
- **Flavor and Nutrient Composition**: Membrane function is critical for the synthesis of secondary metabolites, including sugars, acids, and phenolic compounds that contribute to fruit flavor and nutritional value. Cold stress affects the transport of precursors needed for these compounds, leading to changes in flavor profiles and antioxidant content. For example, reduced sugar transport can result in less sweet fruits, while altered phenolic synthesis can change the bitterness or astringency of certain fruits, impacting consumer acceptance.

3. Ice Crystal Formation and Cellular Dehydration (Freezing Stress)

Freezing temperatures pose a unique challenge to plants, as the formation of ice crystals in and around cells can cause irreversible damage. When temperatures fall below 0°C, ice crystals begin to form in extracellular spaces (the spaces outside of cells), initiating a cascade of physical and osmotic stresses that lead to cellular dehydration and structural damage. The physical impact of freezing can puncture cell membranes and rupture cells, often resulting in cell death. However, plants have evolved various mechanisms to protect themselves from freezing damage, including the production of antifreeze proteins and solutes that help prevent ice formation and preserve cellular integrity.

1. **Extracellular Ice Crystal Formation and Cellular Dehydration**

When temperatures drop below freezing, water in the extracellular spaces solidifies first due to the higher freezing point of extracellular fluids compared to intracellular fluids. This process begins with the formation of ice crystals in the cell wall spaces and intercellular spaces, creating an osmotic gradient that affects cellular water dynamics.

- **Osmotic Gradient and Water Movement**: The formation of ice crystals outside cells reduces the water potential in the extracellular space, creating an osmotic gradient that draws water out of the cell to balance the concentration. As water moves out of the cell, intracellular fluid becomes increasingly concentrated, leading to cellular dehydration. The loss of cellular water causes the cell to shrink, affecting the cytoplasm and organelles, which disrupts cellular function and metabolic processes. Severe dehydration leads to a loss of cell turgor (internal pressure), further compromising the cell's structure and ability to recover from stress.
- **Impact on Cellular Functions**: As cells lose water, enzyme activity slows down due to the reduced availability of water molecules that facilitate enzyme-substrate interactions. Dehydrated cells experience metabolic slowdowns, reduced energy production, and limited ability to repair or produce proteins essential for stress responses. In fruit crops, cellular dehydration during freezing stress can lead to delayed growth, lower fruit set, and increased susceptibility to other stressors like nutrient deficiencies or pathogens.

2. **Physical Damage from Ice Crystal Formation**

Ice crystals can have a destructive physical impact on cellular structures, particularly cell membranes, which are crucial for maintaining the selective permeability and integrity of cells.

- **Puncturing and Rupturing of Cell Membranes**: Ice crystals that form in extracellular spaces can physically press against cell membranes. As ice expands, the crystals can puncture or rupture the membrane, causing irreversible damage. When the membrane integrity is compromised, ions, metabolites, and other essential cellular contents leak out, further depleting the cell's resources and compromising cellular function. Once the membrane is punctured, the cell is unable to maintain osmotic balance or control the influx and efflux of molecules, leading to cell death.
- **Intracellular Ice Formation and Its Consequences**: In extreme cases, ice crystals may also form within the cell itself if the freezing rate is rapid. Intracellular ice formation is especially damaging because it physically disrupts the organelles and cytoskeleton within the cell. The presence of ice inside the cell can fragment organelles like chloroplasts and mitochondria, which are essential for photosynthesis and respiration. This damage is generally irreversible, leading to widespread cell death and tissue necrosis. In fruit crops, intracellular ice formation can affect the quality of fruits by causing texture changes, discoloration, and tissue breakdown.
- **Mechanical Stress on Cell Walls and Tissues**: The expansion of ice in extracellular spaces also puts mechanical pressure on cell walls and tissues. In some cases, this pressure causes tissue cracking, creating additional points of entry for pathogens and increasing

susceptibility to infections. For instance, cold-induced tissue damage can lead to the development of water-soaked lesions on fruits and leaves, which attract microbial pathogens and further compromise the health of the plant.

3. **Adaptive Mechanisms: Antifreeze Proteins and Cryoprotective Solutes**

To survive freezing conditions, many plants have evolved adaptive mechanisms that prevent or mitigate the formation of damaging ice crystals. These adaptations include the production of antifreeze proteins and cryoprotective solutes, which help maintain cellular integrity by controlling ice formation and reducing cellular dehydration.

» **Antifreeze Proteins (AFPs)**: Antifreeze proteins are specialized proteins that bind to ice crystals and inhibit their growth, preventing the formation of large, damaging ice structures. These proteins work by adhering to the surface of small ice crystals, effectively lowering the freezing point of water and limiting further ice formation.

- **Mechanism of Action**: AFPs attach to the tips of growing ice crystals, creating a barrier that blocks additional water molecules from joining the ice structure. By controlling the size and shape of ice crystals, AFPs reduce the likelihood of physical damage to cell membranes and other cellular structures.
- **Role in Cold-Adapted Plants**: AFPs are commonly found in plants that endure extreme cold, such as winter rye and some fruit species like strawberries and apples. The presence of AFPs enables these plants to tolerate freezing conditions and recover more effectively when temperatures rise, protecting buds, leaves, and fruits from cold-induced damage.

» **Cryoprotective Solutes**: Cryoprotective solutes are small molecules, such as sugars, amino acids, and organic acids, that accumulate in plant cells in response to cold stress. These solutes help stabilize cellular structures and prevent ice formation within cells.

- **Role of Sugars and Soluble Carbohydrates**: Sugars, such as sucrose and trehalose, are effective cryoprotectants that protect cellular structures by stabilizing cell membranes and proteins. Sugars lower the freezing point of cellular fluids, reducing the likelihood of intracellular ice formation. They also help retain water in the cytoplasm, preventing severe dehydration and maintaining cell turgor during freezing events.
- **Accumulation of Compatible Osmolytes**: Osmolytes, such as proline and glycine betaine, accumulate in response to cold stress and act as osmoprotectants. These compounds help retain water within the cell by balancing the osmotic pressure caused by extracellular ice formation. By preserving hydration levels, osmolytes prevent cellular collapse and ensure that enzymes and cellular components remain functional during freezing conditions.

» **Role of Cold Acclimation and Gene Expression**: Cold acclimation, a process where plants gradually adapt to lower temperatures, is key to increasing tolerance to freezing stress. During acclimation, plants upregulate genes involved in producing AFPs, cryoprotectants, and other protective proteins. This gene expression leads to biochemical and structural changes that enhance the plant's ability to withstand freezing temperatures. Cold acclimation is particularly beneficial for fruit crops that experience seasonal temperature fluctuations, as it helps protect developing buds, flowers, and fruits from unexpected frosts.

4. Implications for Fruit Crop Quality and Yield

The effects of freezing stress on cellular integrity and plant health have important consequences for fruit crops, particularly those grown in temperate climates where frost and freezing temperatures are common.

- **Delayed Growth and Flowering**: Freezing stress can damage leaf and stem tissues, reducing the plant's photosynthetic capacity and delaying vegetative growth. Damaged leaves and stems may need to be repaired or replaced, diverting resources away from flower and fruit development. In fruit crops like apples, pears, and cherries, delayed growth can lead to a shortened growing season, reducing the time available for fruit ripening and harvest.
- **Impact on Fruit Quality and Storage**: Fruits that experience freezing stress often develop structural and biochemical changes that affect quality. For example, freeze-damaged fruits may exhibit tissue browning, water-soaked lesions, and changes in texture due to cellular rupture. In addition, freezing stress can disrupt the synthesis of sugars, acids, and aroma compounds, leading to fruits with reduced sweetness, altered flavor, and lower nutritional value. These changes impact both fresh consumption and processing quality, as freeze-damaged fruits are more susceptible to bruising, decay, and reduced shelf life.
- **Yield Losses and Economic Impact**: Severe freezing stress can lead to flower and bud mortality, especially in early-blooming fruit crops such as peaches and plums. The loss of flowers or immature fruits due to freezing temperatures significantly reduces potential yield, impacting farmers' income and market availability. For instance, a sudden frost during the blooming period can result in widespread flower loss, leading to a substantial reduction in fruit set and yield.

4. Increased Production of Reactive Oxygen Species (ROS)

Reactive oxygen species (ROS) are chemically reactive molecules derived from oxygen, including superoxide radicals ($O_2^{\bullet-}$), hydrogen peroxide (H_2O_2), and hydroxyl radicals ($^{\bullet}OH$). ROS play dual roles in plant cells: at moderate levels, they function as signaling molecules that help regulate growth and stress responses, but at excessive levels, they cause oxidative damage to cellular structures. Low temperatures disrupt the balance of ROS production and scavenging, leading to an accumulation of ROS that exacerbates the damage caused by cold stress. Cold stress limits the activity of the electron transport chain and reduces the production of antioxidants, increasing ROS levels and their impact on plant cells.

1. Cold-Induced Disruption of Electron Transport Chains and ROS Accumulation

Under optimal conditions, ROS are generated as byproducts of metabolic processes like photosynthesis and respiration, where they are quickly neutralized by antioxidant systems. However, low temperatures interfere with the efficiency of these processes, leading to an imbalance that increases ROS production.

- **Effect on Photosynthetic Electron Transport Chain**: In chloroplasts, the photosynthetic electron transport chain (ETC) is a primary source of ROS. Low temperatures reduce the efficiency of this chain by slowing down electron transfer between the photosystems (Photosystem I and Photosystem II). As electrons become "stalled" within the chain, they are more likely to react with oxygen, forming superoxide radicals. This process is particularly problematic in the chloroplasts of leaves exposed to light, where excess energy that cannot be utilized effectively increases the production of ROS.

- **Impact on Mitochondrial Respiration**: The mitochondrial ETC, responsible for cellular respiration, is another source of ROS under cold stress. Low temperatures slow down respiration, leading to a backlog of electrons in the mitochondria that react with oxygen to form ROS. This increase in mitochondrial ROS production can damage mitochondrial membranes and proteins, affecting ATP production. Energy shortages due to impaired respiration further stress the plant, as the energy required for recovery and repair is limited.

2. Oxidative Damage to Cellular Components

Excessive ROS can damage a wide range of cellular components, including lipids, proteins, and nucleic acids. This oxidative damage worsens the physical damage caused by cold stress, such as membrane rigidity and cellular dehydration, making it even more challenging for the plant to recover.

- **Lipid Peroxidation and Membrane Damage**: ROS readily oxidize the unsaturated fatty acids in cell membranes, a process known as lipid peroxidation. This oxidation leads to the breakdown of the lipid bilayer, increasing membrane permeability and causing leakage of ions and metabolites. Lipid peroxidation produces toxic byproducts, such as malondialdehyde (MDA), which further destabilize the membrane and disrupt cellular homeostasis. In fruit crops, this membrane damage can lead to tissue browning, water-soaked lesions, and reduced firmness, affecting both quality and storage life.
- **Protein Oxidation and Enzyme Inactivation**: ROS can modify proteins by oxidizing amino acid residues, leading to structural changes that inhibit enzyme activity and protein function. For example, oxidative damage to enzymes involved in photosynthesis, such as Rubisco, reduces the efficiency of carbon fixation, limiting carbohydrate production and energy availability. Similarly, oxidative damage to respiratory enzymes impairs ATP synthesis, leading to energy deficits that slow growth, repair, and other metabolic processes essential for stress recovery.
- **DNA Damage and Genomic Instability**: ROS can also react with nucleic acids, causing DNA strand breaks, mutations, and cross-linking. DNA damage affects replication and transcription, leading to errors in gene expression and reduced cell viability. Accumulated DNA damage can lead to genomic instability, affecting cell division and tissue regeneration. In fruit crops, DNA damage can impact growth rates, reduce fruit set, and alter developmental patterns, ultimately reducing yield and quality.

3. Antioxidant Defense Mechanisms in Response to Cold Stress

Plants have evolved complex antioxidant defense systems to mitigate ROS damage under stressful conditions. These systems include both enzymatic and non-enzymatic antioxidants that work together to neutralize ROS, maintain redox balance, and protect cellular structures from oxidative stress.

- **Enzymatic Antioxidants**: Key antioxidant enzymes include superoxide dismutase (SOD), catalase (CAT), and various peroxidases, each of which plays a specific role in detoxifying ROS.
  - ◊ **Superoxide Dismutase (SOD)**: SOD catalyzes the conversion of superoxide radicals (O_2•–) into hydrogen peroxide (H_2O_2), reducing the concentration of superoxide radicals in chloroplasts and mitochondria. This reaction is essential for controlling ROS levels in areas with high metabolic activity, such as the chloroplasts during photosynthesis.
  - ◊ **Catalase (CAT)**: Catalase breaks down hydrogen peroxide into water and oxygen, preventing H_2O_2 from accumulating to toxic levels. This enzyme is particularly active in peroxisomes, where it neutralizes hydrogen peroxide produced during photorespiration, a process that becomes more pronounced under cold stress.
  - ◊ **Peroxidases (POD)**: Peroxidases, such as ascorbate peroxidase (APX), reduce hydrogen peroxide by using electron donors like ascorbate. These enzymes are crucial for detoxifying ROS in chloroplasts, mitochondria, and cytosol, helping protect membranes and proteins from oxidative damage.
- **Non-Enzymatic Antioxidants**: Non-enzymatic antioxidants, including ascorbate (vitamin C), glutathione, carotenoids, and flavonoids, provide additional ROS-scavenging capacity. These molecules directly neutralize ROS and support enzymatic antioxidants in maintaining redox balance under cold stress.
- **Ascorbate and Glutathione**: These antioxidants are involved in the ascorbate-glutathione cycle, a key pathway for detoxifying hydrogen peroxide. This cycle operates in multiple cellular compartments, providing continuous ROS-scavenging support and helping maintain cellular health under cold stress.
- **Carotenoids**: Carotenoids, such as xanthophylls, protect chlorophyll in chloroplasts by quenching singlet oxygen and other ROS. They help dissipate excess light energy as heat, preventing ROS formation and reducing the oxidative load on chloroplasts.
- **Flavonoids**: Flavonoids are secondary metabolites with potent antioxidant properties. They scavenge ROS, particularly in vacuoles and cytoplasm, and play a role in regulating stress responses, enhancing the plant's overall resilience to cold-induced oxidative stress.

4. **Limitations of Antioxidant Defenses Under Extreme or Prolonged Cold Stress**

Although antioxidant defenses are effective in neutralizing ROS under moderate cold stress, prolonged exposure to extreme low temperatures can overwhelm these systems, leading to oxidative stress and cellular damage.

- **Depletion of Antioxidant Reserves**: Antioxidant molecules, especially non-enzymatic antioxidants like ascorbate and glutathione, are consumed in the process of neutralizing ROS. Under prolonged cold stress, these reserves may be depleted faster than they can be replenished, reducing the plant's capacity to scavenge ROS and increasing susceptibility to oxidative damage.

- **Reduced Enzyme Activity in Cold Conditions**: Low temperatures not only increase ROS production but also slow down enzyme activity, including that of antioxidant enzymes. This reduction in enzyme efficiency limits the plant's ability to detoxify ROS, allowing oxidative damage to accumulate in cellular structures. As antioxidant enzyme activity declines, ROS accumulate to toxic levels, leading to cell death and tissue necrosis.
- **Accumulation of Oxidative Damage**: As ROS levels exceed the cell's scavenging capacity, oxidative damage accumulates, compromising membrane integrity, protein function, and DNA stability. This accumulation of damage impairs growth, reduces cellular viability, and weakens the plant's overall health. In fruit crops, such oxidative stress can lead to symptoms like chlorosis (leaf yellowing), browning of tissues, and early senescence, all of which negatively impact crop quality and yield.

5. Implications for Fruit Crop Health and Productivity

The oxidative stress caused by ROS accumulation under low temperatures has significant implications for the growth, quality, and yield of fruit crops, especially those grown in temperate and cold climates.

» **Delayed Growth and Development**: Oxidative damage to cellular components slows down growth, as damaged cells must repair or replace affected structures before resuming normal metabolic activities. In fruit crops, this delay in growth can result in reduced leaf area, which limits photosynthesis and further reduces energy availability for fruit development.

» **Impact on Fruit Quality and Shelf Life**: Oxidative damage to membranes and cellular structures during cold stress affects the quality of fruits, particularly in terms of texture, color, and nutrient content. Freeze-induced tissue browning, water-soaked lesions, and softening are common symptoms of oxidative damage in fruits. Such damage makes fruits more susceptible to bruising and decay, reducing their marketability and shelf life.

» **Yield Losses Due to Premature Senescence**: The cumulative effects of ROS-induced damage can trigger early senescence, or aging, in plant tissues. Premature leaf drop reduces the plant's ability to photosynthesize, leading to energy shortages that limit fruit set and development. In crops like apples, peaches, and berries, early senescence caused by oxidative stress can lead to significant yield losses and reduced crop quality.

5. Cold-Responsive Proteins and Cryoprotectants

Cold stress poses significant challenges to plant cells, impacting their structural integrity, metabolic processes, and overall growth. To cope with low temperatures, plants have developed various adaptive strategies, including the production of cold-regulated (COR) proteins and the accumulation of cryoprotectants. These adaptations help protect cells from freezing damage, stabilize cellular structures, and maintain membrane fluidity. While effective, these protective mechanisms are energy-intensive, often diverting resources from growth and fruit production.

1. Cold-regulated (COR) Proteins and Their Role in Cellular Stability

COR proteins are specialized proteins that plants produce in response to low temperatures. These proteins are crucial for enhancing cold tolerance, as they protect cellular structures from the physical and biochemical stresses associated with cold environments.

- **Stabilizing Membrane Fluidity**: Membrane fluidity is essential for maintaining selective permeability, which allows ions, nutrients, and water to move efficiently across cell membranes. Low temperatures reduce membrane fluidity, causing the lipid bilayer to become rigid and potentially crack, leading to leakage of cellular contents. COR proteins help maintain membrane fluidity by binding to membrane components, reducing the effects of cold-induced rigidity and preserving the membrane's flexibility. This stabilization is critical for cellular homeostasis, as it prevents the excessive loss of ions and metabolites that would otherwise compromise cellular function.
- **Protection Against Protein Denaturation**: At low temperatures, proteins are susceptible to denaturation or misfolding, which can render them inactive. COR proteins help protect other proteins from denaturation by acting as molecular chaperones. By interacting with proteins susceptible to cold damage, COR proteins assist in preserving their functional structure. This support is particularly important for enzymes involved in critical processes like photosynthesis, respiration, and stress signaling, allowing these functions to continue at a baseline level even under cold stress.
- **Gene Expression and Regulation of COR Proteins**: The production of COR proteins is regulated by cold-responsive genes, which are activated through complex signaling pathways in response to temperature drops. These genes are often part of the C-repeat binding factor (CBF) pathway, where transcription factors bind to specific DNA sequences to upregulate COR genes. This regulatory pathway enables plants to respond quickly to sudden cold exposure, producing COR proteins in sufficient quantities to protect cellular structures. The ability to quickly upregulate COR proteins is particularly advantageous for plants in temperate climates where temperatures can fluctuate unpredictably.

2. Cryoprotectants: Compounds That Prevent Freezing Damage

In addition to COR proteins, plants produce a range of small molecules known as cryoprotectants. These compounds, including proline, sucrose, and glycine betaine, accumulate within cells during cold stress, acting as osmoprotectants and stabilizers to protect cellular structures from freezing damage.

- **Role of Proline**: Proline is an amino acid that accumulates in high concentrations during cold stress. It functions as an osmoprotectant by balancing the osmotic pressure between the intracellular and extracellular environments, which helps prevent cellular dehydration. Proline also acts as a molecular chaperone, stabilizing proteins and protecting them from denaturation. Additionally, proline has antioxidant properties, scavenging reactive oxygen species (ROS) that are produced as a byproduct of cold stress, thereby reducing oxidative damage to cellular components.
- **Sucrose as a Cryoprotectant**: Sucrose, a soluble sugar, is another critical cryoprotectant that accumulates in plant cells under cold stress. Sucrose stabilizes cell membranes by inserting itself between lipid molecules, which helps maintain membrane fluidity and prevent phase separation. Additionally, sucrose acts as an energy reserve that plants can mobilize to fuel metabolic processes during extended periods of low temperature. By stabilizing membranes and providing an energy source, sucrose plays a dual role in supporting cell survival under cold stress.

- **Glycine Betaine and Membrane Protection**: Glycine betaine is an osmolyte that accumulates in response to low temperatures, particularly in cold-tolerant plant species. Glycine betaine helps maintain osmotic balance within cells, preventing dehydration by reducing water loss to the extracellular space. Glycine betaine also interacts with lipid molecules in cell membranes, enhancing membrane stability and reducing the risk of rupture due to ice formation. In fruit crops, glycine betaine accumulation can protect cells in developing buds and fruits from cold-induced damage, supporting crop survival and productivity during seasonal frosts.

3. **Mechanisms of Cryoprotectant Function: Reducing Ice Nucleation and Protecting Cellular Integrity**

Cryoprotectants play a crucial role in mitigating the impact of freezing temperatures by inhibiting ice formation and preserving cellular integrity. These compounds prevent damage in multiple ways:

- **Lowering the Freezing Point of Cellular Fluids**: Cryoprotectants accumulate in the cytoplasm and lower the freezing point of intracellular fluids, reducing the likelihood of ice formation within cells. This effect is critical for preventing intracellular ice, which can cause irreparable physical damage by puncturing cell membranes and organelles. By reducing ice nucleation, cryoprotectants help cells survive short periods of freezing temperatures without suffering structural damage.
- **Preventing Cellular Dehydration**: As extracellular ice forms, it creates an osmotic gradient that draws water out of cells, leading to cellular dehydration. Cryoprotectants help retain water in the cytoplasm by balancing osmotic pressure, which minimizes water loss and prevents cellular shrinkage. This osmotic protection is especially beneficial for tissues such as buds and leaves, where cellular dehydration can cause significant structural damage and impair photosynthetic capacity.
- **Preserving Enzyme Activity and Protein Structure**: Cold temperatures can disrupt enzyme activity and protein structure, reducing cellular function and overall metabolic efficiency. Cryoprotectants stabilize proteins by binding to their surfaces, preventing them from unfolding or aggregating. This protection allows enzymes involved in critical pathways, such as photosynthesis and respiration, to maintain activity, albeit at reduced rates. In fruit crops, maintaining some level of metabolic activity is essential for preventing tissue damage and supporting recovery when temperatures rise.

4. **Energy Costs and Trade-Offs Associated with Cold Adaptations**

While COR proteins and cryoprotectants enhance cold tolerance, their production and maintenance require substantial energy and resources. This resource allocation comes with trade-offs that impact plant growth and productivity.

- **Energy Demands of Cryoprotectant Synthesis**: The synthesis of cryoprotectants like proline, sucrose, and glycine betaine is energy-intensive, as these molecules are produced in response to stress signals and require ATP and metabolic precursors. This energy expenditure reduces the amount of ATP available for other processes, such as growth and cell division. For instance, plants divert resources toward cryoprotectant production rather than leaf expansion or root growth, which can lead to stunted growth under prolonged cold conditions.

- **Competition Between Cold Tolerance and Growth**: Plants must prioritize survival over growth under extreme cold, which often results in delayed flowering and reduced fruit set. The resources allocated to COR proteins and cryoprotectants can slow down vegetative and reproductive growth, impacting yield. In fruit crops, these trade-offs can lead to fewer flowers, smaller fruits, and lower overall productivity, particularly if cold stress is prolonged or occurs during key developmental stages such as flowering and fruiting.
- **Reduced Photosynthetic Efficiency**: Cold-stressed plants often exhibit reduced photosynthetic efficiency due to slowed enzyme activity and changes in chloroplast function. The energy-intensive nature of cold tolerance mechanisms, such as COR protein synthesis and cryoprotectant accumulation, exacerbates this limitation by further taxing the plant's energy reserves. Lower photosynthetic output reduces the plant's carbohydrate pool, which affects fruit sweetness, flavor, and storage potential, impacting both fresh and processed produce quality.

5. Implications for Crop Management and Breeding for Cold Tolerance

The roles of COR proteins and cryoprotectants in cold tolerance have significant implications for improving crop resilience to low temperatures.

- **Breeding for Enhanced Cold Tolerance**: Breeding programs focus on selecting crop varieties that naturally produce higher levels of COR proteins and cryoprotectants under cold stress. These cold-tolerant varieties are better equipped to handle seasonal frosts and unexpected temperature drops, maintaining cellular integrity and productivity. For instance, cold-tolerant varieties of apples, grapes, and citrus produce higher levels of cryoprotectants, which enable them to survive early spring frosts that can devastate more sensitive cultivars.
- **Use of Plant Growth Regulators (PGRs)**: Certain plant growth regulators can be applied to stimulate the production of COR proteins and cryoprotectants, enhancing a plant's cold tolerance. For example, applying abscisic acid (ABA) before expected cold events can induce cold-responsive genes, priming the plant to withstand freezing temperatures. This pre-treatment is particularly valuable in commercial fruit production, where even minor frost damage can result in significant economic losses.
- **Optimizing Cultural Practices**: Farmers and growers can use protective practices, such as mulching and frost blankets, to minimize cold exposure and reduce the demand for energy-intensive cold tolerance mechanisms. By maintaining a stable microenvironment, these practices reduce the need for cryoprotectant accumulation, conserving resources that the plant can allocate toward growth and fruit development once temperatures rise.

6. Impact on Photosynthesis and Chloroplast Function

Photosynthesis is the process by which plants convert light energy into chemical energy, stored as carbohydrates, which fuel growth and development. This complex process is highly temperature-sensitive, especially within chloroplasts—the cellular organelles where photosynthesis takes place. Cold stress disrupts chloroplast function by affecting the stability of the photosynthetic apparatus, reducing energy production, and causing chlorophyll degradation. These disruptions lead to decreased photosynthesis, which limits carbohydrate production, causing energy shortages that affect overall plant health, growth, and productivity.

1. Cold-Induced Impairment of Photosynthetic Apparatus and Photosystem II

The photosynthetic machinery in chloroplasts consists of two main complexes, Photosystem I (PSI) and Photosystem II (PSII), along with an electron transport chain that drives the conversion of light energy into chemical energy. PSII, in particular, is sensitive to low temperatures, and its impairment under cold stress has significant downstream effects on photosynthesis.

- **Photosystem II Sensitivity to Cold**: PSII is responsible for the initial capture of light energy and the generation of electrons for the electron transport chain. Low temperatures reduce the efficiency of electron transfer within PSII by affecting the stability of key proteins, especially the D1 protein. The D1 protein in PSII undergoes continuous damage and repair under normal conditions, but low temperatures slow down the repair mechanisms, leading to a backlog of damaged PSII complexes. This damage disrupts the photosynthetic electron flow, reducing the production of ATP and NADPH—two molecules essential for carbon fixation in the Calvin cycle.
- **Slowing of the Electron Transport Chain**: Low temperatures reduce the mobility and energy of electrons, causing "bottlenecks" in the electron transport chain. As electron transfer becomes inefficient, energy production declines, limiting the plant's ability to convert absorbed light into usable energy. This reduction affects ATP synthesis, further impacting the Calvin cycle's efficiency in synthesizing carbohydrates. The slowdown in ATP and NADPH production affects the plant's energy balance, reducing growth rates and impairing reproductive development, especially in fruit crops where energy demand is high.

2. Reduced Carbohydrate Production and Energy Deficits

Photosynthesis is the primary source of carbohydrates, which are essential for cellular energy, growth, and development. When cold temperatures reduce photosynthetic efficiency, carbohydrate production declines, resulting in energy shortages that limit plant growth and physiological functions.

- **Decline in Carbon Fixation and Carbohydrate Availability**: The Calvin cycle, responsible for converting atmospheric CO_2 into carbohydrates, relies on ATP and NADPH produced by the light reactions of photosynthesis. Reduced ATP and NADPH availability limits the rate of carbon fixation, causing a decline in carbohydrate synthesis. This shortage of carbohydrates impacts all aspects of plant metabolism, as sugars are used not only for energy but also as building blocks for cell walls, proteins, and other cellular components. In fruit crops, reduced carbohydrate production directly affects fruit size, sweetness, and overall yield, as fruits require substantial sugar content for flavor and marketability.
- **Energy Deficits and Their Impact on Growth and Development**: Carbohydrates also act as signaling molecules that influence plant development and stress responses. Low

carbohydrate levels signal energy shortages, which trigger the plant to conserve resources and prioritize survival over growth. This shift in resource allocation reduces cell division, cell expansion, and tissue differentiation, leading to stunted growth, delayed flowering, and poor fruit set. For example, in crops like tomatoes, peppers, and strawberries, low-temperature stress can result in smaller plants with fewer flowers and fruits, reducing yield potential.

- **Effect on Storage Reserves**: Perennial plants, such as fruit trees, rely on stored carbohydrates during early growth stages in spring before photosynthesis reaches its full capacity. Low-temperature stress depletes these reserves by reducing photosynthesis and limiting the replenishment of carbohydrate stores, leaving plants vulnerable to additional stressors and affecting growth in subsequent seasons.

3. Chlorophyll Degradation and Chlorosis

Chlorophyll, the green pigment responsible for capturing light energy, is highly susceptible to cold-induced degradation. Loss of chlorophyll leads to **chlorosis** (leaf yellowing), which further limits photosynthetic capacity and reduces the plant's ability to recover from stress.

- **Cold-Induced Chlorophyll Breakdown**: Chlorophyll molecules are housed within chloroplasts, specifically in the thylakoid membranes where the light-dependent reactions of photosynthesis occur. Cold temperatures destabilize these thylakoid membranes, leading to the breakdown of chlorophyll molecules. Enzymes involved in chlorophyll synthesis and degradation become imbalanced under cold stress, often resulting in accelerated chlorophyll degradation. This process is visible as chlorosis, where leaves turn yellow due to the loss of green pigmentation.
- **Impact on Photosynthetic Efficiency**: Chlorosis directly reduces photosynthetic efficiency, as chlorophyll is essential for capturing light energy. Lower chlorophyll levels limit the plant's ability to absorb sunlight, reducing the overall energy available for photosynthesis. In fruit crops, this decrease in photosynthetic capacity impacts carbohydrate accumulation, which is critical for fruit quality and sweetness. Leaf chlorosis can also reduce the plant's appeal, especially in ornamental and leafy vegetable crops where leaf color is a primary quality attribute.
- **Effects on Antioxidant Capacity and Cellular Health**: Chlorophyll degradation also affects the production of antioxidants, such as carotenoids, which play a protective role in detoxifying reactive oxygen species (ROS) generated under stress. As chlorophyll levels decrease, the plant's antioxidant defenses weaken, making it more susceptible to oxidative damage. This increased vulnerability to ROS exacerbates cellular damage, further impairing photosynthesis and overall plant health.

4. Long-Term Implications of Impaired Photosynthesis on Fruit Crop Health and Productivity

The disruptions to chloroplast function and photosynthesis under cold stress have far-reaching implications for fruit crop productivity, quality, and resilience.

- **Delayed Growth and Reproductive Development**: Reduced photosynthesis limits the energy available for growth and reproductive processes, leading to delayed or reduced flowering, fruit set, and fruit development. In fruit crops, low carbohydrate availability due to impaired photosynthesis can delay fruit maturation and affect the quality of the harvest. In crops like grapes, apples, and cherries, energy deficits during early growth stages can lead to smaller fruits with less sugar content, affecting both flavor and marketability.

- **Reduced Fruit Quality and Nutritional Value**: Cold-stressed plants produce fruits with altered biochemical composition due to the lack of sufficient carbohydrates and the disruption of secondary metabolism. Reduced photosynthesis and chlorophyll degradation can lead to lower sugar levels and diminished flavor profiles in fruits, impacting consumer appeal. Additionally, the decline in antioxidant compounds due to chlorophyll loss reduces the nutritional value of fruits, as antioxidants contribute to the health benefits of fruits like berries, apples, and citrus.
- **Increased Susceptibility to Pests and Diseases**: Impaired photosynthesis and chlorophyll degradation weaken the plant's natural defenses, making it more vulnerable to pests and pathogens. The weakened antioxidant defenses and decreased energy reserves compromise the plant's ability to produce defense compounds that deter pests and resist infections. In fruit crops, susceptibility to pathogens can lead to pre- and post-harvest losses, reducing yield and increasing the need for chemical control measures.

5. Adaptive Mechanisms to Preserve Photosynthesis Under Cold Stress

To mitigate the impact of cold stress on photosynthesis, some plants employ adaptive mechanisms that enhance chloroplast stability and protect the photosynthetic apparatus.

- **Production of Cold-Regulated (COR) Proteins**: COR proteins stabilize cellular structures, including thylakoid membranes in chloroplasts. By binding to membrane proteins and lipids, COR proteins reduce cold-induced membrane rigidity, preserving the functionality of the photosynthetic apparatus. This stabilization helps maintain some level of photosynthetic activity even at low temperatures, supporting growth and stress recovery.
- **Accumulation of Cryoprotectants**: Cryoprotectants, such as sucrose and proline, accumulate in chloroplasts and other cellular compartments under cold stress. These molecules stabilize membranes and proteins by reducing ice nucleation and osmotic stress. In chloroplasts, cryoprotectants help prevent chlorophyll degradation and support enzyme activity, allowing photosynthesis to continue at a reduced rate and ensuring a baseline level of energy production.
- **Increase in Antioxidant Production**: Plants under cold stress upregulate antioxidant enzymes and compounds that help detoxify ROS generated during impaired photosynthesis. For instance, carotenoids and flavonoids play crucial roles in protecting chlorophyll from photooxidation, which is particularly important when PSII and other photosynthetic complexes are damaged. By enhancing antioxidant levels, plants can reduce oxidative damage to chloroplasts, improving their resilience to cold.

7. Hormonal Changes

Hormones play a central role in coordinating plant growth, development, and responses to environmental stress. Cold stress disrupts the hormonal balance in plants, leading to increased levels of certain stress-related hormones that help the plant prioritize survival over growth. Among these hormones, abscisic acid (ABA) and ethylene are particularly significant in shaping the plant's response to low temperatures. While these hormonal changes provide short-term protection, they can also reduce photosynthesis, accelerate leaf senescence, and lead to premature fruit drop, ultimately impacting crop yield and fruit quality.

1. Increase in Abscisic Acid (ABA) and Water Conservation

Abscisic acid (ABA) is commonly referred to as the "stress hormone" in plants due to its pivotal role in helping plants respond to various environmental stresses, including cold. Cold stress triggers a rise in ABA levels, which leads to several adaptive responses aimed at minimizing water loss and enhancing stress tolerance.

- **Stomatal Closure and Water Conservation**: One of ABA's primary actions under cold stress is to induce stomatal closure. Stomata are small openings on the leaf surface that allow gas exchange, facilitating CO_2 uptake for photosynthesis and releasing oxygen and water vapor. By closing stomata, ABA reduces water loss through transpiration, which is beneficial for conserving internal water reserves under cold conditions. However, stomatal closure also limits CO_2 uptake, which restricts photosynthesis and reduces carbohydrate production, leading to energy deficits that affect growth and development. This reduced photosynthetic efficiency is particularly detrimental in fruit crops, where energy demands are high to support fruit growth and quality.
- **Inhibition of Growth Processes**: Elevated ABA levels under cold stress also inhibit growth-promoting hormones, such as auxins, cytokinins, and gibberellins. Auxins regulate cell elongation, cytokinins promote cell division, and gibberellins encourage stem and leaf expansion. The increase in ABA effectively suppresses these growth hormones, redirecting resources away from growth and towards stress survival. This inhibition results in stunted growth, delayed flowering, and reduced fruit set, especially in crops exposed to prolonged cold stress. In cold-sensitive fruit crops, such as tomatoes and grapes, this hormone-driven inhibition of growth can significantly reduce yield and marketable quality.
- **Impact on Nutrient Uptake and Transport**: Stomatal closure and reduced water movement also affect nutrient uptake, as the flow of water through the xylem is essential for transporting nutrients from the roots to aboveground tissues. Limited nutrient transport under high ABA levels can lead to deficiencies in essential nutrients like nitrogen, potassium, and calcium. These deficiencies further compromise growth and cellular function, leading to chlorosis (leaf yellowing), poor fruit quality, and reduced overall productivity.

2. Increase in Ethylene Production and Its Effects on Senescence and Fruit Drop

Ethylene is a gaseous hormone involved in various processes, including fruit ripening, senescence, and responses to environmental stress. While ethylene plays positive roles in regulating growth under normal conditions, cold-induced increases in ethylene can lead to detrimental effects, particularly for leaf retention and fruit stability.

- **Promotion of Leaf Senescence**: Under cold stress, ethylene production may increase, which accelerates the process of leaf senescence (aging). Senescence involves chlorophyll breakdown, cell degradation, and nutrient remobilization, processes that allow the plant to conserve resources during adverse conditions. While senescence is an adaptive response, it also reduces photosynthetic capacity by limiting leaf area, which further restricts carbohydrate production and energy availability. In fruit crops, early leaf senescence reduces the plant's ability to produce and allocate carbohydrates to developing fruits, resulting in smaller fruits with reduced sweetness, flavor, and nutritional value.
- **Premature Fruit Drop**: Ethylene promotes fruit ripening, but under cold stress, excessive ethylene levels can lead to premature ripening and fruit drop. This effect is especially problematic for climacteric fruits, such as apples, tomatoes, and bananas, which continue to ripen after harvest. Cold-induced ethylene production can cause fruits to detach prematurely from the plant, resulting in immature, underdeveloped fruits that lack optimal size, sweetness, and flavor. Premature fruit drop reduces yield, affecting both fresh market availability and storage quality, as these fruits are less likely to have a long shelf life.
- **Impact on Reproductive Development**: In addition to affecting leaf and fruit retention, elevated ethylene levels can disrupt reproductive processes, reducing flower formation and limiting successful pollination. Ethylene inhibits the synthesis of floral hormones, reducing the number of flowers produced and decreasing fruit set. In crops like citrus and grapes, cold-induced ethylene production can lead to poor flowering, fewer fruits, and overall lower yield, impacting both productivity and profitability.

3. **Hormonal Trade-Offs: Survival Versus Growth and Productivity**

The increase in ABA and ethylene under cold stress represents a hormonal trade-off where plants prioritize immediate survival over long-term growth and reproductive success. While this shift allows plants to survive low temperatures, it comes at the cost of reduced yield and fruit quality.

- **Resource Allocation and Energy Costs**: High ABA levels signal the plant to conserve resources and limit growth processes. By closing stomata and inhibiting growth-promoting hormones, the plant conserves water and energy but at the expense of growth. The energy diverted to producing ABA and other stress-related responses reduces the resources available for flowering, fruiting, and overall development. In fruit crops, these energy limitations can result in smaller fruits, delayed ripening, and reduced nutritional quality, all of which impact consumer appeal and market value.
- **Reduced Photosynthetic Efficiency and Carbohydrate Shortages**: Stomatal closure under high ABA levels limits CO_2 availability for photosynthesis, leading to reduced carbohydrate production. Carbohydrates are essential for growth and energy storage, and their shortage affects multiple aspects of plant metabolism, from cell division to stress recovery. The energy deficits caused by limited photosynthesis reduce the plant's ability to support fruit development, leading to lower yields and diminished fruit quality. Additionally, carbohydrate shortages weaken the plant's defense systems, increasing vulnerability to secondary stressors, such as pathogens and nutrient deficiencies.

- **Long-Term Implications for Yield and Quality**: Prolonged cold stress with elevated ABA and ethylene levels can lead to a cycle of inhibited growth, reduced photosynthetic capacity, and diminished reproductive output. This cumulative effect is particularly concerning for perennial fruit crops, where cold-induced yield reductions impact multiple growing seasons. In apples, peaches, and grapes, for example, repeated cold stress can lead to smaller fruits, altered flavor profiles, and reduced storage quality, affecting the profitability and marketability of the crop.

4. **Adaptive Mechanisms and Potential Interventions for Cold Stress Management**

Understanding the hormonal changes induced by cold stress offers insights into potential strategies to mitigate the impact of low temperatures on crop productivity.

- **Breeding and Genetic Modification for Hormonal Resilience**: Plant breeders have developed cold-tolerant varieties with modified hormone regulation pathways that maintain lower ABA and ethylene levels under stress. These varieties are more resistant to premature leaf senescence and fruit drop, allowing them to sustain growth and fruit production under low-temperature conditions. For example, cold-tolerant tomato and grape cultivars produce lower levels of ethylene under stress, reducing the risk of premature ripening and fruit drop, and thus maintaining higher yields.
- **Application of Plant Growth Regulators (PGRs)**: The use of plant growth regulators can help modulate hormone levels and enhance cold tolerance in fruit crops. For example, treatments with anti-ethylene compounds, such as silver thiosulfate (STS), can inhibit ethylene effects, delaying leaf senescence and extending fruit retention. Similarly, applying cytokinins before cold events can counteract the effects of ABA, promoting cell division and delaying senescence, thereby enhancing cold resilience. These treatments can be particularly useful in commercial fruit production, where even minor yield losses due to cold stress can have significant economic impacts.
- **Optimized Cultural Practices**: Cultural practices such as mulching, proper irrigation, and using frost protection methods like row covers or wind machines can help mitigate the effects of cold stress by creating a more stable microenvironment. By maintaining warmer soil and reducing the exposure of plants to low temperatures, these practices can help prevent the hormonal imbalances that lead to stomatal closure, reduced photosynthesis, and premature fruit drop.

Table 4: Impact of Low Tmperature on Fruit Crops

S.N	Name of Fruit Crops/Cultivars	Effect of Low Temperature
1.	Almond	Pollination & fertilization is highly reduced
2.	Aonla	Oozing of water from fruits
3.	Banana	Below 10^{0}C leads to impedance of inflorescence and malformation of bunches.
4.	Ber	fruits are shrivelled, brown and turn black
5.	Citrus	Drying up of fruits and twigs
6.	Mango	Floral induction

7.	Pomegranate	Hardening of fruits

c. Radiation Stress

Plants rely on sunlight to drive photosynthesis, the process that produces the energy necessary for growth and development. However, certain types of light, especially high-intensity light and ultraviolet (UV) radiation, can cause **radiation stress**, which disrupts photosynthesis and damages cellular structures. High light intensity and UV exposure impact plants differently, but both forms of radiation stress can inhibit photosynthetic efficiency, affect chlorophyll stability, and lead to oxidative damage, particularly in sensitive crops.

1. Ultraviolet (UV) Radiation and Its Effects on Photosynthesis

Ultraviolet (UV) radiation consists of high-energy wavelengths that are divided into three categories: UVA, UVB, and UVC. UVB (280-315 nm) and UVC (100-280 nm) are particularly harmful to plants, with UVB being the primary concern since UVC is generally absorbed by the Earth's atmosphere and does not reach the surface. Exposure to UV radiation has several detrimental effects on the photosynthetic apparatus in chloroplasts.

- **Damage to Photosystem II (PSII)**: UVB radiation affects the integrity of Photosystem II (PSII), a core component of the photosynthetic apparatus responsible for capturing light and driving electron transport. UVB radiation damages the D1 protein in PSII, which is essential for electron flow. When the D1 protein is damaged, PSII efficiency declines, limiting ATP and NADPH production, which are vital for the Calvin cycle and carbohydrate synthesis. This reduction in PSII activity results in decreased photosynthetic rates, ultimately impacting growth and yield, particularly in fruit crops that rely on high-energy inputs for fruit development.
- **Generation of Reactive Oxygen Species (ROS)**: UV radiation causes the formation of reactive oxygen species (ROS) in chloroplasts, such as singlet oxygen and superoxide radicals. ROS are highly reactive molecules that can oxidize lipids, proteins, and nucleic acids, leading to oxidative stress. Excess ROS can damage chloroplast membranes, impairing electron transport and reducing the plant's capacity to carry out photosynthesis. Additionally, ROS damage chlorophyll molecules, leading to chlorosis (leaf yellowing), which further reduces the plant's ability to absorb light and capture energy.
- **Reduction in Chlorophyll and Carotenoid Content**: UV radiation leads to the breakdown of chlorophyll and carotenoids, pigments essential for capturing light energy and protecting chloroplasts from photooxidative damage. Carotenoids, which help quench excess light energy and protect chlorophyll from ROS, degrade under prolonged UV exposure, leaving chlorophyll molecules more vulnerable to oxidative damage. This reduction in pigment content directly affects the plant's ability to absorb light, reducing photosynthetic efficiency and carbohydrate production.

2. Impact of High Light Intensity on Photosynthesis

While light is essential for photosynthesis, excessive light intensity, especially when combined with other stressors such as heat or drought, can overwhelm the photosynthetic apparatus, leading to a condition known as **photoinhibition**. High light intensity, particularly in sunny environments or at high altitudes, can exceed the plant's ability to utilize the absorbed energy, resulting in damage to photosynthetic components.

- **Photoinhibition of Photosystem II (PSII)**: Under high light conditions, PSII absorbs more light energy than it can process, leading to the overexcitation of chlorophyll molecules. This excess energy is transferred to oxygen, producing ROS that can damage the D1 protein in PSII. Damaged D1 protein disrupts electron transport, reducing the efficiency of PSII and limiting ATP and NADPH production. While plants have repair mechanisms to replace damaged D1 protein, prolonged exposure to high light overwhelms these repair systems, leading to chronic photoinhibition, which significantly reduces photosynthetic rates.
- **Heat Dissipation and Non-Photochemical Quenching (NPQ)**: Plants have protective mechanisms, such as non-photochemical quenching (NPQ), to dissipate excess light energy as heat and prevent damage to the photosynthetic apparatus. However, NPQ has a limited capacity and can become saturated under prolonged high light exposure. When NPQ is overwhelmed, excess light energy generates ROS, causing oxidative damage to chloroplast membranes and photosynthetic proteins. The resulting oxidative stress further inhibits photosynthesis and can lead to cellular injury, leaf bleaching, and necrosis (cell death) in severe cases.
- **Impact on Carbon Fixation and Carbohydrate Production**: High light intensity, by damaging PSII and reducing ATP and NADPH availability, impairs the Calvin cycle, where CO_2 is fixed into carbohydrates. Reduced carbon fixation affects carbohydrate availability for growth, development, and energy storage. In fruit crops, reduced carbohydrate production leads to smaller fruits with lower sugar content, impacting flavor and marketability. Additionally, energy deficits limit flower formation and fruit set, reducing overall yield.

3. Adaptive Mechanisms to Mitigate Radiation Stress

Plants have evolved several adaptive responses to protect themselves from UV radiation and high light intensity, enhancing their resilience to radiation stress and minimizing damage to the photosynthetic apparatus.

- **Production of UV-Absorbing Compounds**: Plants produce UV-absorbing compounds, such as flavonoids and phenolic acids, in response to UV radiation. These compounds accumulate in the epidermal layers of leaves and act as a sunscreen, reducing the penetration of harmful UV rays into deeper tissues. By absorbing UV radiation, these compounds protect chloroplasts from UV-induced oxidative damage, preserving photosynthetic efficiency and limiting cellular injury. In fruit crops, flavonoids and phenolic compounds also contribute to antioxidant properties, which improve nutritional quality and offer health benefits to consumers.
- **Dynamic Adjustment of Chlorophyll and Carotenoid Content**: Plants adjust the levels of chlorophyll and carotenoids in response to high light intensity. Carotenoids, such

as xanthophylls, play a key role in the xanthophyll cycle, a photoprotective mechanism that dissipates excess light energy as heat. This process helps reduce the energy load on PSII and protects chlorophyll molecules from photodamage. By dynamically regulating chlorophyll and carotenoid levels, plants can optimize light capture under varying light conditions while preventing photoinhibition and oxidative damage.

- **Increase in Antioxidant Enzyme Activity**: Antioxidant enzymes, such as superoxide dismutase (SOD), catalase, and peroxidases, are upregulated in response to ROS generated by UV and high light intensity. These enzymes detoxify ROS, protecting cellular structures from oxidative damage. SOD converts superoxide radicals into hydrogen peroxide, which is further broken down by catalase and peroxidases into water and oxygen, reducing the oxidative load in chloroplasts. This antioxidant defense system allows plants to manage ROS levels and mitigate the damaging effects of radiation stress on photosynthesis.

4. **Long-Term Implications of Radiation Stress on Plant Health and Productivity**

The impact of UV radiation and high light intensity on photosynthesis has significant implications for plant growth, development, and productivity, particularly in fruit crops.

- **Reduced Growth and Yield**: Chronic radiation stress reduces photosynthetic efficiency, leading to energy shortages that limit growth and reproductive output. In fruit crops, lower carbohydrate production due to impaired photosynthesis can delay fruit maturation, reduce fruit size, and decrease sugar content, affecting both yield and quality. Fruit crops such as grapes, tomatoes, and peppers are particularly susceptible to yield losses under high light conditions, as they require substantial energy for fruit development.
- **Effects on Fruit Quality and Nutritional Value**: Radiation stress affects fruit quality by reducing the synthesis of carbohydrates, sugars, and secondary metabolites such as antioxidants and flavonoids. In crops like berries, apples, and tomatoes, lower sugar content impacts flavor, while reduced antioxidant levels affect the nutritional value of the fruit. Additionally, high light intensity and UV exposure can cause sunburn or discoloration on the fruit surface, reducing marketability and shelf life.
- **Increased Susceptibility to Secondary Stressors**: Radiation stress can weaken plant defenses, making them more susceptible to additional stressors such as drought, nutrient deficiencies, and pathogens. Damaged chloroplasts and reduced photosynthetic capacity limit the plant's ability to produce energy for stress responses, leaving it vulnerable to infection and environmental fluctuations. This compounded stress increases the risk of crop losses, especially in regions with high UV exposure or intense sunlight.

d. Wind Stress

Wind is a natural environmental factor that can have both positive and negative effects on plants. While gentle winds can promote air circulation and reduce disease incidence, strong or prolonged winds impose **wind stress**, which impacts plant physiology, structure, and water balance. Wind stress influences a plant's ability to grow, take up water and nutrients, and maintain structural integrity, and can lead to reduced yield and quality in sensitive crops, particularly in exposed environments.

1. Physical Impact on Plant Structure and Stability

Wind exerts mechanical forces on plants, challenging their structural integrity and stability. These forces affect all parts of the plant, including stems, branches, and root systems, especially in taller and woody plants like fruit trees. The constant movement from strong winds can lead to bending, breakage, and uprooting, making plants more vulnerable to disease, reducing their access to nutrients, and limiting their photosynthetic capacity.

1. Stem Bending and Lodging

One of the primary impacts of wind on plants is **stem bending** and, in severe cases, **lodging**, where plants are forced to the ground. Lodging occurs when the force of the wind overwhelms a plant's natural structural strength, causing it to fall or bend significantly, which can have immediate and long-term effects on growth and productivity.

- **Horizontal Force and Stem Flexibility**: Wind exerts a horizontal force on plants, causing the stems and branches to bend. Flexible plants can withstand this force to an extent, but young trees and herbaceous crops with weaker stems are highly susceptible. Lodging can be particularly problematic for fruit trees, where heavy canopies add to the wind force experienced by the plant. In crops like corn or wheat, lodging can cause stems to twist and lean, reducing exposure to sunlight and impacting grain fill and yield.
- **Limited Access to Sunlight and Air**: Lodged plants experience a reduction in photosynthetic efficiency due to decreased exposure to sunlight. With limited air circulation around the lodged parts, plants become more susceptible to fungal infections and pest infestations. These issues further reduce growth, leading to stunted plants with lower energy production. In fruit crops, restricted light and air exposure impact fruit development and quality, as insufficient sunlight reduces sugar accumulation, impacting fruit sweetness and marketability.
- **Increased Disease Susceptibility**: The close contact of leaves and stems with the ground creates favorable conditions for disease development. Soil-borne pathogens such as fungi and bacteria can infect lodged plants through surface wounds or leaf contact with soil. Lodging also creates a humid microenvironment, promoting disease spread. For instance, fruit crops like tomatoes and strawberries are highly susceptible to fungal infections, which spread quickly in humid conditions. Lodging in these crops often results in lower yields, poor fruit quality, and increased disease management costs.

2. Branch and Trunk Breakage

Branch and trunk breakage is a common consequence of wind stress, particularly in large trees and fruit crops with dense canopies. Strong winds can break branches, stems, or even the trunk, resulting in a range of negative impacts, from reduced productivity to increased susceptibility to infections.

- **Damage to Fruit-Bearing Branches**: Fruit trees with heavy canopies, especially those laden with maturing fruits, are highly susceptible to branch breakage under wind stress. The added weight of fruits increases the load on branches, making them more vulnerable to the mechanical force of the wind. Breakage of fruit-bearing branches not only reduces current yield but also affects future production, as the tree redirects resources to healing and recovery. This is particularly problematic in apple, pear, and peach orchards, where broken branches lead to direct economic losses due to lost fruits.

- **Increased Susceptibility to Pests and Pathogens**: Branch and trunk breakage creates open wounds that serve as entry points for pests and pathogens. These wounds expose the plant's vascular tissue, increasing the risk of infection from diseases such as bacterial canker, fungal infections, and wood-boring pests. For instance, in citrus and stone fruit trees, wounds from broken branches can allow fungal pathogens like Phytophthora to infect the plant, leading to decay and further weakening of the tree.
- **Economic Losses in Fruit Orchards**: For commercial fruit orchards, branch and trunk breakage can result in significant economic losses. Strong winds, especially during the growing season, can lead to reduced yields and poor-quality fruits, which are less marketable. Additionally, managing wind-damaged trees requires labor-intensive pruning and sometimes replanting, increasing production costs. In severe cases, high winds can damage entire rows of fruit trees, impacting both short-term productivity and the long-term sustainability of the orchard.

Root System Instability

Wind stress also affects the root system, which provides the plant with anchorage and access to water and nutrients. Repeated exposure to strong winds can destabilize the root system, particularly in trees and shrubs with shallow or poorly developed roots, leading to root instability and, in severe cases, uprooting.

- **Uprooting and Reduced Water Absorption**: In regions with high winds, trees with shallow root systems or those growing in loose, sandy soils are more susceptible to uprooting. Uprooted trees lose their ability to absorb water and nutrients, leading to dehydration and nutrient deficiencies. For young trees, uprooting is particularly damaging, as they may lack the established root network necessary for recovery. In fruit orchards, uprooted trees must often be removed and replaced, resulting in additional labor and reduced productivity.
- **Compromised Anchorage in Newly Transplanted Crops**: Young or recently transplanted crops are especially vulnerable to root instability under wind stress. The root systems of newly transplanted plants take time to establish and anchor firmly in the soil. When exposed to wind, these plants experience greater movement, which can damage delicate root hairs and delay root establishment. This phenomenon, known as **transplant shock**, results in reduced growth rates and poor establishment success. In fruit nurseries and newly planted orchards, managing transplant shock is critical for long-term orchard productivity.
- **Soil Erosion and Root Exposure**: Wind stress can also lead to soil erosion, particularly in sandy or loose soils, which can expose roots to the surface. Exposed roots are more susceptible to desiccation, pest infestations, and temperature fluctuations, affecting the plant's nutrient and water uptake. In regions with high wind exposure, soil erosion around the root zone can destabilize the tree, leading to further wind damage and reduced crop

yields. For crops like grapevines, root exposure can also affect root temperature, which influences grape flavor and quality.

3. **Adaptive Mechanisms to Strengthen Plant Stability Against Wind Stress**

Plants have developed various adaptations to improve their resilience to wind stress, enhancing their structural integrity and stability. These adaptations help plants withstand mechanical forces and reduce the likelihood of structural damage.

- **Increased Lignification and Thicker Cell Walls**: Plants exposed to frequent wind stress often develop thicker cell walls and increased lignification, a process that strengthens cell walls by incorporating lignin, a rigid and complex polymer. This adaptation, called **thigmomorphogenesis**, improves structural stability by making stems and branches more resistant to bending and breakage. In fruit trees, enhanced lignification supports the weight of fruit-bearing branches, reducing the risk of breakage under windy conditions.
- **Shorter, More Compact Growth Forms**: Plants that experience consistent wind exposure tend to grow shorter and more compact to reduce wind resistance. Compact growth with thicker stems and smaller leaves helps the plant conserve energy and prevent lodging. In fruit trees, a more compact canopy reduces wind exposure and provides better protection for developing fruits. Growers can encourage compact growth forms through selective pruning and training techniques, which enhance wind resilience while maintaining productivity.
- **Production of Flexible Tissues**: Some plants develop more flexible tissues that allow them to bend rather than break under wind stress. Flexible tissues absorb mechanical forces more effectively, reducing the risk of structural damage. For example, young shoots and branches in fruit trees may have greater flexibility than older, lignified branches, enabling them to sway with the wind and reduce the risk of breakage. This flexibility allows the plant to recover quickly after a wind event and continue normal growth.

4. **Management Practices to Minimize Wind Stress in Crop Production**

Growers can implement various management practices to protect plants from the physical impacts of wind stress, particularly in fruit orchards where strong winds can lead to economic losses.

- **Use of Windbreaks and Shelterbelts**: Windbreaks, such as rows of trees, shrubs, or hedges, help reduce wind speed and create a more stable microenvironment for crops. Shelterbelts not only protect individual plants from direct wind exposure but also reduce soil erosion, improve moisture retention, and promote biodiversity. In fruit orchards, windbreaks minimize fruit drop and prevent branch breakage, preserving both yield and quality.

Installation of Support Structures: Staking, trellising, and other support systems provide mechanical stability to plants, reducing the risk of lodging and breakage. In fruit crops like tomatoes, grapes, and apples, support structures help maintain canopy structure, protect fruit-bearing branches, and facilitate harvest operations. Trellises and stakes are particularly valuable for young trees or vining plants, providing the support needed for stable growth under windy conditions.

- **Soil Management and Mulching**: To prevent soil erosion and root exposure, growers can use soil management practices such as mulching, which helps stabilize the soil around the root zone. Mulching not only reduces erosion but also retains moisture and moderates soil temperature, which benefits root stability and nutrient uptake. In wind-exposed areas, applying a thick layer of organic mulch around the base of fruit trees and crops can help protect roots, reduce water loss, and improve overall plant health.

2. Water Loss Through Increased Transpiration

Wind stress plays a major role in increasing water loss from plants through **transpiration**, the process by which water vapor exits the plant through stomata on leaf surfaces. While transpiration is essential for nutrient transport and cooling, excessive water loss under windy conditions can create significant water deficits. These deficits impact cellular function, growth, and the plant's resilience to stress, often leading to reduced productivity and quality in fruit crops.

1. Stomatal Regulation and Water Conservation

Plants regulate water loss by controlling the opening and closing of stomata, which are small pores on the leaf surface that allow gas exchange. In windy conditions, transpiration rates increase as wind removes the layer of humid air around leaves, creating a steeper water vapor gradient. This gradient accelerates water loss, forcing the plant to adapt by closing its stomata to conserve moisture.

- **Stomatal Closure and CO_2 Limitation**: In response to wind-induced water loss, plants may partially or fully close their stomata to reduce transpiration. While this response conserves water, it also limits CO_2 uptake, as stomata are necessary for gas exchange. Without sufficient CO_2, photosynthetic efficiency declines, reducing the plant's ability to produce the carbohydrates needed for growth and development. In fruit crops, such as grapes, apples, and tomatoes, reduced photosynthesis due to stomatal closure can affect fruit development, leading to smaller fruits with lower sugar content, impacting flavor and marketability.
- **Energy Shortages and Reduced Growth**: When CO_2 intake is restricted due to stomatal closure, the Calvin cycle, which is responsible for carbon fixation, slows down. This reduction in photosynthetic activity causes an energy shortage, limiting the plant's capacity for growth and repair. Fruit trees and vines, which require high energy input for flowering, fruit set, and fruit expansion, are particularly affected by these shortages. Stomatal regulation under wind stress can thus lead to slower growth, delayed fruit maturation, and compromised quality, as lower carbohydrate levels result in fruits that are less sweet and flavorful.

2. Risk of Desiccation and Water Deficits

Prolonged exposure to wind can lead to **desiccation**, or extreme water loss, which affects leaves, stems, and young shoots. This desiccat\n becomes particularly harmful in dry regions or during drought periods when soil moisture levels are already low.

- **Extreme Water Loss and Leaf Damage**: Wind accelerates water evaporation from leaf surfaces, causing leaves to lose more water than they can replenish. As a result, leaves and young shoots may dry out, turn brown, and become brittle. This desiccation is especially damaging to fruit crops, where young, tender tissues such as flowers and new leaves are vital for future fruit development. The drying and browning of leaves also reduce photosynthetic capacity, further limiting energy production and reducing the plant's resilience to other stressors.

- **Impact on Sensitive Tissues**: In fruit trees, young leaves and flowers are highly sensitive to water stress. Desiccation of these tissues can lead to flower drop, which directly reduces fruit set and yield. For example, in crops like peaches, citrus, and cherries, flower drop due to water stress can result in significantly lower yields. The loss of young leaves, which are responsible for producing energy and carbohydrates, further compromises the plant's growth and reproductive capacity, impacting both current and future fruit production.
- **Increased Vulnerability in Drought-Prone Regions**: In areas with limited water availability, the combination of wind stress and low soil moisture exacerbates the risk of desiccation. Plants that are already struggling with limited water resources are more likely to experience wilting and severe dehydration under windy conditions. Prolonged exposure to wind stress in such environments can lead to permanent wilting, tissue necrosis, and, in severe cases, plant death.

3. **Imbalance Between Water Uptake and Loss**

Wind-driven water loss often exceeds a plant's ability to take up water from the soil, especially if root growth and stability are compromised by mechanical forces. This imbalance between water uptake and loss creates significant challenges for plant health and productivity, particularly in fruit crops that require a stable water supply for fruit development.

- **Reduced Water Uptake Due to Root Instability**: Mechanical instability caused by wind stress can limit root growth and anchorage, affecting the plant's ability to absorb water efficiently. When roots are unable to maintain a steady water supply, the plant cannot compensate for the high transpiration rates caused by wind, resulting in water deficits. In fruit trees, such as apple and pear trees, water deficits can reduce fruit size, affect flavor, and decrease overall yield, as the plant prioritizes survival over fruit development.
- **Water Imbalance and Fruit Quality**: In fruit crops, water imbalance affects fruit expansion and development, as water is crucial for cell enlargement and nutrient transport to the fruit. When water availability is limited, fruits may not reach their full size and may develop a tougher texture. In addition, reduced water content in fruits affects sugar transport and accumulation, leading to less sweet and flavorful produce. For instance, in grapes, water stress under windy conditions can result in smaller berries with lower sugar content, affecting the quality of wine grapes and table grapes alike.
- **Increased Susceptibility to Secondary Stressors**: Plants experiencing water stress from wind exposure are more vulnerable to additional stressors, such as nutrient deficiencies and pathogens. Water plays a vital role in nutrient transport, and water shortages can lead to nutrient imbalances that affect metabolic functions. For instance, calcium deficiencies, which commonly occur in water-stressed plants, can lead to disorders such as blossom

end rot in tomatoes and bitter pit in apples, reducing fruit quality. Additionally, water-stressed plants are more susceptible to pests and diseases, as weakened tissues and stressed immune responses make it easier for pathogens to infect the plant.

4. Adaptive Mechanisms to Minimize Water Loss and Enhance Resilience

Plants have developed various adaptive mechanisms to reduce water loss and increase resilience to wind-induced water stress. These adaptations help plants conserve moisture and protect sensitive tissues from desiccation.

- **Thicker Cuticles and Waxy Coatings**: Many plants exposed to frequent wind stress develop thicker cuticles and waxy coatings on their leaves to reduce water loss. The cuticle is a waxy layer on the leaf surface that acts as a barrier to water vapor, slowing down transpiration. This adaptation is particularly beneficial in arid and windy environments, where water conservation is critical for survival. In fruit crops like apples and citrus, a thicker cuticle protects leaves and fruits from excessive water loss, enhancing the plant's resilience to wind stress.
- **Smaller and Rolled Leaves**: Plants in windy areas often develop smaller leaves, which reduce the surface area exposed to wind and thereby lower transpiration rates. Additionally, some plants can roll their leaves inward during periods of intense wind, reducing the exposure of stomata to dry air and conserving water. This adaptation helps minimize water loss and protect internal tissues from dehydration, particularly in crops grown in arid regions or windy climates.
- **Deeper Root Systems**: Some plants respond to wind-induced water stress by developing deeper root systems that can access moisture from deeper soil layers. This adaptation improves the plant's ability to absorb water even when surface moisture is limited due to wind-driven evaporation. For instance, fruit trees like grapevines, with deep-rooting abilities, can access water from deeper soil layers, helping them withstand wind stress in arid environments.

5. Management Practices to Mitigate Wind-Induced Water Stress in Crop Production

Growers can employ several management strategies to reduce the effects of wind-induced water stress, particularly in fruit crops that are sensitive to water shortages.

- **Use of Windbreaks**: Installing windbreaks, such as rows of trees or hedges, can help reduce wind speed around crop plants, creating a more sheltered environment. Windbreaks lower transpiration rates, allowing plants to conserve water and reducing the risk of desiccation. In fruit orchards, windbreaks protect young trees and developing fruits from wind damage and water loss, helping maintain fruit size, flavor, and quality.
- **Optimized Irrigation Practices**: Providing supplemental irrigation during windy periods helps counteract the increased water loss through transpiration. Drip irrigation, for example, delivers water directly to the root zone, improving water use efficiency and ensuring that plants receive adequate moisture even during windy conditions. Regular irrigation helps prevent water deficits, which is essential for maintaining growth and fruit quality in crops like tomatoes, berries, and citrus.

- **Mulching to Conserve Soil Moisture**: Mulching around the base of plants helps retain soil moisture by reducing evaporation from the soil surface. Organic mulches, such as straw or wood chips, create a protective layer that limits water loss and insulates the soil, keeping it cooler. In windy conditions, mulch helps maintain stable soil moisture levels, which is beneficial for young trees and crops with shallow roots. For fruit crops, mulching helps sustain root moisture levels, supporting fruit growth and reducing the risk of yield loss.

3. Nutrient Uptake and Transport Limitations

Nutrient uptake and transport within plants depend on the stability and functionality of the root system, as well as sufficient water availability for nutrient transport. Wind stress disrupts these factors by destabilizing roots and increasing water loss through transpiration, leading to reduced nutrient absorption, nutrient imbalances, and altered growth patterns. For fruit crops, these limitations impact growth, fruit quality, and overall productivity.

1. Reduced Nutrient Absorption

Plants rely on a steady supply of water and a stable root system to absorb nutrients from the soil. Wind stress disrupts these processes by increasing water loss and reducing root functionality, limiting the plant's ability to absorb essential nutrients.

- **Effect of Water Loss on Nutrient Transport**: Water is a critical medium for transporting nutrients from the soil to the roots and throughout the plant. Wind stress increases transpiration rates, causing rapid water loss and leading to water deficits that hinder nutrient transport. When plants lose water faster than they can absorb it, nutrient uptake is disrupted, as the movement of nutrients in the xylem (water-conducting tissue) depends on sufficient water flow. Without adequate nutrient transport, plants experience deficiencies that affect growth and development.
- **Impact of Nutrient Deficiencies on Plant Health**: Nutrients like nitrogen, potassium, and calcium are vital for cellular function, growth, and stress tolerance. Nitrogen is necessary for chlorophyll production and protein synthesis, while potassium regulates stomatal function and water balance, and calcium strengthens cell walls and enhances stress resilience. Under wind-induced water stress, the limited uptake of these nutrients affects overall plant health, leading to stunted growth, reduced leaf development, and diminished stress tolerance. In fruit crops, potassium and calcium are especially important for fruit development and firmness; deficiencies in these nutrients can lead to smaller fruits with poor flavor and reduced shelf life.
- **Long-Term Consequences for Fruit Crops**: In fruit-bearing plants, reduced nutrient absorption due to wind stress impacts fruit quality and yield. For example, potassium deficiency affects sugar transport and accumulation in fruits, resulting in less sweet and flavorful produce. Calcium deficiency compromises cell wall stability, leading to softer, less resilient fruits that are more susceptible to post-harvest diseases. The combination of smaller size, lower sugar content, and shorter shelf life diminishes the market value of fruit crops like apples, grapes, and strawberries.

2. **Nutrient Imbalance and Leaf Scorching**

Wind stress can lead to nutrient imbalances, where certain nutrients are deficient while others accumulate, causing visible symptoms like leaf scorching or chlorosis. These imbalances impact the plant's aesthetic appeal, productivity, and photosynthetic efficiency, especially in crops that require balanced nutrition for optimal growth.

- **Leaf Scorching Due to Water and Nutrient Shortages**: Leaf scorching is a common symptom of nutrient imbalances caused by wind stress, appearing as browning or burning at the edges and tips of leaves. Wind stress-induced water loss disrupts nutrient transport, limiting the plant's ability to move minerals and essential nutrients to the leaves. The resulting nutrient shortage, particularly of potassium and calcium, leads to browning and drying of leaf edges, reducing photosynthetic surface area and limiting energy production. For fruit crops, leaf scorching can affect fruit yield by reducing the plant's capacity to produce the carbohydrates necessary for fruit development.
- **Chlorosis from Nitrogen and Magnesium Deficiencies**: Chlorosis, or leaf yellowing, is another common sign of nutrient imbalance due to wind stress. Nitrogen and magnesium are essential for chlorophyll production, and deficiencies in these nutrients hinder photosynthesis. Wind-stressed plants often struggle to absorb adequate nitrogen and magnesium, resulting in reduced chlorophyll content and yellowing of leaves. Chlorosis reduces the plant's aesthetic appeal, which is especially critical for leafy vegetables and ornamental plants, and decreases photosynthetic efficiency, impacting growth and productivity.
- **Impaired Growth and Reduced Quality**: Nutrient imbalances due to wind stress can lead to uneven growth patterns and lower overall crop quality. In fruit trees, imbalances in nutrients like potassium, calcium, and magnesium can lead to poor fruit development, smaller fruits, and lower sugar accumulation. For instance, in grapevines, magnesium deficiency can reduce berry size and affect flavor, impacting the quality of wine grapes. The aesthetic and nutritional quality of crops like leafy greens, tomatoes, and peppers is also compromised, reducing their market value.

3. **Impact on Root Growth and Development**

Wind stress not only affects aboveground plant parts but also alters root growth patterns. Plants exposed to strong winds often prioritize root growth for anchorage and stability, shifting resources away from shoot growth and nutrient uptake.

- **Increased Root Growth for Stabilization**: Under wind stress, plants may divert resources to root growth to enhance stability. This shift in resource allocation helps plants anchor more firmly in the soil, reducing the risk of uprooting in high winds. However, the increased focus on root growth comes at the expense of shoot growth, resulting in fewer leaves and branches. In fruit crops, limited shoot growth means reduced canopy size, which affects the plant's photosynthetic capacity and limits fruit-bearing potential.
- **Reduced Aboveground Growth and Photosynthetic Capacity**: As resources are directed to the roots, the plant may produce fewer leaves and branches. This reduction in leaf

area limits photosynthesis, which directly impacts energy production and carbohydrate availability. For fruit crops, the reduction in canopy size decreases fruit-bearing sites and restricts fruit production. The lower carbohydrate levels also limit fruit growth and sugar accumulation, resulting in smaller fruits with lower sugar content, affecting both flavor and quality.

- **Reduced Nutrient Uptake Efficiency**: Wind stress can impair nutrient uptake efficiency, as nutrient absorption depends on a balance between root and shoot growth. When resources are diverted to stabilizing roots, nutrient uptake efficiency declines, as fewer nutrients are available to support aboveground growth and fruit development. Additionally, if roots are damaged by wind-induced instability or soil erosion, nutrient absorption is further compromised, leading to deficiencies that affect growth and crop quality. In fruit trees, reduced nutrient uptake can result in fewer and lower-quality fruits, impacting both yield and marketability.

4. Adaptive Mechanisms to Improve Nutrient Uptake Under Wind Stress

Plants have evolved adaptive responses to improve nutrient uptake and maintain resilience under wind-induced nutrient stress. These adaptations help plants balance root stability with nutrient absorption, enhancing their ability to cope with wind stress.

- **Enhanced Root Depth and Distribution**: Wind-stressed plants often develop deeper and more widespread root systems to enhance anchorage and improve access to nutrients in the soil. Deeper roots allow plants to tap into nutrient-rich layers of soil, providing a steady supply of essential nutrients even under windy conditions. In fruit crops, deeper root systems improve water and nutrient uptake, helping sustain fruit development and quality even in challenging environments.
- **Altered Leaf Morphology for Improved Nutrient Retention**: Some plants exposed to wind stress develop smaller leaves or thicker cuticles, which help reduce water loss and conserve nutrients. Smaller leaves have less surface area exposed to wind, reducing transpiration rates and minimizing nutrient loss. In crops like apples and grapes, smaller or more compact canopies can help conserve nutrients, supporting healthy fruit development without depleting nutrient resources.
- **Increased Production of Secondary Metabolites**: Wind-stressed plants may increase the production of secondary metabolites, such as flavonoids and phenolics, which play protective roles and help retain nutrients within cells. These compounds reduce oxidative stress, enhance nutrient retention, and support the plant's overall stress tolerance. In fruit crops, secondary metabolites contribute to fruit quality, flavor, and nutritional value, improving resilience to wind-induced nutrient limitations.

5. Management Strategies to Mitigate Nutrient Limitations Due to Wind Stress

To mitigate nutrient uptake limitations caused by wind stress, growers can implement various management practices that enhance nutrient availability, root stability, and overall plant health.

- **Supplemental Fertilization**: Providing supplemental fertilization can help offset nutrient deficiencies caused by wind stress. Foliar feeding, where nutrients are applied directly to the leaves, can supply nitrogen, potassium, and calcium to plants experiencing nutrient uptake challenges. This approach is particularly beneficial for fruit crops, as it directly supports fruit development, firmness, and shelf life, maintaining crop quality even in windy environments.

- **Use of Windbreaks to Reduce Transpiration and Soil Erosion**: Installing windbreaks, such as rows of trees or hedges, around crop fields helps reduce wind speed, minimizing water and nutrient loss through transpiration. Windbreaks also reduce soil erosion, which protects the root zone and improves nutrient retention. In fruit orchards, windbreaks help prevent nutrient deficiencies and maintain stable soil conditions, supporting long-term growth and productivity.
- **Soil Management and Mulching**: Applying organic mulches around the base of plants helps retain soil moisture, reduce temperature fluctuations, and enhance nutrient availability. Mulching also prevents soil erosion and protects the root zone, reducing the risk of nutrient depletion due to wind exposure. In fruit crops, mulching supports root stability and nutrient uptake, sustaining healthy growth and improving fruit quality.

4. Impact on Photosynthesis and Carbohydrate Production

Photosynthesis is essential for plant growth, as it provides the energy and carbohydrates necessary for various physiological processes, including flowering, fruit set, and ripening in fruit crops. Wind stress indirectly affects photosynthesis by promoting water loss, which leads to stomatal closure, reduces CO_2 intake, and increases oxidative stress. Additionally, wind alters leaf orientation, which limits sunlight capture. These factors collectively reduce photosynthetic efficiency and carbohydrate production, impacting crop yield and quality, especially in fruit-bearing plants.

1. **Reduction in CO_2 Uptake**

Wind stress causes an increase in transpiration, leading plants to close their stomata to conserve water. While this helps prevent water loss, it also restricts CO_2 intake, which is essential for the Calvin cycle in photosynthesis. Reduced CO_2 availability limits carbohydrate production, impacting energy-demanding processes in plants, particularly in fruit crops.

- **Impact on Photosynthesis and Carbohydrate Production**: When stomata close, the supply of CO_2 for photosynthesis decreases, limiting the Calvin cycle's ability to fix carbon into carbohydrates. Reduced photosynthesis affects the plant's overall energy balance, as fewer carbohydrates are available for growth and reproductive processes. Fruit crops like grapes, apples, and berries, which rely on high carbohydrate levels for flavor and sweetness, are particularly affected. Lower photosynthetic rates result in less sugar accumulation in fruits, diminishing flavor quality, which is a key factor in marketability and consumer preference.
- **Effects on Flowering and Fruit Development**: Carbohydrates are not only essential for fruit sweetness but also for other energy-intensive processes like flowering and fruit set. Reduced carbohydrate availability can delay flowering or lead to poor fruit set, reducing yield. In crops like citrus, peaches, and cherries, insufficient energy can result in smaller fruits or lower-quality produce. Reduced carbohydrate availability can also

lead to premature fruit drop, as the plant prioritizes survival over reproduction under stress conditions.

- **Long-Term Implications for Crop Quality and Yield**: Over time, the reduction in CO_2 uptake and carbohydrate production limits the plant's growth potential, affecting both current and future productivity. In perennial fruit crops, repeated cycles of limited photosynthesis due to wind stress can weaken the plant, leading to smaller, lower-quality fruits, reduced flowering, and compromised long-term productivity.

2. **Oxidative Stress and Cellular Damage**

High wind speeds can increase the production of reactive oxygen species (ROS), which cause oxidative damage to various cellular components, including those in the chloroplasts where photosynthesis occurs. ROS damage chlorophyll molecules and disrupt the photosynthetic machinery, leading to reduced energy production and photosynthetic efficiency.

- **ROS Production and Damage to Photosynthetic Machinery**: Wind stress-induced water loss and increased metabolic activity can generate excessive ROS, including superoxide radicals, hydrogen peroxide, and hydroxyl radicals. These ROS cause oxidative damage to chloroplasts, particularly to chlorophyll molecules, lipids, and proteins involved in the photosynthetic electron transport chain. Damage to chlorophyll reduces the plant's ability to capture light energy, while protein and lipid oxidation disrupt electron flow and ATP synthesis, both of which are crucial for energy production in photosynthesis.
- **Impact on Antioxidant Systems**: Plants have antioxidant systems, including enzymes like superoxide dismutase (SOD) and catalase, that neutralize ROS. However, when wind stress is prolonged or severe, ROS production can overwhelm these defenses, leading to oxidative stress. As a result, chloroplasts experience structural damage, which not only reduces photosynthetic efficiency but can also lead to cell death if oxidative stress becomes severe. In fruit crops, this damage affects leaf retention and quality, as oxidative stress can cause leaf drop and necrosis, further reducing the plant's photosynthetic capacity and impacting yield.
- **Consequences for Growth and Productivity**: If the plant's antioxidant systems cannot keep up with ROS production, oxidative stress can lead to cell damage and even cell death. This damage reduces overall productivity, as fewer functional cells remain for photosynthesis. In crops like apples and grapes, where high energy levels are required for fruit ripening, reduced photosynthesis due to oxidative stress can delay ripening and reduce fruit size, affecting both quality and market value. Additionally, oxidative damage impacts leaf health, resulting in chlorosis and necrosis, which reduce the plant's ability to perform photosynthesis effectively.

3. **Impact on Leaf Orientation and Sunlight Capture**

Wind stress can alter the physical orientation of leaves, as plants adjust their leaf position to reduce wind resistance. This change in leaf orientation reduces the surface area exposed to sunlight, which limits light capture and further impairs photosynthetic efficiency.

- **Leaf Angle Adjustment and Reduced Sunlight Capture**: Under strong winds, plants may angle or reposition their leaves to reduce exposure to the wind. While this adjustment helps minimize physical damage and water loss, it also decreases the amount of sunlight absorbed by the leaves, limiting the energy available for photosynthesis. In crops with large or dense canopies, such as fruit trees, the reduced light interception can significantly impact photosynthetic rates and carbohydrate production.

- **Compromised Energy Production and Carbohydrate Accumulation**: Sunlight is essential for the light-dependent reactions in photosynthesis, where light energy is converted into chemical energy. When leaf orientation changes reduce sunlight capture, it decreases the energy input for these reactions, reducing ATP and NADPH production. This reduction limits the Calvin cycle's ability to fix carbon and produce carbohydrates, resulting in lower energy availability. In fruit-bearing plants, lower carbohydrate accumulation directly impacts fruit size, flavor, and texture, as fruits fail to develop the sugars and structural components necessary for high-quality produce.
- **Long-Term Effects on Growth and Yield**: Over time, limited light capture due to altered leaf orientation reduces overall growth and yield, especially in fruit trees and other tall crops with extensive foliage. Reduced energy production limits vegetative growth, affecting leaf, branch, and root development, which are essential for sustained productivity. In fruit crops, a smaller canopy due to limited growth reduces the plant's ability to produce sufficient fruits, impacting both yield and quality. Additionally, reduced carbohydrate production can lead to weaker plants that are less resilient to other stressors, further compounding the effects of wind stress on productivity.

4. Adaptive Mechanisms to Mitigate Wind-Induced Photosynthetic Limitations

Plants have developed adaptive mechanisms to help mitigate the impacts of wind-induced limitations on photosynthesis and carbohydrate production. These adaptations enable plants to maintain photosynthetic efficiency and enhance resilience to wind stress.

- **Adjustment of Leaf Morphology**: Some plants adapt to windy environments by developing smaller, thicker, or waxier leaves that reduce water loss and increase resilience to physical damage. Smaller leaves have less surface area for wind to affect, and thicker leaves can better withstand mechanical stress. In fruit crops, adapting leaf morphology helps maintain a functional canopy and preserves photosynthetic capacity, supporting healthy fruit development even in windy conditions.
- **Enhanced Antioxidant Production**: Plants exposed to wind stress often increase the production of antioxidants, such as flavonoids and carotenoids, which help neutralize ROS and protect chloroplasts from oxidative damage. These antioxidants support the plant's defense against oxidative stress, preserving chloroplast function and reducing cell damage. For fruit crops, higher antioxidant levels in leaves and fruits can improve quality, as these compounds contribute to flavor, color, and nutritional value.
- **Flexible Leaf Petioles and Stems**: Some plants develop flexible leaf petioles and stems, which allow leaves to sway with the wind rather than resist it. This flexibility helps prevent damage to the photosynthetic machinery and ensures that leaves continue capturing

light efficiently. In crops like grapevines, flexible stems and petioles help leaves remain exposed to sunlight, supporting sustained photosynthesis and energy production despite windy conditions.

5. **Management Practices to Reduce the Impact of Wind Stress on Photosynthesis and Carbohydrate Production**

Growers can employ various management practices to minimize the effects of wind stress on photosynthesis, supporting carbohydrate production and improving crop resilience.

- **Installing Windbreaks for Protection**: Windbreaks, such as rows of trees, hedges, or fences, can reduce wind speed around crop fields, creating a more stable environment for plants. By reducing the intensity of wind exposure, windbreaks help maintain normal stomatal function, reduce water loss, and improve CO_2 uptake. For fruit crops, windbreaks protect the canopy and help sustain photosynthesis, improving carbohydrate production and fruit quality.
- **Supplemental Irrigation to Counteract Water Loss**: Providing additional irrigation during windy conditions helps compensate for increased water loss and reduces the need for stomatal closure. Drip irrigation, which delivers water directly to the root zone, is especially effective in windy environments, as it minimizes water loss and ensures plants have adequate moisture for nutrient and CO_2 uptake. In fruit orchards, supplemental irrigation supports carbohydrate accumulation, enhancing fruit sweetness and size.
- **Optimized Nutrient Management**: Ensuring that plants have access to essential nutrients, particularly potassium, magnesium, and nitrogen, supports photosynthesis and increases resilience to oxidative stress. Potassium helps regulate stomatal function, magnesium is essential for chlorophyll production, and nitrogen is necessary for protein synthesis. By optimizing nutrient availability, growers can improve photosynthetic efficiency and carbohydrate production, supporting healthy growth and fruit quality even under wind stress.

5.Adaptive Mechanisms to Mitigate Wind Stress

Wind stress imposes significant mechanical and physiological challenges on plants, including water loss, nutrient imbalances, and oxidative stress. To cope with these challenges, plants have developed several adaptive responses, which range from strengthening their structural components to altering growth patterns and increasing the production of protective compounds. These adaptations help plants withstand frequent wind exposure and maintain their growth and productivity, especially in crops that are regularly exposed to windy conditions.

1. **Strengthening of Cell Walls and Tissues**

One of the primary responses of plants to wind stress is the **strengthening of cell walls and tissues**, which enhances their mechanical stability. When plants are exposed to frequent wind, they often develop thicker cell walls and more lignified tissues. This response, known as **thigmomorphogenesis**, helps plants withstand mechanical forces, reducing the likelihood of stem breakage, lodging, and other physical damage.

- **Thicker Cell Walls for Increased Rigidity**: Wind-stressed plants may deposit additional cellulose, hemicellulose, and lignin in their cell walls, making them thicker and more rigid. This increased rigidity provides better structural support and makes plant stems, branches, and leaves more resilient to bending and breakage under wind pressure. In fruit trees, thicker cell walls in branches help reduce the risk of branch breakage, particularly when branches are loaded with developing fruits. For instance, apple and pear trees with thicker cell walls are less likely to lose branches under high wind conditions, which helps prevent yield loss.
- **Lignification for Enhanced Structural Support**: Lignin is a complex polymer that strengthens cell walls by providing rigidity and resistance to compressive forces. Plants exposed to regular wind stress often increase lignin production in stems and branches, making them sturdier and less prone to lodging. This lignification is particularly beneficial for fruit crops with heavy canopies, as it reduces the risk of trunk and branch breakage under strong winds. For example, in grapevines and citrus trees, lignification of the main stems and branches supports the plant's overall stability, minimizing the chances of structural damage that could impact yield.
- **Reduced Risk of Lodging and Fruit Loss**: Strengthened cell walls and increased lignification make wind-stressed plants more resilient to lodging, a condition where plants fall over completely. In crops like wheat or young fruit trees, lodging can limit sunlight exposure and increase disease susceptibility. By reinforcing their structural components, plants reduce the likelihood of lodging, allowing them to maintain photosynthesis and energy production. In fruit crops, these adaptations help prevent fruit loss, as stronger branches are less likely to break, reducing the risk of premature fruit drop and yield reduction.

2. **Growth Modifications**

Plants may also alter their growth patterns in response to wind stress, prioritizing stability over height and size. By modifying their growth, plants reduce wind resistance and improve their resilience to physical damage.

- **Shorter, More Compact Growth Form**: Wind-exposed plants often grow shorter and more compact, which lowers their center of gravity and reduces the surface area exposed to wind. This growth pattern reduces wind resistance and the likelihood of bending or breaking. For fruit trees and other tall crops, a compact canopy provides stability under windy conditions, reducing the risk of damage to branches and leaves. In wind-exposed areas, fruit trees like apple and peach may develop a shorter stature with thicker trunks, providing greater resistance to mechanical forces and reducing the risk of branch loss.
- **Thicker Stem Base for Stability**: To further stabilize themselves, wind-stressed plants may allocate more resources to strengthening the base of their stems. A thicker stem base increases the plant's structural stability, helping it withstand the mechanical forces of wind. In fruit crops, a thicker stem base supports the weight of the canopy and reduces the risk of uprooting. This adaptation is especially beneficial for young trees and newly transplanted crops, which are more susceptible to wind-induced instability.

- **Smaller Leaves to Reduce Wind Resistance**: Some plants exposed to wind stress develop smaller leaves, which reduce the surface area available for wind to exert force on. Smaller leaves not only decrease wind resistance but also conserve water by reducing the transpiration surface area. In fruit trees, smaller leaves help maintain a stable canopy structure while protecting the plant from excessive water loss. This adaptation is common in crops like citrus and olive trees, where smaller leaves help the plant withstand windy conditions without compromising its ability to produce fruit.

3. **Increased Production of Protective Compounds**

Wind stress can lead to the generation of reactive oxygen species (ROS), which cause oxidative stress and damage cellular structures. To counteract this, plants may increase the production of protective compounds, such as antioxidants and secondary metabolites, which help detoxify ROS and protect cells from damage.

- **Antioxidants to Combat Oxidative Stress**: Plants produce a variety of antioxidants, including enzymes like superoxide dismutase (SOD), catalase, and non-enzymatic compounds like flavonoids and carotenoids, to neutralize ROS. Under wind stress, these antioxidants help protect chloroplasts and other cellular structures from oxidative damage, preserving the plant's ability to perform photosynthesis and maintain cellular function. In fruit crops, higher antioxidant levels in leaves and fruits contribute to improved stress resilience and, in some cases, enhanced fruit quality. For instance, increased flavonoid content in grapevines not only improves resistance to oxidative stress but also contributes to grape flavor and color, enhancing the quality of wine grapes.
- **Secondary Metabolites for Cellular Protection**: Secondary metabolites, such as phenolic compounds and alkaloids, serve as protective agents against environmental stress. These compounds can stabilize cellular membranes, protect chlorophyll molecules, and reduce the impact of oxidative damage on photosynthetic efficiency. In fruit crops, secondary metabolites contribute to fruit flavor, color, and nutritional value, making them beneficial for both stress resilience and quality. In tomatoes and berries, for example, higher phenolic levels improve the plant's tolerance to wind stress while enhancing antioxidant content in the fruits, which is a desirable trait for consumers.
- **Improved Nutritional and Flavor Quality in Fruits**: In some fruit crops, increased production of antioxidants and secondary metabolites enhances fruit quality, as these compounds are responsible for flavor, color, and health benefits. For example, anthocyanins in berries and flavonoids in citrus fruits improve the plant's resistance to oxidative stress while contributing to the color and flavor of the fruit. This dual benefit of increased resilience and quality makes protective compounds especially valuable in fruit crops grown in wind-exposed regions.

4. **Other Adaptive Responses to Wind Stress**

In addition to structural reinforcement, growth modifications, and increased production of protective compounds, plants employ other adaptive strategies that improve resilience to wind stress and enhance their overall stability.

- **Alteration of Leaf Orientation**: Wind-stressed plants may alter leaf orientation to reduce wind exposure. By angling or repositioning their leaves, plants reduce the surface area exposed to wind, minimizing water loss and protecting sensitive tissues from physical damage. This adaptation is particularly beneficial for crops with large canopies, as adjusted leaf orientation helps maintain photosynthesis and energy production despite wind exposure.
- **Deeper Root Systems for Anchorage and Nutrient Uptake**: To improve anchorage and maintain nutrient uptake under windy conditions, plants may develop deeper and more extensive root systems. Deep roots anchor the plant firmly in the soil, reducing the risk of uprooting and improving access to water and nutrients from deeper soil layers. In fruit trees, a deeper root system supports stable growth and allows the plant to better withstand the mechanical forces of wind, minimizing stress on the canopy and branches.
- **Production of Flexible Tissues**: Some plants develop more flexible tissues that can sway with the wind rather than resist it. This flexibility helps prevent damage to stems, branches, and leaves, allowing the plant to maintain its structural integrity without risking breakage. In grapevines and other crops with flexible growth forms, this adaptation enables leaves and shoots to move with the wind, reducing physical stress and maintaining photosynthetic efficiency.

5. **Implications for Crop Management in Windy Environments**

Understanding these adaptive responses to wind stress can inform management practices that support crop resilience, reduce physical damage, and improve overall productivity in windy conditions.

- **Plant Selection and Breeding for Wind Resistance**: Breeding programs can focus on developing wind-resistant crop varieties with enhanced structural stability, thicker cell walls, and higher levels of antioxidants. Selecting fruit crop varieties with adaptive traits, such as compact growth forms and flexible stems, improves resilience to wind stress, allowing farmers to maintain productivity in wind-prone areas.
- **Use of Windbreaks and Shelterbelts**: Planting windbreaks, such as rows of trees or shrubs, around crop fields reduces wind exposure and helps prevent water loss, nutrient deficiencies, and physical damage. By creating a more sheltered environment, windbreaks allow plants to retain moisture, improve nutrient uptake, and reduce oxidative stress, supporting healthy growth and fruit quality.
- **Optimized Irrigation and Nutrient Management**: Wind-stressed plants require additional water and nutrients to counteract the effects of increased transpiration and limited nutrient uptake. Providing supplemental irrigation and balanced fertilization helps plants maintain growth and resilience, especially during periods of high wind exposure. In fruit orchards, optimized irrigation and nutrient management improve carbohydrate accumulation, supporting energy-intensive processes like flowering and fruiting.

Implications for Crop Management and Yield in Windy Environments

Wind stress poses multiple challenges to plant growth and productivity, affecting photosynthesis, water balance, nutrient uptake, and structural integrity. These impacts are especially critical in fruit production, where the quality and quantity of yield are essential to economic success. By understanding the effects of wind stress and implementing effective management strategies, growers can improve resilience, reduce crop losses, and maintain productivity in windy environments.

1. **Use of Windbreaks and Shelterbelts**

Windbreaks and **shelterbelts** are effective structures for protecting crops from wind stress by reducing wind speed and creating a more stable microenvironment. Windbreaks can be composed of rows of trees, shrubs, or hedges planted strategically around crop fields, while shelterbelts may include multiple rows of different plant species to maximize protection.

- **Reduction in Wind Speed and Water Loss**: Windbreaks reduce wind velocity, which in turn decreases the rate of transpiration and water loss from plants. This is particularly beneficial for fruit crops, which are sensitive to dehydration. By lowering the rate of water loss, windbreaks help maintain adequate moisture levels in the soil and in the plant itself, reducing the likelihood of water deficits. For crops like apple, citrus, and avocado trees, maintaining a stable water balance is crucial for fruit size and quality.
- **Minimizing Physical Damage and Fruit Drop**: By buffering crops from strong winds, windbreaks help minimize physical damage such as leaf tearing, branch breakage, and fruit drop. In fruit orchards, where mature trees may bear heavy fruit loads, reduced wind exposure prevents premature fruit drop, which is critical for maintaining yield. In crops like grapes and cherries, fruit drop can lead to significant yield losses, making windbreaks a valuable asset for protecting harvest quality and quantity.
- **Improving Microclimate and Soil Health**: Windbreaks create a stable microclimate that not only protects plants from direct wind exposure but also reduces soil erosion, preserving soil structure and nutrient content. Shelterbelts and windbreaks trap moisture and promote a cooler, more humid microenvironment, which benefits crops that thrive in moderate humidity. The stability provided by windbreaks also supports beneficial insect populations, contributing to pollination and pest control in fruit orchards, improving long-term crop health and sustainability.

2. **Support Systems for Wind-Sensitive Crops**

In windy environments, the structural stability of crops is crucial to prevent lodging, stem breakage, and yield loss. **Support systems** such as trellises, staking, and netting are often used to stabilize plants, particularly for wind-sensitive crops like tomatoes, grapes, and berry bushes.

- **Preventing Lodging and Stem Breakage**: Support systems help stabilize plants by providing physical support against wind forces, preventing them from bending, lodging, or falling over. For tall or heavy-yielding fruit crops like tomatoes, peppers, and grapes, trellises or stakes offer vertical support, reducing the risk of stem breakage under windy conditions. In grapevines, for example, trellises prevent vine collapse, allowing plants to maintain access to sunlight and air circulation, which are crucial for fruit quality.
- **Enhanced Fruit Retention and Improved Harvesting**: Support systems also aid in fruit retention by minimizing movement during windy conditions. Fruits that are well-

supported are less likely to fall prematurely, which is especially important for crops like cherries and blueberries, where fruit loss can impact yield and marketable quality. Support systems further assist in efficient harvesting, as organized plant structures are easier to access, reducing damage to fruits during collection and enhancing post-harvest quality.

- **Customized Support Solutions for Different Crops**: Each fruit crop has unique structural needs, and support systems can be customized to accommodate these requirements. For sprawling plants like strawberries and squash, ground-level supports such as low trellises and netting help protect fruits from direct wind contact. For taller crops like tomatoes, adjustable stakes provide flexibility to accommodate plant growth while ensuring stability in high winds. By tailoring support structures to specific crops, growers can maximize plant stability, improve growth, and maintain yield quality in windy conditions.

3. Optimized Irrigation and Nutrient Management

Wind-stressed plants experience increased transpiration and nutrient transport limitations, which require adjustments in **irrigation** and **nutrient management** to ensure plant health and productivity. These adjustments help counteract the effects of wind-induced water loss and nutrient deficiencies, improving resilience and supporting optimal growth.

- **Supplemental Irrigation to Counteract Water Loss**: Increased transpiration due to wind stress requires additional water to maintain plant hydration and reduce stress. By adjusting irrigation schedules, growers can prevent water deficits, particularly in fruit trees with high water demands. Drip irrigation systems, which deliver water directly to the root zone, are especially effective for minimizing evaporation losses in windy conditions. For fruit crops like apples, peaches, and citrus, supplemental irrigation supports continuous growth, fruit development, and flavor enhancement, as water availability is directly linked to sugar accumulation in fruits.
- **Foliar Sprays and Targeted Nutrient Application**: Wind-stressed plants often require additional nutrients to maintain cellular function and mitigate oxidative stress. Foliar sprays provide a direct nutrient source to leaves, helping counteract deficiencies in potassium, calcium, and magnesium, which are critical for cell wall strength, stomatal regulation, and photosynthesis. For instance, potassium and calcium foliar sprays strengthen cell walls, reducing leaf scorch and physical damage under windy conditions. In fruit crops, these nutrients improve fruit firmness, shelf life, and disease resistance, supporting better marketable quality.
- **Enhanced Fertilization Practices**: Adjusting fertilization practices to accommodate the nutrient needs of wind-stressed plants is essential for maintaining growth and productivity. Applying slow-release fertilizers ensures a steady supply of nutrients to plants, reducing the risk of nutrient leaching and maximizing uptake efficiency. For crops like strawberries, grapes, and tomatoes, which require a balanced nutrient profile for optimal growth, enhanced fertilization helps mitigate the impacts of wind stress on nutrient uptake and transport, improving yield and fruit quality.

4. Additional Strategies to Enhance Resilience to Wind Stress

In addition to windbreaks, support systems, and optimized irrigation, there are other effective practices that can help plants withstand wind stress and maintain productivity in challenging environments.

- **Pruning and Canopy Management**: Pruning helps maintain a compact canopy structure, which reduces wind resistance and the risk of branch breakage. By removing excess foliage and creating a well-balanced canopy, plants are better able to withstand wind stress without sacrificing photosynthesis and fruit production. In fruit trees like apple, cherry, and citrus, pruning improves air circulation, reduces disease susceptibility, and minimizes the likelihood of wind-induced fruit drop.
- **Mulching for Soil Stability and Moisture Retention**: Mulching around the base of plants helps conserve soil moisture by reducing evaporation and insulating the soil from temperature fluctuations. In windy environments, mulch stabilizes the soil and prevents erosion, protecting the root zone and enhancing nutrient availability. Mulching is particularly beneficial for shallow-rooted crops like strawberries and blueberries, as it helps maintain stable moisture levels and reduces the risk of root exposure in high winds.
- **Crop Rotation and Companion Planting**: Certain companion plants can act as natural windbreaks, providing additional shelter and reducing wind exposure for sensitive crops. Tall, sturdy plants like sunflowers or corn can be strategically planted to shield shorter, wind-sensitive crops. Crop rotation also benefits soil structure and fertility, ensuring that plants have access to essential nutrients even in windy conditions, which supports healthier growth and improved yield.

5. Impacts on Crop Yield and Long-Term Productivity

Implementing these management strategies can have significant positive effects on crop yield and productivity, particularly in fruit production where quality and yield are essential to economic viability.

- **Improved Fruit Quality and Market Value**: By reducing wind-induced stress through effective crop management, growers can improve fruit size, sweetness, firmness, and overall quality. In fruit crops, stable water and nutrient availability support optimal carbohydrate accumulation, which enhances flavor, aroma, and color. In crops like grapes, apples, and berries, improved quality translates to higher market value, as consumers prioritize fruits with superior taste, texture, and nutritional content.
- **Enhanced Resilience to Secondary Stressors**: Wind stress can weaken plants, making them more susceptible to secondary stressors like pest infestations and diseases. By implementing protective measures such as windbreaks, support systems, and optimized irrigation, growers can reduce plant vulnerability and enhance resilience. Healthy, well-nourished plants are better equipped to resist pathogens and recover from damage, which helps maintain productivity and reduce crop losses over time.
- **Sustainable Long-Term Yield**: Effective wind management practices contribute to sustainable long-term productivity by preserving soil health, improving nutrient retention, and supporting stable growth. For perennial crops, such as fruit trees, consistent application of these practices enhances resilience over multiple growing seasons, ensuring

that the orchard remains productive and profitable. By reducing wind-related damage and improving overall crop health, growers can achieve consistent yields and reduce the need for costly interventions, supporting sustainable agricultural practices in windy environments.

Table 5: Effect of Wind Stress on Fruit Crops

S. No	Name of Crop	Effect of wind Stress
1.	Banana	Uprooting, tearing and removal of whole leaf lamina, susceptible to toppling
2.	Litchi	Fruit skin cracking and sun burn
3.	Mango	Fruit Drop
4.	Papaya	Crinkled leaves, reduced growth, fruit set, fruit quality, and productivity, susceptible to toppling
5.	Strawberry	Increased symptoms of bacterial or fungal diseases

e. Soil Stress Conditions (Salinity, Alkalinity, Ion Toxicity, Fertilizer Toxicity, etc.)

Soil stress conditions significantly influence plant growth and productivity. Stress from salinity, alkalinity, ion toxicity, and excessive fertilizer application disrupt water balance, nutrient absorption, and cellular processes, affecting crop yield and quality. Understanding these conditions is essential for managing soil health and supporting sustainable agriculture, especially in fruit crops that are sensitive to soil composition.

1. Salinity Stress

Soil salinity arises when the soil contains high levels of soluble salts, such as sodium chloride (NaCl) and calcium sulfate ($CaSO_4$). Salinity is particularly problematic in arid and semi-arid regions, where high evaporation rates cause salts to accumulate on the soil surface. This stress imposes severe challenges for plant growth by affecting water uptake, nutrient availability, and metabolic processes, which are essential for healthy growth and productivity, especially in sensitive fruit crops.

1. **Osmotic Stress and Reduced Water Uptake**

High salt concentrations in soil increase the osmotic potential, making it challenging for plants to absorb water. This **osmotic stress** limits water availability to the plant, even if there is sufficient water in the soil. The reduced water uptake creates a water deficit within the plant, leading to dehydration and a cascade of physiological responses aimed at conserving water.

- **Dehydration and Stomatal Closure**: To conserve water, plants under osmotic stress may close their stomata, the small pores on the leaf surface that regulate gas exchange. Stomatal closure reduces water loss through transpiration but also restricts CO_2 intake, which is essential for photosynthesis. This reduction in photosynthesis limits the plant's ability to produce carbohydrates and energy needed for growth, flowering, and fruit production. In fruit crops such as tomatoes and grapes, reduced photosynthetic rates can lead to smaller, less flavorful fruits and delayed ripening.

- **Impact on Growth and Development**: Osmotic stress impacts cellular turgor, the pressure that helps cells maintain their shape and structure. Reduced turgor leads to wilted, stunted growth as the cells lose water and become flaccid. For fruit crops, maintaining cell turgor is essential for proper fruit expansion and development. Under salinity-induced osmotic stress, fruit growth is limited, resulting in smaller fruits with lower market value. Additionally, reduced energy production due to limited photosynthesis impacts flowering and fruit set, lowering overall crop yield.
- **Adaptations to Conserve Water**: Some plants respond to osmotic stress by producing osmoprotectants, compounds that help maintain cellular water potential. These compounds, such as proline and glycine betaine, accumulate in plant cells to counteract dehydration and stabilize proteins and membranes. However, producing osmoprotectants requires significant energy, which further reduces growth and productivity in fruit crops that rely heavily on high energy levels for fruit development.

2. **Ion Toxicity and Nutrient Imbalance**

In addition to osmotic stress, high levels of sodium (Na^+) and chloride (Cl^-) ions can accumulate in plant tissues, leading to **ion toxicity**. These ions compete with essential nutrients for absorption sites on the root cell membranes, disrupting nutrient balance and causing deficiencies that impact cellular processes and growth.

- **Sodium Interference with Nutrient Uptake**: Sodium ions (Na^+) can interfere with the absorption of essential nutrients such as potassium (K^+), calcium (Ca^{2+}), and magnesium (Mg^{2+}). Potassium is critical for stomatal regulation, as it helps control water movement and transpiration rates. A deficiency in potassium disrupts water balance, affecting the plant's ability to cope with salinity stress. Calcium, which is crucial for cell wall stability, becomes deficient under high sodium conditions, weakening cell walls and making the plant more susceptible to diseases and pest attacks. In fruit crops, these nutrient imbalances reduce fruit quality and storage life, as weaker cell walls result in softer fruits that are prone to decay.
- **Chloride Toxicity and Leaf Damage**: Chloride ions (Cl^-), though essential in small amounts, become toxic when accumulated in large quantities. High chloride levels disrupt photosynthesis by affecting chloroplast function and inhibiting enzymes involved in the process. Chloride toxicity often leads to leaf burn, characterized by browning at the edges and tips of leaves, and ultimately to leaf drop. In fruit trees like citrus and avocado, chloride toxicity reduces leaf area and photosynthetic capacity, which directly impacts fruit size, yield, and flavor.
- **Effect on Fruit Quality and Yield**: Nutrient imbalances due to salinity affect various aspects of fruit quality, such as size, texture, and flavor. Potassium deficiency, for example, leads to lower sugar accumulation in fruits, reducing sweetness and market appeal. Calcium deficiency weakens cell structure, resulting in softer fruits with reduced shelf life. For high-value crops like grapes and strawberries, the loss of fruit quality due to salinity stress translates into significant economic losses.

3. Oxidative Stress and Cellular Damage

Salinity stress disrupts metabolic processes in plants, leading to an overproduction of **reactive oxygen species (ROS)**. ROS, such as superoxide radicals, hydrogen peroxide, and hydroxyl ions, are highly reactive molecules that damage cellular components, including lipids, proteins, and DNA. The accumulation of ROS creates oxidative stress, which exacerbates cellular damage and impairs plant health.

- **Membrane Damage and Loss of Integrity**: ROS can oxidize lipids in cell membranes, leading to lipid peroxidation. This process damages the structure and integrity of cell membranes, making them more permeable and causing leakage of ions and other cellular contents. In fruit crops, this loss of membrane integrity affects water and nutrient transport, leading to reduced growth and yield. Membrane damage also affects fruit quality, as leaky cells contribute to softening and susceptibility to post-harvest decay.
- **Damage to Photosynthetic Machinery**: Salinity-induced oxidative stress damages chloroplasts, the organelles responsible for photosynthesis. ROS can oxidize chlorophyll and other components of the photosynthetic machinery, reducing the plant's ability to capture and convert light energy. Chloroplast damage leads to chlorosis, where leaves turn yellow due to reduced chlorophyll levels, limiting photosynthetic efficiency. In sensitive crops like strawberries and peppers, reduced photosynthesis from oxidative stress diminishes energy availability for growth and fruit development, impacting yield and fruit quality.
- **Antioxidant Defense and Energy Cost**: Plants have antioxidant defense systems, including enzymes like superoxide dismutase (SOD) and catalase, that detoxify ROS and protect cells from oxidative damage. However, producing antioxidants requires energy and resources that could otherwise support growth and reproduction. In fruit crops, the energy diverted to antioxidant production reduces the availability of resources for fruit development, leading to smaller, lower-quality fruits. In some cases, prolonged oxidative stress overwhelms the plant's defense systems, resulting in cell death and tissue necrosis, further reducing crop productivity.

4. Management Strategies to Mitigate Salinity Stress

Addressing salinity stress requires a combination of soil management, water management, and the use of tolerant crop varieties. These strategies help reduce salt accumulation, improve water and nutrient availability, and support plant resilience.

- **Leaching and Improved Irrigation Practices**: Leaching involves applying sufficient water to flush excess salts below the root zone, reducing salt concentration in the soil. Drip irrigation, which delivers water directly to the roots, minimizes evaporation and salt accumulation on the soil surface, making it an effective practice for managing salinity. High-quality irrigation water with low salt content is also essential, as saline water can exacerbate salt build-up and increase osmotic stress. In regions with limited freshwater availability, growers may use rainwater harvesting or mix low-salinity water with irrigation water to dilute salt levels.

- **Application of Soil Amendments**: Soil amendments like gypsum (calcium sulfate) help displace sodium ions in the soil, allowing them to be leached away. Gypsum provides calcium, which improves soil structure and enhances nutrient availability by reducing sodium toxicity. Organic matter, such as compost or mulch, can also improve soil texture, increase water retention, and enhance microbial activity, promoting nutrient availability and root health. For fruit orchards, these soil amendments improve root functionality, supporting healthier trees and reducing the impact of salinity on fruit quality.
- **Use of Salt-Tolerant Crop Varieties**: Breeding and selecting salt-tolerant crop varieties are effective strategies for reducing the impact of salinity stress. Salt-tolerant plants have adaptations, such as enhanced ion transport mechanisms and increased production of osmoprotectants, that allow them to survive and produce in saline soils. For example, salt-tolerant grapevine and citrus varieties can maintain productivity in moderate salinity, providing farmers with a sustainable option for managing salinity stress.
- **Balanced Fertilization and Foliar Feeding**: Applying balanced fertilizers that provide adequate potassium and calcium can mitigate nutrient imbalances caused by salinity. Potassium and calcium are essential for cell wall strength and water regulation, and their availability improves plant resilience to osmotic and ionic stress. In cases where root absorption is limited by salinity, foliar sprays with potassium and calcium provide a direct source of these nutrients, supporting healthy growth and fruit development in crops like tomatoes and strawberries.

2. Alkalinity Stress

Alkalinity stress occurs in soils with high pH levels, typically above 7.5, which impacts nutrient solubility and the activity of beneficial soil microbes. Alkaline soils are common in areas with poor drainage, lime buildup, or high bicarbonate levels. High pH affects root function and restricts nutrient uptake, which is especially problematic for fruit crops that prefer slightly acidic to neutral soil. As a result, plants growing in alkaline soils often experience stunted growth, nutrient deficiencies, and reduced productivity.

1. **Reduced Nutrient Solubility and Uptake**

High pH in alkaline soils affects the solubility and availability of essential nutrients, particularly iron (Fe), manganese (Mn), zinc (Zn), and phosphorus (P). These nutrients are vital for plant metabolism, chlorophyll production, and enzyme function, but they become less accessible to plants in alkaline conditions, leading to deficiencies that impair growth and development.

- **Iron Deficiency and Chlorosis**: Iron is a crucial nutrient for chlorophyll synthesis and various enzymatic functions, but in alkaline soils, it becomes insoluble and unavailable to plants. Iron deficiency, also known as iron chlorosis, is commonly observed in young leaves, where a yellowing or pale color appears while veins remain green. This chlorosis reduces photosynthetic efficiency, leading to poor energy production, stunted growth, and reduced fruit development. In fruit crops such as citrus and grapes, iron chlorosis can significantly affect fruit quality, size, and yield.
- **Zinc and Manganese Deficiency**: Like iron, zinc and manganese are also critical for

chlorophyll production and enzyme activation, but they become insoluble at high pH levels. Zinc deficiency causes interveinal chlorosis (yellowing between veins) and leaf deformities, while manganese deficiency results in pale green or yellow leaves with small, irregular spots. These deficiencies reduce the plant's photosynthetic capacity, limiting energy production and growth. In crops like apples and cherries, zinc and manganese deficiencies due to alkalinity stress lead to smaller, poor-quality fruits with reduced market value.

- **Phosphorus Imbalance**: Phosphorus, essential for energy transfer and root development, also becomes less available in alkaline soils, as it forms insoluble compounds that plants cannot absorb. Low phosphorus availability affects root growth, flowering, and fruit set, leading to reduced productivity. For fruit crops, insufficient phosphorus impacts root expansion, limiting access to water and nutrients and reducing overall yield and fruit quality.

2. **Impact on Root Growth and Microbial Activity**

Alkaline conditions impact root growth directly and indirectly by limiting microbial activity in the soil. Both root function and soil microbes are essential for effective nutrient cycling and uptake, and disruptions in these processes lead to further nutrient limitations, affecting plant health and productivity.

- **Inhibited Root Growth and Water Uptake**: High pH levels restrict root growth, as the altered chemical environment affects root cell membrane permeability and enzyme activity. Reduced root growth limits the plant's ability to absorb water and nutrients, increasing the risk of dehydration and nutrient deficiencies. In fruit crops, root growth is essential for stability and access to soil resources, so restricted root development leads to poor growth and reduced productivity.
- **Decline in Beneficial Soil Microorganisms**: Soil microorganisms, such as mycorrhizal fungi and nitrogen-fixing bacteria, play a crucial role in nutrient cycling and uptake. Mycorrhizal fungi enhance phosphorus availability and water absorption, while nitrogen-fixing bacteria convert atmospheric nitrogen into a form plants can use. Alkaline conditions inhibit the activity of these microorganisms, reducing their populations and limiting nutrient availability. In alkaline soils, crops like apples and cherries are particularly affected by reduced nitrogen availability, leading to poor leaf and fruit development, smaller fruit size, and diminished flavor.
- **Effects on Nitrogen and Phosphorus Cycling**: Soil microbes contribute to nutrient cycling, breaking down organic matter and releasing nutrients in plant-available forms. In alkaline soils, microbial activity declines, slowing down nitrogen and phosphorus cycling and leading to nutrient imbalances that impact plant growth. Nitrogen deficiency affects vegetative growth, while low phosphorus availability impairs root function and fruit development. This nutrient imbalance reduces growth and yield, particularly in fruit crops, which require consistent nutrient levels for high-quality production.

3. **Increased Risk of Secondary Stresses**

Alkaline soils are often associated with poor drainage, compaction, and soil aeration issues, which increase the risk of secondary stresses such as root diseases and oxygen deficiency. These conditions further restrict nutrient uptake and make plants more susceptible to pathogens, impacting long-term crop productivity.

- **Soil Compaction and Restricted Root Aeration**: Alkaline soils often suffer from compaction, which reduces the space for air within the soil structure. Compacted soils limit root aeration, increasing the likelihood of oxygen-deficient conditions. Without adequate oxygen, roots struggle to respire, leading to poor energy production and reduced nutrient uptake. In fruit orchards, restricted root aeration weakens tree stability and impacts fruit production, making trees more susceptible to stress and reducing orchard productivity over time.
- **Increased Susceptibility to Root Diseases**: Poor drainage, commonly associated with alkaline soils, creates anaerobic (oxygen-poor) conditions that promote the growth of pathogenic fungi and bacteria. Pathogens such as Phytophthora thrive in waterlogged, anaerobic soils, leading to root rot and further impairing nutrient absorption. For fruit crops, root diseases weaken trees and increase the likelihood of premature fruit drop, reducing both yield and quality. In severe cases, root rot can lead to plant death, significantly impacting orchard productivity.
- **Long-Term Weakening of Trees in Fruit Orchards**: In fruit orchards, the combination of alkaline soil, poor drainage, and compaction leads to weakened trees that struggle to maintain adequate growth and yield. Weakened trees are more vulnerable to environmental stresses, pests, and pathogens, which further reduce productivity. Over time, this accumulation of stressors impacts orchard sustainability, leading to decreased yield and lower fruit quality, as trees cannot access the nutrients needed for robust fruit production.

4. Management Strategies to Alleviate Alkalinity Stress

Addressing alkalinity stress requires careful soil management, including pH adjustments, improved drainage, and soil amendment practices. These strategies aim to improve nutrient availability, enhance root function, and support beneficial soil microorganisms, particularly in fruit crops sensitive to high pH.

- **pH Adjustment Using Soil Amendments**: Adding soil amendments such as sulfur or organic matter can help lower soil pH and improve nutrient availability. Sulfur, when added to the soil, is oxidized by soil bacteria into sulfuric acid, which lowers pH and increases the availability of micronutrients like iron and zinc. Organic matter, such as compost or peat, improves soil structure, supports microbial activity, and enhances nutrient retention. In fruit orchards, these pH adjustments allow for better root development and nutrient uptake, supporting healthy growth and productivity.
- **Gypsum Application for Improved Soil Structure**: Gypsum (calcium sulfate) is often applied to alkaline soils to improve soil structure and reduce compaction. Gypsum displaces sodium ions, which contribute to alkalinity, and improves soil texture, enhancing drainage and root aeration. For fruit crops, gypsum application supports deeper root growth, which improves access to water and nutrients and reduces the impact of alkalinity on productivity.

- **Enhanced Irrigation Practices and Drainage Management**: Improving drainage is essential in alkaline soils to prevent waterlogging and reduce the risk of root diseases. Installing drainage systems or using raised beds can help manage water flow and reduce soil compaction. Additionally, using high-quality irrigation water with low salt content helps avoid further pH increases due to bicarbonate build-up. For fruit trees, proper drainage and irrigation practices are critical for maintaining healthy root systems and supporting high-quality fruit production.
- **Incorporating Organic Matter for Microbial Health**: Organic matter supports soil microbial communities by providing nutrients and improving soil structure. Mycorrhizal fungi and nitrogen-fixing bacteria thrive in soils rich in organic matter, which enhances nutrient cycling and availability. By incorporating compost or other organic materials, growers can improve soil fertility, enhance root health, and support beneficial microbes. In alkaline soils, organic matter additions improve nitrogen and phosphorus availability, supporting better growth and productivity in fruit crops like apples, grapes, and cherries.

5. Management Strategies to Mitigate Soil Stress Conditions

Addressing soil stress conditions requires a combination of soil management, irrigation adjustments, and the use of tolerant crop varieties to minimize the impact of salinity, alkalinity, ion toxicity, and fertilizer toxicity on plant health.

- **Soil Amendments and pH Adjustments**: Applying soil amendments like gypsum can help displace sodium ions and reduce soil salinity, while sulfur can be used to lower soil pH in alkaline soils. Organic matter additions, such as compost, improve soil structure and enhance microbial activity, promoting better nutrient availability and root health. In fruit orchards, pH adjustments improve root function and nutrient uptake, supporting healthy growth and productivity.
- **Irrigation Management**: Using high-quality irrigation water with low salt content and avoiding excessive irrigation reduce the accumulation of toxic ions in the soil. Leaching, which involves flushing excess salts below the root zone with ample water, helps reduce salinity. Additionally, drip irrigation minimizes water waste and salt build-up, providing steady moisture to fruit crops without overloading the soil with salts.
- **Use of Salt-Tolerant Crop Varieties**: Selecting salt-tolerant or alkaline-tolerant varieties helps reduce yield losses in areas prone to soil stress. For example, salt-tolerant grapevine or citrus varieties can withstand moderate salinity, allowing growers to maintain productivity even in challenging conditions.
- **Controlled Fertilizer Application**: Applying fertilizers in balanced amounts and using slow-release formulations reduce the risk of fertilizer toxicity. Soil testing before fertilization helps identify existing nutrient levels and prevents over-application. In fruit production, balanced fertilization ensures optimal growth, fruit size, and flavor, while minimizing the risk of nutrient imbalances and toxicity.

3. Ion Toxicity

Ion toxicity occurs when certain ions, particularly sodium (Na^+), chloride (Cl^-), and boron (B), accumulate to harmful levels in the soil or plant tissue, disrupting normal cellular functions. Excessive ion concentrations result from conditions such as high soil salinity, poor soil management, or irrigation with saline water. High ion levels interfere with cellular homeostasis, enzyme activity, and nutrient uptake, leading to growth inhibition, leaf necrosis, and diminished fruit quality. This stress is particularly detrimental in fruit crops, where even moderate ion toxicity can impact yield, fruit set, and overall productivity.

1. Sodium and Chloride Toxicity

Sodium (Na^+) and chloride (Cl^-) ions are prevalent in saline soils and often accumulate through irrigation with saline water. While plants can tolerate low levels of these ions, excessive sodium and chloride disrupt cellular balance, interfere with vital nutrient functions, and negatively impact growth. Sodium and chloride toxicity is especially concerning for salt-sensitive fruit crops such as strawberries, peppers, grapes, and citrus, as these crops exhibit visible stress symptoms that reduce their commercial value and productivity

- **Disruption of Cellular Homeostasis and Ion Imbalance**: Sodium toxicity primarily affects cellular homeostasis by disrupting the delicate balance of ions within plant cells. Potassium (K^+) is an essential nutrient that plays a critical role in maintaining osmotic balance and activating various enzymes required for photosynthesis, protein synthesis, and other metabolic processes. However, sodium competes with potassium for uptake and transport within the cell, displacing potassium ions and disrupting potassium-dependent functions. For example, potassium is crucial for stomatal regulation, allowing plants to control water loss and gas exchange. When sodium displaces potassium, the plant loses its ability to regulate stomata effectively, leading to excessive water loss, reduced CO_2 intake, and impaired photosynthesis.
- **Membrane Instability and Increased Susceptibility to Dehydration**: Sodium ions alter the structural integrity of cell membranes, affecting their permeability and stability. This alteration increases membrane leakage, causing essential nutrients and ions to escape from cells, further disrupting cellular function. When cell membranes become less stable, cells lose their ability to retain water effectively, causing dehydration. In fruit crops, this dehydration effect is particularly damaging to roots and leaves, as it limits the plant's ability to absorb water and nutrients, leading to visible symptoms such as wilting, leaf browning, and reduced growth. For instance, in strawberries, sodium toxicity impairs root function, restricting water and nutrient absorption, which ultimately reduces fruit size and quality.
- **Impact on Enzyme Activity and Metabolism**: Sodium interferes with enzyme activity by altering the ionic environment within cells. Enzymes require specific ion concentrations to function optimally, and sodium disrupts this balance, reducing the efficiency of metabolic reactions involved in growth, energy production, and stress response. As metabolic processes slow down, energy production declines, limiting the plant's capacity for new growth and reducing the formation of vital compounds needed for fruit development.

This disruption is critical in fruit crops, where energy is needed for flowering, fruit set, and ripening.

- **Chloride Toxicity and Photosynthetic Inhibition**: While chloride ions play a minor role in photosynthesis and osmotic regulation, they become toxic at high concentrations. Excess chloride accumulates in plant tissues, particularly in chloroplasts, where it interferes with photosynthetic enzymes. Chloride toxicity impairs photosynthesis by inhibiting chlorophyll production and damaging the photosynthetic machinery, leading to chlorosis (yellowing of leaves) and reduced energy production. In salt-sensitive fruit crops like grapes, citrus, and avocado, the accumulation of chloride ions causes leaf burn, characterized by browning and drying at the edges of leaves, which limits the plant's ability to perform photosynthesis efficiently. This reduction in photosynthetic activity decreases energy availability, which is essential for growth and fruit development, leading to lower yields and poor fruit quality.

2. Boron Toxicity

Boron is an essential micronutrient involved in cell wall formation, membrane integrity, and reproductive processes in plants. However, when present in excessive amounts, boron becomes toxic, especially in crops with low boron tolerance. Boron toxicity is often observed in soils with high boron content or where irrigation water contains elevated boron levels. This condition can be particularly harmful in fruit crops like citrus, avocado, and stone fruits, which are sensitive to boron levels.

- **Accumulation in Leaves and Cellular Disruption**: Boron toxicity generally manifests as yellowing or browning at the edges of older leaves, progressing to necrosis (cell death) as toxicity levels increase. This leaf burn is caused by the accumulation of boron in leaf tissues, where it interferes with cellular functions and damages membranes. High boron concentrations disrupt cell wall structure, affecting cell stability and transport processes within tissues. In severe cases, boron toxicity leads to cell death, which weakens overall plant health and reduces photosynthetic capacity.
- **Impact on Flower Formation and Fruit Set**: Boron plays a critical role in the reproductive development of plants, including pollen tube formation and fertilization. However, excessive boron disrupts these reproductive processes, leading to poor flower formation and reduced fruit set. In crops like citrus and avocado, boron toxicity affects the number and quality of flowers, leading to fewer fruits and reduced yields. Additionally, poor pollination due to boron toxicity results in incomplete or misshapen fruits, which reduces market value and limits the economic viability of affected orchards.
- **Long-Term Effects on Fruit Quality and Shelf Life**: Excessive boron levels not only reduce fruit set but also impact fruit quality by causing structural weaknesses and lowering shelf life. Fruits that develop under boron toxicity are often smaller, less flavorful, and more prone to post-harvest decay. In crops like grapes, apples, and stone fruits, these quality reductions affect both yield and profitability, as lower-quality fruits are less desirable in the marketplace. Boron toxicity thus has a lasting impact on crop performance, particularly in high-value fruit crops where appearance, flavor, and storage life are key factors.

3. Inhibition of Growth and Reproductive Development

Ion toxicity significantly inhibits plant growth by disrupting nutrient uptake and essential metabolic processes. In fruit crops, reduced growth and impaired reproductive development have direct consequences for yield, fruit quality, and economic returns.

- **Growth Suppression and Leaf Necrosis**: High levels of sodium, chloride, or boron led to a general reduction in growth as cellular functions are compromised. Toxic ion accumulation causes leaf necrosis, where tissues die due to ion-induced damage, limiting photosynthetic area and reducing energy production. In fruit crops, leaf necrosis weakens the plant's overall health and resilience, making it more susceptible to other stresses. Growth suppression is particularly detrimental in perennial crops like grapes and citrus, where consistent leaf area and photosynthetic capacity are essential for long-term productivity and fruit quality.
- **Interference with Nutrient Uptake and Water Transport**: Toxic ions compete with essential nutrients like potassium, calcium, and magnesium, impairing the plant's ability to absorb and transport these nutrients. Potassium is critical for photosynthesis and stomatal regulation, calcium stabilizes cell walls, and magnesium is essential for chlorophyll synthesis. When toxic ions displace these nutrients, the plant experiences deficiencies that affect cellular function, water regulation, and overall vitality. In crops like strawberries, grapes, and tomatoes, nutrient deficiencies lead to smaller fruits, altered flavor, and reduced yield, as the plant cannot meet the high nutrient demands for optimal fruit development.
- **Reduced Flowering, Fruit Set, and Fruit Quality**: The reproductive stages of flowering, fruit set, and fruit development are highly sensitive to ion toxicity, as these processes require substantial energy and nutrient support. High sodium or chloride levels reduce flowering and fruit set by limiting energy production and disrupting reproductive development. For instance, in grapevines, sodium toxicity has been shown to reduce the number of flower clusters, limiting potential yield. The fruits that do form under ion stress are often of lower quality, with reduced size, sugar content, and flavor, which are crucial factors in fruit crops destined for fresh consumption or processing.

Management Strategies to Mitigate Ion Toxicity

Ion toxicity is a significant challenge in agriculture, particularly in soils with high sodium, chloride, or boron levels, where these ions can inhibit plant growth and productivity. Addressing ion toxicity requires integrated soil and water management strategies to reduce toxic ion concentrations, support nutrient availability, and enhance plant resilience. These practices are essential in fruit orchards and other high-value crops, where ion toxicity can impact yield, fruit quality, and economic returns.

1. Leaching Excess Salts from Soil

Leaching is the process of applying large volumes of water to flush excessive salts below the root zone, reducing the concentration of harmful ions like sodium and chloride. By moving these ions away from the roots, leaching prevents the build-up of toxic ions around plant roots and improves soil health.

- **Drip and Flood Irrigation Techniques**: Drip and flood irrigation methods are commonly used to leach salts, as they provide sufficient water to wash away excessive ions without causing waterlogging. Drip irrigation delivers water directly to the root zone, minimizing water wastage and preventing salt accumulation on the soil surface. In flood irrigation, water is spread across the soil to saturate the top layer and flush salts downward. However, with flood irrigation, careful management is required to avoid creating standing water, which can lead to root suffocation and soil erosion.
- **Low-Salinity Irrigation Water**: The quality of irrigation water is crucial for effective leaching. Using low-salinity water prevents further accumulation of sodium and chloride ions in the soil. Where freshwater is limited, growers can mix saline water with freshwater to reduce salt content or use rainwater harvesting to provide an alternative low-salinity water source. For fruit orchards, using low-salinity water ensures that ion concentrations remain within manageable limits, reducing the risk of toxicity and supporting healthy root development and nutrient uptake.
- **Regular Leaching for Consistent Growth and Yield**: Regular leaching prevents ion buildup in the soil over time, supporting healthier root function and consistent crop growth. In fruit orchards, periodic leaching improves soil structure and reduces osmotic stress, enabling plants to absorb water and nutrients efficiently. This practice is particularly beneficial in salt-sensitive fruit crops like strawberries, apples, and grapes, where excessive ion concentrations can significantly reduce yield and fruit quality. Regular leaching helps maintain an optimal growing environment, ensuring sustained productivity and high-quality fruit.

2. **Application of Soil Amendments**

Soil amendments like gypsum (calcium sulfate) and organic matter help displace toxic ions from soil particles, improving soil structure and enhancing nutrient availability. Soil amendments play a critical role in reducing ion toxicity by supporting better drainage, aeration, and root health.

- **Gypsum for Sodium Displacement**: Gypsum is a widely used amendment for managing sodium toxicity. Calcium in gypsum displaces sodium ions from soil particles, allowing them to be leached away with irrigation. This displacement improves soil structure, reducing compaction and enhancing water infiltration and aeration. In sodium-sensitive crops like strawberries and tomatoes, gypsum application reduces the risk of sodium toxicity, promotes healthier root growth, and supports nutrient uptake, improving overall growth and fruit quality.
- **Improved Soil Drainage and Aeration**: By displacing sodium ions and improving soil structure, gypsum and other soil amendments enhance drainage and aeration, which are essential for healthy root function. Improved drainage prevents waterlogging, which can exacerbate ion toxicity and lead to root suffocation. Enhanced aeration also supports beneficial soil microbes, which play a role in nutrient cycling and availability. In fruit crops, well-aerated soil allows roots to access oxygen, supporting root health, nutrient absorption, and sustained productivity.

- **Organic Matter to Improve Soil Health**: Adding organic matter, such as compost or manure, improves soil texture, water retention, and microbial activity. Organic matter increases soil organic carbon levels, which enhance nutrient availability and root growth, creating a more resilient soil environment. The organic acids in compost also help to chelate toxic ions, reducing their availability to plants. For fruit orchards, organic matter additions improve soil fertility and root health, reducing the negative impacts of ion toxicity on growth and fruit production.

3. **Use of Salt-Tolerant Crop Varieties**

Salt-tolerant crop varieties possess genetic adaptations that allow them to exclude or compartmentalize toxic ions, improving resilience in saline or ion-toxic soils. These varieties provide a viable option for maintaining productivity in areas with high salinity or ion toxicity.

- **Mechanisms of Salt Tolerance**: Salt-tolerant varieties exclude toxic ions from roots or compartmentalize them in vacuoles to prevent ion buildup in vital tissues. By limiting the movement of harmful ions to leaves and other sensitive parts, these plants maintain cellular homeostasis and reduce the physiological stress associated with high ion concentrations. Salt-tolerant varieties also produce osmoprotectants like proline, which help cells retain water and prevent dehydration under saline conditions.
- **Improved Growth and Reproductive Success**: Salt-tolerant varieties exhibit stable growth and reproductive development even in saline soils, as their adaptations protect essential metabolic processes from ion-induced disruptions. For example, salt-tolerant grape or citrus varieties can grow and produce fruit with minimal impact from sodium or chloride toxicity, making them valuable options in regions with saline soils or irrigation water. These varieties offer a sustainable way to maintain productivity and quality, even in challenging environments.
- **Regional Suitability and Variety Selection**: Choosing the right salt-tolerant varieties is essential for achieving high yields in saline-prone regions. Breeders have developed salt-tolerant varieties of several fruit crops, including grapes, citrus, and tomatoes, which can withstand moderate salinity levels without compromising growth or fruit quality. By selecting salt-tolerant varieties suited to local soil and climate conditions, growers can reduce the impact of ion toxicity, ensuring stable yields and high-quality produce.

4. **Nutrient Management through Foliar Feeding**

Foliar feeding is the practice of applying essential nutrients directly to the leaves, bypassing soil uptake limitations. This method is particularly useful in soils affected by ion toxicity, where nutrient competition and uptake inhibition reduce nutrient availability.

- **Targeted Nutrient Supply**: Foliar feeding supplies nutrients like potassium, calcium, and magnesium directly to the leaves, ensuring they reach critical plant tissues without interference from toxic ions in the soil. Potassium and calcium are especially beneficial for combating the effects of sodium toxicity, as they support stomatal regulation, water balance, and cell wall strength. In fruit crops, foliar feeding provides a direct nutrient source, promoting healthy growth, flower development, and fruit formation.

- **Improving Stress Tolerance and Growth**: Foliar feeding helps improve stress tolerance by enhancing the plant's nutrient status, which supports critical physiological functions under ion-toxic conditions. Potassium aids in osmotic regulation, allowing plants to maintain hydration and prevent wilting, while calcium strengthens cell walls, making plants more resilient to environmental stress. In fruit crops like strawberries, tomatoes, and apples, foliar feeding reduces symptoms of ion toxicity, such as leaf burn and chlorosis, and promotes sustained growth and productivity.
- **Enhancing Fruit Quality and Yield**: By providing essential nutrients during key growth stages, foliar feeding supports better fruit quality and yield in crops affected by ion toxicity. For example, calcium foliar sprays improve fruit firmness, preventing issues like blossom-end rot in tomatoes and increasing shelf life in apples. Potassium application enhances sugar transport and accumulation, improving flavor and sweetness in fruits. In fruit orchards, regular foliar feeding ensures that nutrient levels remain adequate, supporting high-quality fruit production even in soils with elevated sodium or chloride levels.

5. **Additional Management Practices**

Besides leaching, soil amendments, salt-tolerant varieties, and foliar feeding, other management practices can help mitigate ion toxicity and improve plant resilience.

- **Strategic Crop Rotation and Soil Cover**: Planting salt-tolerant cover crops between growing seasons can help prevent salt buildup by reducing evaporation and salt movement to the soil surface. Cover crops like barley or ryegrass absorb excess ions and improve soil structure, reducing ion toxicity for subsequent crops. Crop rotation with salt-tolerant species helps maintain soil health and prevents excessive ion accumulation in soils.
- **Mulching to Retain Soil Moisture**: Mulching helps retain soil moisture, reducing evaporation rates and limiting the upward movement of salts to the root zone. Organic mulches, such as straw or bark, improve soil texture, retain water, and support beneficial microbes. In fruit orchards, mulching reduces water stress and promotes root health, minimizing the impact of ion toxicity on tree stability, growth, and yield.
- **Precision Irrigation to Minimize Salt Accumulation**: Precision irrigation methods, such as drip and micro-sprinkler systems, provide water directly to the root zone, reducing evaporation and salt buildup on the soil surface. This targeted approach ensures that plants receive adequate moisture without contributing to salinity or ion buildup. For fruit crops, precision irrigation supports healthy growth, prevents osmotic stress, and optimizes water use efficiency.

4. Fertilizer Toxicity

Fertilizer toxicity occurs when plants receive excessive nutrients from fertilizers, particularly nitrogen, potassium, or phosphorus-based compounds, which can disrupt the nutrient balance, increase soil salinity, and lead to toxicity symptoms. Fertilizer toxicity is most common in intensively managed agricultural systems where high-yielding crop varieties are grown. While fertilizers play a crucial role in boosting productivity, over-fertilization can hinder plant growth, reduce fruit quality, and cause long-term soil health issues.

1. **Excess Nitrogen and Ammonium Toxicity**

Nitrogen is one of the most important nutrients for plant growth, as it promotes leaf and stem development. However, excessive nitrogen application, especially in the form of ammonium-based fertilizers, can lead to **ammonium toxicity** and unbalanced growth patterns. While nitrogen stimulates vegetative growth, an overabundance often comes at the expense of reproductive growth, impacting fruit production and yield.

- **Ammonium Toxicity and Cellular pH Disruption**: Excess ammonium ions (NH_4^+) in the soil can lead to toxicity by disrupting the pH levels within plant cells. Ammonium uptake decreases the pH in the root zone and in plant tissues, causing an acidic environment that affects enzyme function and metabolic processes. This pH imbalance disrupts normal cell function, which can inhibit nutrient transport and lead to growth suppression. In fruit crops, ammonium toxicity manifests as yellowing or necrotic leaf tips, stunted growth, and poor root development, reducing the plant's ability to absorb water and nutrients.
- **Excessive Vegetative Growth at the Expense of Reproductive Growth**: High nitrogen levels stimulate lush vegetative growth, resulting in larger leaves and stems. While this growth may appear beneficial, it often comes at the expense of flowering and fruit production. Excess nitrogen diverts the plant's resources toward producing leaves rather than flowers or fruit, reducing yield in fruit crops such as tomatoes, peppers, and cucumbers. For instance, tomatoes receiving too much nitrogen may produce dense foliage, which shades the flowers and developing fruits, further reducing fruit set and quality. In addition, the dense canopy can trap humidity, increasing susceptibility to fungal diseases.
- **Delayed Maturity and Reduced Fruit Quality**: In fruit crops, delayed flowering and fruiting due to excess nitrogen can affect fruit ripening, resulting in fruits that are smaller, less flavorful, and prone to physiological disorders. Nitrogen-rich plants tend to have delayed maturity, which can be problematic in crops with specific harvest windows. Additionally, the fruits produced may have a lower sugar content, reducing sweetness and overall quality. For example, in grapes, excessive nitrogen can dilute sugars, leading to poor flavor and lower quality for wine production.

2. **Phosphorus Imbalance and Micronutrient Deficiency**

Phosphorus is essential for energy transfer and root development, but over-application can lead to **micronutrient deficiencies** by interfering with the uptake of iron, zinc, and copper. These deficiencies are particularly problematic in fruit crops that rely on balanced nutrition for optimal fruit quality and flavor.

- **Interference with Iron, Zinc, and Copper Absorption**: High phosphorus levels in the soil can create nutrient imbalances by blocking the uptake of essential micronutrients, especially iron, zinc, and copper. These nutrients are critical for chlorophyll synthesis, enzyme activation, and stress tolerance. When phosphorus interferes with their availability, plants experience deficiencies that limit photosynthesis and metabolic activity. In fruit crops like citrus and apples, iron deficiency caused by phosphorus excess results in iron chlorosis—a condition where leaves turn yellow while veins remain green, reducing photosynthetic efficiency and leading to smaller, less flavorful fruits.

- **Impact on Fruit Quality and Nutritional Content**: Iron, zinc, and copper play significant roles in the development of fruit flavor, color, and nutritional content. Iron is essential for chlorophyll production, which supports photosynthesis and energy availability. Zinc is involved in hormone production and stress resilience, while copper aids in disease resistance and cellular function. Phosphorus-induced deficiencies in these nutrients reduce fruit sweetness, acidity, and color, impacting both quality and marketability. In citrus, iron chlorosis reduces the accumulation of sugars and acids, leading to bland-tasting fruits with pale color and lower nutritional value.
- **Long-Term Effects on Soil Health**: Continuous phosphorus over-application can lead to phosphorus buildup in the soil, which not only hinders nutrient uptake in plants but also contributes to environmental issues like eutrophication. High phosphorus levels in runoff can pollute water bodies, leading to algal blooms that disrupt aquatic ecosystems. In the long run, excessive phosphorus application degrades soil quality and limits the soil's ability to support healthy crop growth.

3. **Increased Soil Salinity from Fertilizer Salts**

Many fertilizers, especially those based on potassium, contain salts that increase soil salinity when applied in high concentrations. **Fertilizer salts** contribute to osmotic stress, which restricts water uptake and impacts plant growth, particularly in fruit crops that require consistent moisture for fruit development and quality.

- **Osmotic Stress and Reduced Water Uptake**: Fertilizer salts raise the osmotic potential of the soil, making it more challenging for plants to absorb water. When salt levels are high, plants lose water to the soil due to osmotic imbalance, resulting in dehydration and wilting even when soil moisture is adequate. In fruit crops like strawberries, grapes, and blueberries, which are highly sensitive to soil moisture, reduced water uptake due to salt stress can lead to smaller fruits, increased susceptibility to wilting, and lower yield.
- **Impact on Root Functionality and Nutrient Absorption**: High salt concentrations affect root cell membranes, reducing root efficiency and impairing the plant's ability to take up essential nutrients. Salt stress damages root cells, leading to nutrient deficiencies that affect growth and productivity. For example, potassium toxicity from over-fertilization can lead to calcium and magnesium deficiencies, as these ions compete for uptake sites on root cell membranes. Calcium deficiency weakens cell walls, making plants more susceptible to diseases, while magnesium deficiency impairs chlorophyll production, leading to chlorosis and reduced photosynthesis. In fruit crops, these nutrient imbalances reduce fruit size, flavor, and structural integrity.
- **Reduced Fruit Quality and Shelf Life**: Excessive soil salinity from fertilizers can lead to poor fruit quality by affecting sugar and acid content, reducing flavor, and shortening shelf life. High salt concentrations around roots limit water availability, leading to fruits

with lower water content and a more fibrous texture. Additionally, salt-stressed plants often produce fruits with diminished sweetness and reduced acidity, affecting overall taste. For example, strawberries grown under saline conditions are often smaller, less sweet, and have a shorter shelf life, making them less marketable and desirable to consumers.

4. Management Strategies to Prevent Fertilizer Toxicity

Fertilizer toxicity, often caused by excessive application of nutrients like nitrogen, phosphorus, or potassium, disrupts nutrient balance and increases soil salinity, leading to issues such as nutrient imbalances, reduced fruit quality, and long-term soil degradation. Preventing fertilizer toxicity requires integrated nutrient management strategies that involve careful nutrient assessment, balanced application, and targeted delivery methods to meet plant needs without causing excess. These strategies are essential in fruit orchards and other high-value crop systems, where fertilizer toxicity can lead to significant economic losses and degrade soil health.

1. Soil Testing and Balanced Fertilizer Application

Regular **soil testing** is fundamental to effective nutrient management, as it allows growers to determine the existing nutrient levels and pH in the soil, helping to identify specific deficiencies or excesses. By assessing soil health through testing, growers can apply only the nutrients that are deficient, reducing the risk of over-application and minimizing fertilizer toxicity.

- **Assessing Nutrient Requirements through Soil Testing**: Soil testing provides detailed information on the levels of nitrogen, phosphorus, potassium, and micronutrients such as iron, zinc, and manganese. This data helps growers tailor fertilizer applications to address only the nutrients needed, avoiding the buildup of non-limiting nutrients that could lead to toxicity. For example, in fruit orchards, soil testing may reveal a surplus of phosphorus, indicating that only nitrogen and potassium need to be applied to support plant health and fruit development.
- **Customized Fertilizer Programs for Fruit Orchards**: In high-value crops like grapes, apples, and citrus, balanced fertilization based on soil test results supports optimal growth, fruit size, and quality. By applying the correct nutrient ratios, growers can enhance photosynthesis, flowering, and fruit set while avoiding nutrient imbalances. For example, excessive nitrogen can promote leafy growth at the expense of fruit production, while phosphorus excess may induce micronutrient deficiencies like iron chlorosis. Balanced fertilizer programs ensure that plants receive the right nutrient quantities for stable growth and high-quality yields.
- **Preventing Environmental Impact and Long-Term Soil Health Issues**: Excess fertilizers not only cause toxicity but can also leach into water bodies, contributing to nutrient pollution and eutrophication. By applying fertilizers based on soil test recommendations, growers can reduce the environmental impact of farming practices, improve long-term soil fertility, and prevent degradation. This approach also minimizes financial waste, as growers avoid unnecessary fertilizer costs and use resources more efficiently.

2. **Controlled-Release and Slow-Release Fertilizers**

Controlled-release fertilizers and **slow-release formulations** are designed to release nutrients gradually, aligning nutrient availability with plant demands and reducing the risk of fertilizer burn and nutrient imbalances. These fertilizers are particularly beneficial in fruit crops, where consistent nutrient supply is crucial for optimal growth and high-quality fruit production.

- **Nutrient Release Aligned with Plant Demand**: Controlled-release fertilizers release nutrients in response to soil temperature and moisture, providing plants with a steady nutrient supply throughout the growing season. This gradual release reduces the likelihood of nutrient leaching or buildup in the soil, ensuring that plants receive nutrients only when they are actively growing and able to utilize them. In fruit crops, where nutrient demands vary with growth stages, controlled-release fertilizers prevent nutrient spikes that could lead to toxicity and support sustained growth, flowering, and fruiting.
- **Reduced Salt Buildup and Osmotic Stress**: Many conventional fertilizers contain salts that increase soil salinity when applied in excess. Controlled-release fertilizers help prevent salt buildup by providing a measured release, reducing the risk of osmotic stress, which limits water uptake and affects growth. In crops like strawberries, tomatoes, and grapes, where soil salinity can impact fruit flavor and texture, slow-release fertilizers maintain a balanced nutrient supply without causing osmotic stress or affecting soil health.
- **Lower Labor Costs and Environmental Impact**: Controlled-release fertilizers reduce the frequency of applications, which can lower labor costs and minimize soil disturbance. By reducing runoff and leaching, these fertilizers also lessen the environmental impact of fertilization practices. For high-value fruit crops, controlled-release fertilizers not only support sustainable growth but also help growers achieve consistent yields and maintain soil fertility over the long term.

3. **Foliar Feeding for Targeted Nutrient Delivery**

Foliar feeding involves applying nutrients directly to plant leaves, bypassing the soil and allowing for immediate nutrient uptake. This targeted method is particularly useful for correcting micronutrient deficiencies, especially those caused by soil conditions that inhibit nutrient availability.

- **Direct Correction of Micronutrient Deficiencies**: Foliar feeding supplies nutrients such as iron, zinc, manganese, and copper directly to the leaves, ensuring that they reach critical tissues without interference from soil chemistry. This is especially beneficial in cases of phosphorus-induced deficiencies, where high phosphorus levels reduce the availability of micronutrients. For instance, iron chlorosis, commonly seen in citrus and apples due to excess phosphorus, can be addressed with foliar iron sprays, helping to restore chlorophyll production, leaf health, and photosynthetic efficiency.
- **Enhanced Plant Resilience and Growth**: Foliar feeding provides immediate relief from nutrient deficiencies, supporting vital physiological functions like photosynthesis, enzyme activity, and cell wall strength. Nutrients like calcium and potassium, when applied as foliar sprays, enhance cell wall stability and osmotic regulation, allowing plants to maintain water balance and prevent wilting. In fruit crops, foliar feeding

supports flowering, fruit set, and quality, promoting larger, more flavorful fruits and ensuring high yield.

- **Precision Application for Optimal Fruit Development**: Foliar feeding can be timed with key growth stages, such as pre-flowering or early fruit development, to provide nutrients precisely when they are most needed. For example, calcium sprays during fruit formation in tomatoes and apples prevent disorders like blossom-end rot and improve fruit firmness and shelf life. Potassium foliar feeding enhances sugar accumulation in fruits, improving sweetness and overall flavor, which is especially beneficial for crops like grapes and strawberries.

4. **Water Management to Reduce Salt Accumulation**

Proper **irrigation practices** play a critical role in preventing salt accumulation from fertilizer applications. Managing water effectively helps dilute and flush excess salts from the root zone, reducing osmotic stress and supporting nutrient uptake.

- **Drip Irrigation for Direct Nutrient and Water Delivery**: Drip irrigation delivers water directly to the root zone, minimizing evaporation and preventing salt accumulation on the soil surface. This method helps maintain consistent soil moisture levels and reduces salt stress on roots, particularly in crops that are sensitive to salinity. For fruit crops, drip irrigation prevents the formation of salt crusts around roots, supporting healthy root function, efficient nutrient uptake, and optimal growth.
- Irrigation Scheduling to Support Fertilizer Application: Scheduling irrigation to coincide with fertilizer application helps dissolve and transport salts below the root zone, reducing the risk of salt buildup. By flushing excess salts away from the root area, irrigation prevents osmotic stress, allowing plants to absorb water more efficiently. In fruit orchards, where soil salinity can affect fruit quality, precise irrigation scheduling supports root health, enhances fruit flavor, and prevents yield losses.
- Maintaining Soil Moisture Levels for Consistent Growth: Soil moisture levels play a crucial role in nutrient absorption and transport within plants. Ensuring adequate moisture levels through consistent irrigation prevents salt-induced dehydration and supports healthy root function. In fruit crops, maintaining appropriate soil moisture helps prevent nutrient imbalances, reduce stress during critical growth stages, and support high-quality fruit production. For example, regular irrigation in strawberry fields prevents salt stress, which can otherwise lead to smaller, less flavorful fruits with reduced shelf life.

5. **Additional Practices to Enhance Fertilizer Efficiency and Prevent Toxicity**

In addition to soil testing, controlled-release fertilizers, foliar feeding, and water management, other practices can enhance fertilizer efficiency and prevent toxicity, improving soil health and supporting sustainable crop production.

- **Use of Organic Fertilizers for Soil Health**: Organic fertilizers release nutrients more slowly than synthetic options, reducing the risk of nutrient buildup and toxicity. They also add organic matter to the soil, improving soil structure, water retention, and microbial activity. For fruit crops, organic fertilizers support long-term soil health and create a balanced nutrient environment that reduces the risk of toxicity.

- **Cover Crops and Mulching to Retain Nutrients**: Cover crops absorb excess nutrients and prevent leaching, while mulching helps retain soil moisture and reduces evaporation. Together, these practices support a balanced nutrient profile and prevent nutrient loss, reducing the need for additional fertilizer applications. For example, planting leguminous cover crops like clover or vetch in fruit orchards adds nitrogen to the soil, reducing the need for synthetic nitrogen fertilizers and minimizing the risk of nitrogen toxicity.
- **Integrated Nutrient Management (INM) for Sustainable Fertilization**: INM combines organic and synthetic fertilizers based on crop requirements, soil conditions, and environmental factors. This approach optimizes fertilizer use, supporting crop health while minimizing environmental impact. In fruit production, INM provides a balanced nutrient supply that enhances growth, prevents toxicity, and promotes long-term soil fertility.

Table 6: Effect of Salt Stress

S.No	Name of Crop	Effect of Salt Stress
1.	Mango	Scorching of leaf tips and margins, Leaf curling, reduces growth, causes abscission of leaves and death of trees
2.	Pineapple	Leaf length and mass, plant fresh and dry weight, leaf elongation and leaf water potentials decline and thus fruit yield and quality.
3.	Grapes	Leaf burn and Necrotic symptoms.

Chapter - 2

Pollution Impact on Fruit Crop Production

I. Pollution – Increased Level of CO_2

The global concentration of carbon dioxide (CO_2) in the atmosphere has been rising due to human activities such as industrial processes, deforestation, and fossil fuel combustion. While CO_2 is a critical component of photosynthesis, and increased levels can enhance plant growth under certain conditions, excessive CO_2 in the atmosphere has complex effects on agricultural productivity, plant physiology, and the environment. The impact of elevated CO_2 is especially relevant to fruit crop production, where changes in growth patterns, nutritional content, and stress tolerance can influence yield and quality. The rise in **global atmospheric CO_2** levels is a direct consequence of industrialization, deforestation, and the burning of fossil fuels. Although CO_2 is integral to plant photosynthesis and can promote growth under certain conditions, the increase in atmospheric CO_2 has complex and far-reaching effects on agriculture, plant physiology, and ecosystem balance. For fruit crop production, in particular, elevated CO_2 levels bring about shifts in growth, nutrient content, pest resistance, and stress tolerance, all of which significantly influence crop yield and quality.

1. Role of CO_2 in Photosynthesis and Plant Growth

CO_2 is essential for photosynthesis, the process by which plants convert light energy into chemical energy, producing carbohydrates that fuel growth. Higher CO_2 levels theoretically enhance the rate of photosynthesis (a phenomenon known as the "CO_2 fertilization effect"), especially in C3 plants, which include most fruit crops. Under elevated CO_2, these plants tend to increase their photosynthetic rate, leading to faster growth and potentially higher yields.

- **Increased Biomass and Yield Potential**: When CO_2 levels are elevated, plants can produce more biomass due to increased photosynthetic rates. For fruit crops like apples, strawberries, and grapes, this can mean more substantial plants with greater leaf area, enhanced flowering, and potentially higher fruit yields. Additionally, CO_2 enrichment in controlled environments (like greenhouses) often results in faster crop cycles, allowing growers to maximize productivity.
- **Enhanced Water-Use Efficiency**: Higher CO_2 often leads to partial stomatal closure in plants, reducing the rate of water loss through transpiration. This reduced water requirement per unit of photosynthesis means that plants can use water more efficiently,

DOI: 10.1201/9781003717409-2

which is beneficial in arid regions where water resources are limited. For water-sensitive crops, improved water-use efficiency under high CO_2 conditions can increase resilience to drought stress.

2. **Impact of Elevated CO_2 on Fruit Nutritional Content**

While increased CO_2 can promote growth, it also leads to changes in the nutrient composition of fruits. Rapid growth under high CO_2 can dilute essential nutrients, reducing the overall nutritional density of fruit crops.

- **Reduced Nutrient Density (Nutrient Dilution Effect)**: Rapid growth spurred by higher CO_2 can result in lower concentrations of vital nutrients like nitrogen, iron, and zinc in plant tissues. This "dilution effect" compromises the nutritional quality of fruits, as they may contain fewer proteins, vitamins, and minerals per unit of weight. For crops like citrus and berries, where nutrient content is a significant quality factor, this reduction can impact marketability and consumer health benefits.
- **Altered Sugar and Acid Content in Fruits**: Elevated CO_2 can influence the sugar-to-acid ratio in fruits, affecting their flavor profile. Increased carbohydrate accumulation can lead to higher sugar levels, often making fruits sweeter, though potentially less complex in flavor due to a reduction in acidity. In wine grapes, for instance, a shift toward higher sugar levels with less acidity can alter the quality of wine, impacting the wine industry where balanced flavor profiles are essential.

3. **Effects on Plant Defense Mechanisms and Pest Pressure**

Elevated CO_2 influences not only plant growth but also its interactions with pests and diseases. Higher CO_2 levels can reduce the concentration of defense-related compounds, making plants more susceptible to pests and certain pathogens.

- **Reduced Production of Defense Compounds**: Some studies have shown that plants grown under elevated CO_2 produce fewer secondary metabolites, compounds involved in natural pest resistance. This reduction in defense compounds may increase the plant's vulnerability to herbivorous pests like aphids and caterpillars. In fruit crops, which are already susceptible to pest damage, lower natural defense levels could lead to higher dependency on pesticides, raising production costs and potential environmental impact.
- **Changes in Pest Populations and Feeding Behavior**: The lower nitrogen content in plants grown under elevated CO_2 may encourage pests to consume more plant material to meet their nutritional needs, leading to more significant damage. For instance, pests feeding on nitrogen-deficient leaves may increase their feeding intensity, affecting plant health and yield. In fruit crops, where pest pressure directly impacts marketable yield, these shifts in pest behavior may require more intensive management practices.

4. **Implications for Soil Microbial Communities and Nutrient Cycling**

Higher CO_2 levels also influence the rhizosphere (root-soil interface), where complex interactions between roots and soil microbes take place. Increased root exudation under elevated CO_2 can enhance microbial activity, which in turn affects nutrient availability and soil health.

- **Enhanced Root Exudation and Microbial Activity**: Elevated CO_2 levels often lead to increased carbon fixation, resulting in more carbon-rich compounds being exuded by roots into the soil. This promotes microbial activity around the roots, enhancing processes like nitrogen fixation and phosphorus solubilization. While beneficial in terms of nutrient cycling, increased microbial activity may also deplete soil nitrogen more quickly, necessitating adjustments in fertilization practices for sustained crop health.
- **Potential Shifts in Soil Nutrient Demand**: With increased microbial competition for soil nutrients, especially nitrogen, plants may face nutrient limitations even in fertilized soils. Fruit crops, which rely on a balanced nutrient supply for high-quality yields, may require more precise nutrient management to compensate for microbial nutrient demand. This effect could drive up fertilizer requirements, particularly for nitrogen, increasing both production costs and environmental impact.

5. **Environmental and Long-Term Agricultural Impacts**

The increase in atmospheric CO_2 presents broader environmental and agricultural challenges, necessitating adaptations in crop management to maintain productivity and quality.

- **Sustainable Crop Management Practices**: The complex effects of elevated CO_2 require crop management practices that enhance sustainability, such as precision fertilization, integrated pest management (IPM), and soil health practices. By optimizing nutrient application and reducing reliance on chemical inputs, these strategies can help mitigate the negative impacts of increased CO_2 on fruit crop production.
- **Potential for Carbon Sequestration**: Increased CO_2 can stimulate greater biomass production, which in turn could enhance carbon sequestration in agricultural soils. Integrating practices like cover cropping, reduced tillage, and organic matter additions can improve soil carbon storage and help offset some of the CO_2 emissions from agricultural activities.

A. Positive Effects of Increased CO_2 on Plant Growth and Photosynthesis

Rising levels of atmospheric CO_2, while concerning in terms of climate change, can have some beneficial effects on plant growth, particularly in C3 plants. C3 plants, which include most fruit crops like apples, peaches, strawberries, and grapes, respond positively to elevated CO_2 through increased rates of photosynthesis and improved water-use efficiency. This response, known as the "**CO_2 fertilization effect**," can lead to more substantial plant growth, higher fruit production, and better adaptation to certain environmental stresses. However, these benefits are contingent on favorable growing conditions, as other stress factors (e.g., nutrient deficiencies, and extreme temperatures) can limit the extent of CO_2-related growth enhancements.

1. **Enhanced Photosynthesis (CO_2 Fertilization Effect)**

Photosynthesis is the fundamental process by which plants convert light energy, water, and CO_2 into glucose and oxygen, fueling growth and metabolic activities. Increased CO_2 concentrations directly impact the Calvin cycle, where CO_2 is fixed into glucose, driving the photosynthetic rate.

- **The CO_2 Fertilization Effect in C3 Plants**: C3 plants are particularly responsive to elevated CO_2 because of their unique photosynthetic pathway, where CO_2 directly influences the Calvin cycle. With more available CO_2, the enzyme Rubisco operates more efficiently, leading to an increase in glucose production. This enhanced photosynthesis means that plants have more energy to allocate toward growth, development, and reproduction. Research has shown that fruit crops like apples, peaches, and strawberries can exhibit increased photosynthetic rates under elevated CO_2, leading to greater energy production that supports improved growth and yield.

- **Potential for Larger Plants and Higher Biomass**: Enhanced photosynthesis enables C3 plants to produce more carbohydrates, which can then be directed toward vegetative and reproductive growth. Fruit crops may develop larger leaves, stronger stems, and a more extensive canopy, which enhances their ability to capture sunlight, further driving photosynthesis. Increased canopy coverage is beneficial in fruit crops, as it supports more flowers and potential fruit sites, positively impacting yield. The larger biomass also contributes to more resilient plants that can withstand certain environmental stresses more effectively.
- **Increase in Flower Set and Fruit Production**: With additional energy from increased photosynthesis, many fruit crops under elevated CO_2 produce more flowers, which can lead to a higher potential fruit set. For example, studies have indicated that apples and strawberries grown under elevated CO_2 conditions show an increase in both flower and fruit numbers, potentially translating to higher yields. This response is particularly advantageous for fruit growers looking to maximize productivity and profitability under greenhouse or controlled-environment agriculture where CO_2 levels can be managed.

2. **Improved Water-Use Efficiency**

Water-use efficiency (WUE) refers to the amount of water used by plants per unit of carbon gained through photosynthesis. Elevated CO_2 can positively impact water-use efficiency by reducing water loss through transpiration while maintaining, or even enhancing, photosynthetic rates.

- **Stomatal Regulation Under Elevated CO_2**: Plants have tiny pores on their leaves called stomata, which regulate gas exchange by opening and closing. Elevated CO_2 often causes plants to partially close their stomata, reducing the rate of water loss through transpiration. While CO_2 enters the leaves for photosynthesis, reduced transpiration means that less water is lost, making plants more efficient in their water use. For fruit crops grown in arid and semi-arid regions, this is particularly advantageous, as water availability is often a limiting factor. Improved water-use efficiency allows these crops to grow with less water, supporting sustainable production.
- **Enhanced Drought Tolerance**: The partial stomatal closure under elevated CO_2 improves water retention within the plant, which helps maintain cellular hydration and turgor pressure, both of which are essential for maintaining plant structure and function. In drought-prone areas, fruit crops such as grapes, peaches, and olives may benefit from this water-saving effect, as it enhances their ability to withstand periods of limited water availability. This improved drought tolerance is crucial for maintaining productivity under challenging environmental conditions.

- **Reduction in Irrigation Needs for Fruit Crops**: With improved water-use efficiency, fruit crops may require less frequent irrigation, which is beneficial both economically and environmentally. Reduced irrigation needs can lower water costs for farmers and lessen the strain on water resources in regions where agriculture competes with other water demands. In crops like strawberries and blueberries, which are typically grown in water-intensive systems, elevated CO_2 can help growers reduce water usage without compromising yield or fruit quality.

3. **Increased Biomass and Fruit Yield**

Alongside enhanced photosynthesis and improved water-use efficiency, elevated CO_2 often leads to an increase in overall plant biomass and, consequently, greater fruit yields. This effect can be especially pronounced in high-value fruit crops, where yield and fruit quality directly impact profitability.

- **Stimulated Cell Division and Expansion**: Elevated CO_2 levels promote cell division and cell expansion, processes that are fundamental for vegetative growth and fruit development. As CO_2 stimulates photosynthesis, the increased production of glucose supports cellular growth and division. In fruit crops, this can lead to larger plants with more extensive root systems and leaf coverage. A larger root system enhances nutrient and water uptake, while a more robust canopy supports greater photosynthetic capacity, further driving growth.
- **Increased Fruit Number and Size**: Many fruit crops respond to elevated CO_2 by producing a greater number of fruits, as well as larger individual fruits. For example, in grapes and apples, the increase in vegetative growth under elevated CO_2 supports a greater number of fruit-bearing branches, leading to more fruit sites and potentially larger yields. Additionally, the larger fruits produced under high CO_2 conditions are often more visually appealing and marketable, which is a crucial factor in commercial fruit production.
- **Enhanced Overall Yield and Economic Benefits**: The increase in biomass and yield associated with elevated CO_2 provides substantial economic benefits for growers. Higher yields mean greater profitability, especially for high-value crops like apples, peaches, and berries, which are valued for their quality and size. In controlled environments, such as greenhouses, where CO_2 levels can be regulated, growers often enrich CO_2 to maximize productivity and obtain consistent, high-quality yields throughout the year. This increase in yield under elevated CO_2 contributes to more efficient land use, allowing for higher output within limited acreage.

4. **Long-Term Productivity and Resource Efficiency**

The positive effects of elevated CO_2 on photosynthesis, water-use efficiency, and yield can contribute to more sustainable and resource-efficient fruit crop production over time. However, these benefits depend on optimal conditions, as other environmental stressors can limit the extent of CO_2-driven growth enhancements.

- **Sustainability in Water-Limited Environments**: In regions with limited water resources, the improved water-use efficiency conferred by elevated CO_2 can play a significant role in supporting sustainable agriculture. By reducing water demand, crops grown under higher CO_2 levels can be cultivated more efficiently, using fewer resources without sacrificing productivity. This is particularly advantageous in areas where water scarcity poses a challenge to agricultural expansion.

- **Potential for Reduced Fertilizer Use**: The increase in photosynthetic efficiency under elevated CO_2 may reduce the need for certain fertilizers, as plants become more efficient at converting CO_2 into biomass. However, this depends on adequate availability of other nutrients, as CO_2 enrichment alone cannot substitute for balanced fertilization. In systems where nutrient management is optimized alongside CO_2 enrichment, growers may see reduced reliance on chemical inputs, leading to cost savings and reduced environmental impact.
- **Challenges and Limitations**: While elevated CO_2 can enhance growth, these benefits are not always fully realized in field conditions, where factors such as nutrient availability, temperature, and pest pressure also play critical roles in plant productivity. Under nutrient-limited conditions, the potential growth benefits of increased CO_2 may be constrained, as plants cannot convert additional CO_2 into biomass without sufficient nitrogen, phosphorus, and other essential elements. Therefore, the positive effects of elevated CO_2 are maximized when combined with well-managed nutrient and water inputs.

B. Negative Effects of Elevated CO_2 on Nutrient Composition and Quality

While elevated CO_2 levels can stimulate plant growth, this increase in photosynthetic activity often leads to undesirable effects on fruit composition and quality. Rapid growth can dilute essential nutrients, alter taste and flavor balance, and compromise fruit texture, all of which are crucial for consumer satisfaction and marketability. These negative effects present challenges for growers and the food industry, especially in high-value fruit crops where quality is directly linked to economic returns.

1. Reduced Nutrient Density

One of the primary concerns of elevated CO_2 is the "dilution effect," where rapid growth driven by increased CO_2 reduces the nutrient concentration in plant tissues. This decrease in nutrient density impacts both the nutritional quality of fruits and their market value.

- **The Dilution Effect and Nutrient Imbalances**: Elevated CO_2 promotes increased carbon fixation, leading to greater production of carbohydrates. However, as plants focus on rapid growth, their uptake of essential minerals such as nitrogen, zinc, iron, and magnesium does not keep pace, resulting in a lower concentration of these nutrients in plant tissues. This effect can reduce the protein content (which is nitrogen-dependent) and mineral density in fruits, making them less nutritious for consumers. For example, apples and berries grown under elevated CO_2 conditions may contain lower levels of iron and zinc, essential minerals for human health. This reduction in nutrient density could affect the nutritional value of fruit-based diets, particularly in communities that rely heavily on fruit for essential vitamins and minerals.
- **Impact on Plant Health and Crop Quality**: Nutrient deficiencies not only affect consumer nutrition but also compromise plant health. Minerals like iron and magnesium

are critical for chlorophyll production and photosynthesis, while nitrogen is essential for protein synthesis and growth. When these nutrients are diluted within plant tissues, overall plant vigor may decline, increasing susceptibility to environmental stressors and diseases. For fruit crops, this can result in reduced resilience, impacting both yield and quality over time.

- **Challenges for Growers and Marketability**: Reduced nutrient density impacts the marketability of fruits, as consumers increasingly seek produce with high nutritional value. Growers and retailers may face challenges in meeting consumer expectations for nutrient-rich fruits, and the economic value of crops could be affected as nutrient-dense produce gains a premium in the market. Nutrient-depleted fruits could require adjustments in fertilization practices to compensate for mineral imbalances, adding costs and labor to the production process.

2. Impact on Flavor and Sugar-Acid Balance

Elevated CO_2 levels can alter the sugar-to-acid ratio in fruits, which is a key determinant of taste and flavor. While higher CO_2 often increases carbohydrate accumulation, the change in sugar and acid balance may result in sweeter but less complex flavors, which can be a drawback for flavor-sensitive fruits and products like wine.

- **Increased Sugar Content and Reduced Acidity**: Elevated CO_2 often results in increased photosynthetic rates, leading to higher carbohydrate synthesis and sugar accumulation in fruits. While sweetness is generally desirable, excessive sugar can make fruits overly sweet and reduce the complexity of flavors. At the same time, elevated CO_2 can lead to a reduction in organic acid levels, which are crucial for the balanced flavor profile of fruits. This reduced acidity, combined with higher sugar content, alters the traditional taste consumers expect, leading to sweeter but "flatter" flavors. In crops like grapes, which are used for winemaking, this shift can disrupt the delicate sugar-acid balance necessary for producing wines with rich, balanced flavors.
- **Effect on Flavor Profile in High-Value Crops**: For fruits like apples, berries, and citrus, flavor is a critical quality attribute that influences consumer choice and market price. The altered sugar-acid balance under elevated CO_2 may make fruits taste less refreshing or complex, affecting consumer satisfaction. In citrus fruits, for instance, a reduction in acidity can result in a less tangy taste, which is an important flavor characteristic. Changes in taste profiles due to elevated CO_2 could impact the food industry, as flavor consistency is essential in both fresh fruit sales and processed products like juices and jams.
- **Reduced Storability and Shelf Life**: The increased sugar content and reduced acidity may also make fruits more susceptible to spoilage, as high sugar levels can promote microbial growth. Acidity acts as a natural preservative, and lower acid levels can decrease the shelf life and storability of fruits, increasing the likelihood of post-harvest losses. For growers and distributors, this means shorter marketing windows and potential economic losses due to increased waste and quality degradation during transport.

3. Changes in Texture and Structural Integrity

Elevated CO_2 levels can influence the cellular structure of plants, affecting fruit firmness and texture. This effect is particularly significant for soft fruits, where texture plays a crucial role in consumer preference and shelf life.

- **Impact on Cell Wall Integrity and Fruit Firmness**: Elevated CO_2 can alter cell wall composition and structure, leading to changes in fruit firmness and texture. Under high CO_2, the production of certain enzymes involved in cell wall remodeling may be affected, leading to softer cell walls that are more prone to damage. Fruits with reduced cell wall integrity are more susceptible to bruising, especially during harvesting, packaging, and transportation. For soft fruits like strawberries, blueberries, and peaches, a loss of firmness can reduce consumer appeal, as these fruits are valued for their texture. In the retail environment, softer fruits are more likely to be rejected or discarded, increasing waste and reducing profitability.
- **Shortened Shelf Life Due to Texture Changes**: The softer texture resulting from elevated CO_2 may reduce the shelf life of certain fruits. Firmness is a key factor for storage and transport, and fruits with compromised structural integrity are more vulnerable to mechanical damage and spoilage. This is particularly relevant for fruits intended for export, where long transport times require durable fruits that can withstand handling and prolonged storage. In peaches and plums, for example, reduced firmness can lead to rapid ripening and softening, limiting the marketability of these fruits and increasing post-harvest losses.
- **Increased Susceptibility to Post-Harvest Pathogens**: Changes in texture and cell wall structure may make fruits more vulnerable to pathogens such as fungi and bacteria, which exploit weakened tissues. Fruits with softer cell walls are more prone to infection by post-harvest pathogens like Botrytis cinerea (graymold) and Penicillium species, which cause spoilage. This susceptibility not only impacts shelf life but also increases the risk of waste throughout the supply chain. For growers and retailers, this may necessitate additional post-harvest treatments or increased use of packaging and refrigeration to extend shelf life, which can add to production and distribution costs.

4. Economic and Agricultural Implications

The negative effects of elevated CO_2 on nutrient density, flavor, and texture in fruit crops have broader implications for agriculture and the economy. As consumers become more health-conscious and demand high-quality produce, these changes could influence market dynamics and production practices.

- **Increased Costs for Quality Maintenance**: The need to maintain nutrient quality and flavor in fruits grown under elevated CO_2 conditions may drive up production costs. Growers may need to adopt additional fertilization strategies to counter nutrient dilution, increasing the cost and environmental impact of fruit production. For soft fruits with reduced texture quality, additional investments in packaging and cold storage may be necessary to minimize damage and prolong shelf life, further raising production and handling costs.

- **Market Demand and Consumer Perception**: With increasing consumer awareness of nutritional quality, the reduction in nutrient density due to elevated CO_2 could influence demand. Fruits with lower nutritional value may become less appealing to health-conscious consumers, impacting marketability and pricing. For high-value crops like berries and apples, which are marketed for their health benefits, reduced nutrient density could lead to a decrease in perceived value, affecting sales and profitability.
- **Need for Adapted Cultivation Practices**: As the agricultural industry adapts to elevated CO_2 levels, there may be a need for breeding programs focused on selecting varieties with improved nutrient retention, flavor stability, and texture resilience. Additionally, precision agriculture practices that optimize nutrient management and irrigation could help mitigate some of the negative effects of elevated CO_2, though these adaptations will require investment in technology and research.

1. Increased CO_2 and Plant-Microbe Interactions

Elevated CO_2 levels affect not only plant physiology and growth directly but also influence the complex interactions between plants and soil microbes, along with the plant's susceptibility to diseases and pests. These indirect effects of increased CO_2 have significant implications for soil health, nutrient availability, pest and disease resistance, and, ultimately, crop productivity. Understanding these interactions is essential for developing effective management strategies, especially in fruit crops, where quality and yield are directly affected by microbial and pest pressures.

1. Changes in Soil Microbial Activity

Elevated CO_2 increases the amount of carbon fixed during photosynthesis, leading to more carbon-rich compounds (such as sugars and organic acids) being exuded by roots into the rhizosphere. These root exudates stimulate microbial activity around the roots, impacting nutrient availability, soil health, and plant growth.

- **Enhanced Microbial Activity in the Rhizosphere**: Root exudates, which are organic compounds secreted by plants into the soil, serve as food sources for soil microbes. Under elevated CO_2, the increased exudation promotes microbial activity, enhancing nutrient cycling processes, such as nitrogen fixation and phosphorus solubilization. Beneficial microbes, such as nitrogen-fixing bacteria and mycorrhizal fungi, play critical roles in improving nutrient availability to plants, making these interactions beneficial in many respects. For fruit crops like strawberries, grapes, and blueberries, which require balanced nutrition for optimal growth and fruit quality, active microbial communities can contribute to healthier, more fertile soils, supporting robust plant growth.
- **Increased Nutrient Competition and Potential Limitations**: While enhanced microbial activity boosts nutrient cycling, it also intensifies competition for nutrients, especially nitrogen. Soil microbes and plants often compete for available nitrogen, and elevated microbial activity can result in nutrient immobilization, where microbes absorb nitrogen and other nutrients before they are accessible to plants. This competition may lead to nutrient limitations, requiring growers to adjust fertilization practices to compensate. In fruit crops, where nitrogen is essential for flowering, fruit set, and fruit development,

managing nutrient competition is crucial to ensure high-quality yields.

- **Effects on Soil Health and Fertilization Needs**: The increased demand for nutrients due to heightened microbial activity may necessitate more frequent or targeted fertilization, particularly in fruit orchards where nutrient requirements are high. Regular soil testing and precise nutrient management can help maintain soil health while addressing nutrient competition. Fertilization practices that incorporate organic amendments, such as compost, may support microbial diversity and provide slow-release nutrients, ensuring a stable supply to both plants and soil microbes over time. This approach enhances long-term soil fertility, supporting sustainable fruit crop production under elevated CO_2 conditions.

2. Impact on Disease Resistance

Elevated CO_2 can influence a plant's defense mechanisms, often altering the production of secondary metabolites and defense-related compounds. While some plants may increase their defensive capacity under stress, others may become more susceptible to diseases, particularly fungal pathogens.

- **Reduced Production of Defense Compounds**: Under elevated CO_2, some plants produce fewer secondary metabolites, such as phenolic compounds, flavonoids, and phytoalexins, which play critical roles in defending against pathogens. These compounds are essential in fruit crops for resisting fungal infections that can cause leaf, stem, or fruit diseases. For instance, strawberries and grapes rely on these natural defense compounds to protect against pathogens like powdery mildew and Botrytis cinerea (graymold). With reduced defense compound production, plants become more vulnerable to these diseases, potentially leading to significant yield losses.
- **Increased Disease Susceptibility in Fruit Crops**: In high-value fruit crops, disease susceptibility under elevated CO_2 can have serious economic implications. Fungal diseases like Botrytis cinerea thrive under conditions of high humidity and can cause rapid decay in fruits post-harvest, resulting in substantial losses during storage and transport. Additionally, powdery mildew, a common fungal disease in grapes, can reduce photosynthetic efficiency and lead to poor fruit quality. Increased susceptibility to these pathogens may require more intensive disease management practices, including regular fungicide applications, which raise production costs and potentially impact the environment.
- **Enhanced Disease Management Strategies**: To counter the reduced natural disease resistance under elevated CO_2, integrated disease management practices become essential. Regular monitoring, cultural practices (such as proper spacing and pruning to improve air circulation), and targeted fungicide applications can help mitigate disease pressure. For organic and sustainable fruit production, alternatives like biological fungicides and resistance-inducing compounds may offer eco-friendly solutions. Implementing a multi-faceted approach to disease management will be critical for maintaining fruit quality and yield in the face of increased disease susceptibility.

3. Changes in Pest Dynamics

Elevated CO_2 levels can influence pest populations and their feeding behaviors by affecting the nutritional quality of plant tissues. Changes in nitrogen content, leaf structure, and secondary metabolites in response to elevated CO_2 can alter how pests interact with crops, often leading to increased pest pressure.

- **Lower Nitrogen Content and Increased Feeding by Herbivores**: Elevated CO_2 often reduces the nitrogen content in plant tissues due to the dilution effect, as rapid growth driven by CO_2 results in lower protein levels. Herbivorous pests, like aphids, caterpillars, and leafhoppers, may respond to this nutrient dilution by consuming more plant material to meet their nitrogen needs. This increased feeding can lead to more significant damage to leaves, stems, and fruits. For fruit crops, where appearance and quality are critical, increased pest feeding can reduce marketable yield, leading to economic losses.
- **Altered Plant Chemistry and Pest Behavior**: The reduction in certain defense compounds under elevated CO_2 can also make plants more attractive to pests. For instance, some plants produce lower levels of tannins and other defensive metabolites, which are natural deterrents against herbivores. Without these deterrents, pests may be more inclined to feed on the plants, increasing the extent of pest damage. In crops like strawberries and apples, where pest resistance is crucial for quality control, this increase in pest pressure may require more frequent intervention and pest control measures.
- **Need for Intensified Pest Management**: As pest pressure increases under elevated CO_2, more robust pest management strategies become necessary. Integrated pest management (IPM) practices, such as using biological control agents (e.g., predatory insects or parasitoids), habitat management, and selective pesticide applications, can help manage pest populations while minimizing environmental impact. For high-value fruit crops, precision agriculture technologies that monitor pest populations in real-time may help optimize pest control efforts, reducing pesticide use while effectively controlling pest infestations.

4. Implications for Crop Management and Sustainable Production

The indirect effects of elevated CO_2 on plant-microbe interactions, disease resistance, and pest dynamics highlight the need for adaptive crop management practices. Growers will need to consider these changes to maintain productivity, fruit quality, and sustainable practices.

- **Enhanced Soil and Nutrient Management**: Given the increased competition for nutrients in soils with active microbial communities, regular soil testing and balanced fertilization become essential to avoid nutrient deficiencies. Practices such as crop rotation, cover cropping, and the use of organic amendments can enhance soil health, promoting a resilient microbial community that supports nutrient cycling without depleting essential elements. In fruit orchards, integrating slow-release fertilizers and compost can provide a steady nutrient supply, supporting both plants and beneficial soil microbes.
- **Integrated Pest and Disease Management (IPDM)**: As both pest and disease pressures are likely to increase under elevated CO_2, adopting a holistic approach to pest and disease control will be essential. Integrated pest and disease management combines biological control, habitat manipulation, crop monitoring, and targeted pesticide use to minimize

crop damage while reducing environmental impacts. For fruit crops, IPDM strategies help maintain high-quality produce with minimal chemical input, supporting sustainable and eco-friendly production.

- **Breeding and Selection for Resilient Varieties**: To address the increased vulnerability to pests and diseases, breeding programs may focus on developing fruit crop varieties with enhanced natural resistance to pathogens and pests, even under elevated CO_2 conditions. Selecting varieties that retain high nutrient density, flavor, and disease resistance under elevated CO_2 will be crucial for maintaining quality in commercial fruit production. These resilient varieties could reduce the need for chemical interventions and improve sustainability in fruit crop systems.

2. Long-Term Implications of Elevated CO_2 on Fruit Crop Adaptation and Management

As atmospheric CO_2 levels continue to rise, fruit crop production systems will need to adapt to maintain productivity, quality, and resilience. The long-term impacts of elevated CO_2 are complex, affecting plant physiology, nutrient composition, pest and disease dynamics, and interactions with the surrounding environment. Farmers, researchers, and agricultural policymakers will need to consider adjustments in crop breeding, fertilization practices, and pest and disease management to meet the challenges posed by elevated CO_2. By implementing adaptive strategies, growers can mitigate the potential negative effects of elevated CO_2 while taking advantage of its positive impacts on growth and productivity.

1. Selection of CO_2-Responsive Crop Varieties

Some crop varieties may respond more favorably to elevated CO_2, maintaining higher yield and quality compared to others. Identifying or developing CO_2-responsive varieties through breeding programs can help ensure fruit crops are well-adapted to the changing atmosphere.

- **Breeding for CO_2-Resilient Traits**: As CO_2 levels rise, breeding programs may focus on selecting varieties that not only show enhanced growth under high CO_2 but also retain nutritional quality and disease resistance. For fruit crops like apples, peaches, and grapes, maintaining nutrient density and flavor quality is essential for meeting consumer expectations. Breeders may prioritize traits such as high nutrient retention, balanced sugar-acid ratios, and strong structural integrity to ensure that fruits remain nutritious, flavorful, and marketable even under elevated CO_2. Additionally, breeding programs could target improved tolerance to drought and nutrient-poor soils, traits that complement the water-use efficiency gains seen under elevated CO_2.
- **Maintaining Nutritional Quality**: With CO_2-driven nutrient dilution becoming a concern, varieties that maintain or concentrate higher levels of essential minerals like nitrogen, iron, and zinc will be valuable for producing fruits that meet dietary requirements. This is particularly important for fruits consumed primarily for their health benefits, such as berries, which are rich in antioxidants, vitamins, and minerals. Breeding efforts focused on maintaining nutrient density in the face of rapid growth induced by elevated CO_2 will ensure that fruits remain nutritious and continue to meet consumer demands for health-promoting produce.

- **Selection for Structural and Disease-Resistant Traits**: Elevated CO_2 can impact plant structure and increase susceptibility to pathogens. Breeding varieties with resilient cell wall structure and robust natural disease resistance can help mitigate the effects of CO_2-driven changes. Varieties that maintain firmness and resist softening will have longer shelf lives and be less susceptible to transport damage, ensuring fruits reach consumers in optimal condition. Additionally, selecting varieties with strong defenses against common pests and diseases will reduce dependency on chemical interventions, supporting sustainable fruit crop production.

2. **Adjustments in Fertilization Practices**

Elevated CO_2 often leads to nutrient dilution, which can reduce the nutritional quality of fruit crops. To counteract this effect and support optimal plant health, adjustments in fertilization practices may be necessary to maintain nutrient levels in fruit tissues.

- **Increased Nitrogen Fertilization to Counteract Nutrient Dilution**: Higher photosynthetic rates under elevated CO_2 can lead to rapid plant growth, which often dilutes nitrogen and other essential nutrients in plant tissues. As a result, increased nitrogen fertilization may be needed to maintain adequate protein levels and support metabolic functions. However, excessive nitrogen application can have negative environmental impacts, including soil degradation, groundwater contamination, and greenhouse gas emissions from nitrogen volatilization. Therefore, nitrogen fertilization must be carefully managed to avoid environmental harm while supporting crop growth.
- **Balanced Fertilization Strategies with Micronutrients**: Alongside nitrogen, elevated CO_2 can lead to deficiencies in micronutrients such as iron, zinc, and magnesium, which are essential for plant development and human health. Fertilizer formulations that include a balance of micronutrients can help counteract the dilution effect and ensure fruits remain rich in essential minerals. For example, iron and zinc are crucial for chlorophyll synthesis and enzyme function, and their presence is important for both plant vitality and nutritional quality. Combining nitrogen with micronutrients in fertilization programs can improve plant health, yield, and the nutrient profile of fruits.
- **Precision Fertilization and Soil Testing**: Regular soil testing will become increasingly important for monitoring nutrient availability in the face of elevated CO_2. Precision fertilization techniques, such as variable-rate applications based on soil nutrient profiles, can optimize nutrient availability, reduce excess application, and minimize environmental impact. In fruit orchards, soil and tissue testing can inform targeted fertilization strategies, ensuring that nutrient levels support optimal fruit quality without risking over-fertilization. By integrating precision agriculture practices, growers can respond to nutrient dilution while maintaining a sustainable approach to soil health and resource use.

3. **Enhanced Pest and Disease Management**

Elevated CO_2 can indirectly influence pest and disease dynamics by altering plant nutritional quality and defense mechanisms. Long-term pest and disease management will need to adapt to these changes to protect yield, quality, and plant health.

- **Increased Pest Pressure and Adjusted IPM Strategies**: Elevated CO_2 often results in lower nitrogen content in plant tissues, which can drive herbivorous pests to consume more plant material to meet their nutritional needs. In fruit crops, this may lead to increased feeding pressure from pests such as aphids, caterpillars, and beetles, which can damage leaves, stems, and fruit. Integrated pest management (IPM) strategies will become more critical in managing pest populations under elevated CO_2 conditions. IPM practices, such as using biological control agents (e.g., predatory insects), habitat management, and targeted pesticide applications, can help control pest populations while minimizing environmental impacts. For high-value fruit crops, IPM ensures that pest pressure is managed without compromising fruit quality or sustainability.
- **Enhanced Disease Monitoring and Biological Control**: Elevated CO_2 can reduce a plant's natural disease resistance by altering the production of defense compounds, making crops more susceptible to fungal infections and other pathogens. Regular disease monitoring and early detection will be crucial for preventing disease outbreaks. Biological control measures, including beneficial microbes that inhibit pathogens, can play a role in reducing disease incidence and severity. For example, the use of Trichoderma species in the rhizosphere has been shown to reduce soil-borne diseases in fruit crops. By incorporating biological control agents, growers can reduce dependency on chemical fungicides, promoting a more sustainable approach to disease management.
- **Selective Breeding for Disease Resistance**: To address the increased vulnerability to diseases under elevated CO_2, breeding programs may focus on developing varieties with enhanced resistance to pathogens. For example, in grapes, breeding for resistance to powdery mildew and Botrytis cinerea (graymold) can reduce losses due to these common fungal diseases. Varieties that maintain high levels of defense compounds and structural resilience under elevated CO_2 conditions will be valuable for reducing disease pressure and minimizing the need for chemical treatments. These resilient varieties will allow for a more sustainable, low-input approach to fruit crop production.

4. **Environmental and Economic Considerations**

Adapting to elevated CO_2 involves balancing the environmental impact of agricultural practices with the economic needs of fruit crop production. Sustainable practices that address nutrient management, pest and disease control, and crop variety selection are essential for long-term profitability and environmental stewardship.

- **Optimizing Resource Use and Reducing Environmental Impact**: Precision agriculture and resource-efficient practices are critical to managing the long-term implications of elevated CO_2. Precision fertilization, targeted irrigation, and IPM not only optimize input use but also reduce environmental footprints. By minimizing excess nitrogen application and reducing reliance on chemical pesticides, growers can maintain soil and water quality while meeting productivity goals. In fruit crops, optimizing resource use through precision farming will support sustainable production and preserve soil health, benefiting both current and future generations of farmers.
- **Balancing Yield with Quality and Nutritional Value**: As elevated CO_2 can increase yield

while diluting nutrient content, finding a balance between productivity and nutritional quality will be crucial for consumer satisfaction and marketability. High-quality produce with retained nutrient density, flavor, and texture will be essential to meet consumer demands. Implementing nutrient management practices that address nutrient dilution and investing in resilient crop varieties will enable growers to maintain this balance, ensuring that fruits remain nutritious, flavorful, and marketable.

- **Economic Impact and Market Adaptation**: The economic implications of adapting to elevated CO_2 are significant, as adjustments in fertilization, pest control, and breeding programs require investment. However, these practices are necessary to maintain productivity and quality, particularly in high-value crops where consumer expectations are high. By investing in sustainable practices, growers can protect their market share and profitability in the face of changing CO_2 levels. Resilient, high-quality fruit production will command higher prices, supporting long-term economic viability and food security.

3. Environmental and Economic Considerations

The implications of elevated CO_2 levels extend beyond immediate crop growth and productivity, influencing both environmental sustainability and the economic landscape of agriculture. While higher CO_2 can stimulate growth in certain crops, particularly those with C3 photosynthetic pathways, this boost in productivity is accompanied by environmental and economic trade-offs. Factors such as soil health, carbon sequestration, input costs, and market demand for quality produce will all play a role in shaping the future of fruit crop production under elevated CO_2 conditions. Adapting agricultural practices to maintain both productivity and sustainability will be critical in managing the complex challenges posed by changing atmospheric CO_2 levels.

1. **Carbon Sequestration Potential**

Elevated CO_2 levels can enhance biomass production, which in turn can contribute to carbon sequestration in agricultural soils and plant tissues. Carbon sequestration is the process by which CO_2 is absorbed by plants through photosynthesis and stored as carbon in roots, leaves, stems, and soil organic matter. While this process offers potential benefits for mitigating climate change, the extent of carbon sequestration depends on crop type, soil health, and management practices.

- **Increased Biomass and Soil Carbon Storage**: Under elevated CO_2, fruit crops may produce more biomass due to increased photosynthesis. This additional biomass, particularly from root systems, can contribute to soil organic matter when plants decompose. Organic matter in the soil is a critical component of soil carbon storage, as it helps trap carbon in stable forms, reducing atmospheric CO_2. In fruit orchards, for example, the larger root systems and greater canopy coverage under elevated CO_2 can increase the carbon stored both above and below ground, contributing to carbon sequestration.

- **Role of Cover Crops and Organic Amendments**: Incorporating cover crops and organic amendments, such as compost or manure, can enhance soil carbon sequestration in orchards and other fruit-growing systems. Cover crops, like legumes or grasses, add organic matter to the soil as they decompose, improving soil structure and nutrient availability while capturing carbon. Organic amendments increase soil organic carbon levels and improve the soil's capacity to hold water and nutrients. By fostering soil health through these practices, growers can enhance carbon sequestration while creating a more resilient growing environment that supports long-term productivity.

- **Reduced Tillage for Soil Carbon Preservation**: Tillage disrupts soil structure, accelerating the decomposition of organic matter and releasing stored carbon into the atmosphere. By reducing tillage, growers can help retain soil carbon and improve soil stability. Reduced tillage practices, such as minimal or no-till, are particularly beneficial in orchards where soil disturbance is minimal. In fruit crop systems, reducing tillage not only preserves soil carbon but also supports beneficial soil microbial communities, which play a vital role in nutrient cycling and plant health.

2. **Economic Impact on Crop Production**

While elevated CO_2 may increase yields for certain crops, including many fruit crops, the economic impact is influenced by additional input costs and market dynamics. Higher yields may boost profitability in the short term, but potential reductions in quality, increased management costs, and changing consumer preferences add complexity to the economic equation.

- **Short-Term Profit Gains from Higher Yields**: The initial increase in yield under elevated CO_2 can lead to higher profits, especially for high-value fruit crops where market prices are linked to volume. For instance, apples, grapes, and berries may produce more fruit per plant under higher CO_2 conditions, providing a direct boost to revenue. Increased yield potential is particularly advantageous in commercial fruit production, where maximizing output per hectare is critical for profitability. However, these gains are highly dependent on the ability to maintain quality, as consumers are increasingly demanding fruits that are nutritious and flavorful.
- **Increased Input Costs for Fertilizers and Pest Control**: To sustain the quality and nutrient density of fruits under elevated CO_2, growers may need to invest more in fertilizers and pest management. Nutrient dilution resulting from CO_2-driven growth requires additional fertilization, particularly with nitrogen and essential micronutrients like iron and zinc. Higher CO_2 can also affect pest dynamics, often increasing pest pressure due to altered plant chemistry. These factors may require more intensive pest management, adding to input costs. The cost of these additional inputs can offset short-term profit gains from higher yields, creating a more complex economic landscape for growers.
- **Balancing Productivity with Quality and Market Demand**: While increased yield may offer economic benefits, reductions in fruit quality and nutrient density can impact market demand. Consumers increasingly prioritize nutritious, flavorful, and visually appealing produce, especially in the premium fruit market. Lower nutrient density, altered sugar-acid balance, and reduced firmness may affect the acceptability and price of fruit crops like apples, strawberries, and grapes. For crops like wine grapes, where flavor complexity and acidity are key, changes in fruit chemistry due to elevated CO_2 can affect product quality, potentially reducing market value. Therefore, maintaining a balance between productivity and quality will be crucial for long-term profitability in fruit crop production.

3. Sustainability and Environmental Impact

The sustainability of agricultural practices under elevated CO_2 will be vital for addressing both environmental and economic challenges. Sustainable practices that focus on resource efficiency, soil health, and ecosystem balance will play a critical role in mitigating the negative impacts of elevated CO_2 on the environment.

- **Optimizing Resource Use with Precision Agriculture**: Precision agriculture techniques, such as site-specific fertilization, targeted irrigation, and real-time monitoring, help growers optimize resource use and reduce environmental impacts. By using sensors and data analysis, precision agriculture enables growers to apply nutrients and water only where they are needed, reducing excess and minimizing nutrient runoff. In fruit orchards, precision agriculture can improve fertilizer and water-use efficiency, supporting both productivity and environmental sustainability. This approach helps growers manage nutrient and water resources effectively, reducing costs and limiting pollution.
- **Reducing Greenhouse Gas Emissions**: While elevated CO_2 can enhance plant growth, additional nitrogen fertilization required to maintain nutrient density can lead to increased emissions of nitrous oxide, a potent greenhouse gas, if not managed carefully. By employing controlled-release fertilizers and integrating organic amendments, growers can reduce nitrogen losses and lower greenhouse gas emissions. Sustainable practices that minimize synthetic fertilizer use also improve soil health, supporting a more resilient and productive agricultural system.
- **Long-Term Soil Health and Biodiversity**: Maintaining soil health and biodiversity is essential for the sustainability of fruit crop production under elevated CO_2. Healthy soils with high organic matter content and diverse microbial communities are better equipped to support productive crops while retaining carbon and nutrients. Practices such as crop rotation, cover cropping, and reduced tillage help maintain soil health and foster biodiversity in the soil ecosystem. In fruit crop systems, promoting biodiversity and soil health improves resilience to pests and diseases, reduces the need for chemical inputs, and supports sustainable yields over the long term.

4. Market Adaptation and Consumer Perception

The effects of elevated CO_2 on fruit quality and nutrient composition may influence consumer perception and market trends, particularly as consumers become more health-conscious and environmentally aware. Market adaptation strategies that address these changes can help maintain demand and profitability for fruit producers.

- **Marketing for Nutritional and Quality Attributes**: To adapt to potential changes in fruit nutrient density and flavor, producers may need to focus on marketing high-quality and nutrient-rich varieties to meet consumer demand. Varieties that maintain higher nutrient density and appealing flavors under elevated CO_2 will be critical for differentiating products in a competitive market. Marketing strategies that emphasize quality, freshness, and nutritional value will appeal to consumers who prioritize health benefits in their food choices.

- **Premium Pricing for Sustainable Practices**: As consumers increasingly value sustainably produced foods, adopting environmentally responsible practices can be a valuable marketing point. Fruit producers who adopt practices like reduced tillage, organic fertilization, and IPM can position their products as environmentally friendly, justifying premium pricing. For high-value crops like grapes and apples, certifications for sustainable practices can attract environmentally conscious consumers, enhancing profitability while supporting environmental goals.
- **Investment in Research and Development**: The long-term adaptation to elevated CO_2 conditions may require investment in research and development to support sustainable practices, crop breeding, and precision agriculture technologies. Research efforts focused on developing CO_2-responsive varieties, optimizing nutrient management, and improving pest control strategies will help fruit producers navigate the challenges of elevated CO_2. Collaboration between researchers, government agencies, and the agricultural industry can accelerate innovation and make sustainable practices more accessible to growers.

II. Industrial Waste

Industrial waste, generated by a variety of manufacturing and production processes, poses a significant environmental threat due to its potential to contaminate air, water, and soil. The implications of industrial waste are particularly concerning for agriculture, as waste byproducts can introduce harmful pollutants, alter soil chemistry, and impact crop productivity and quality. In the context of fruit crop production, exposure to industrial waste can lead to a range of challenges, including nutrient imbalances, accumulation of toxic heavy metals, and alterations in soil microbial activity—all of which influence crop resilience, yield, and marketability. Managing the impacts of industrial waste requires a comprehensive understanding of its sources, types, and effects, along with strategies for mitigation and adaptation.

1. Sources and Types of Industrial Waste

Industrial waste is generated from a wide range of manufacturing processes and industries, including mining, chemical manufacturing, textile production, metal processing, electronics, and more. These wastes consist of various materials, from organic and inorganic compounds to hazardous chemicals and metals, each with unique properties and distinct environmental and agricultural impacts. The sources and types of industrial waste are essential for assessing its effects on agriculture, particularly in fruit crop production, where exposure to these wastes can affect soil health, plant growth, and crop quality.

Solid Waste

Solid industrial waste comprises a variety of materials, including both non-hazardous and hazardous substances, which can accumulate in soils and persist for long periods. These materials originate from manufacturing processes, packaging, and other industrial activities, and they pose unique environmental risks, especially in agricultural areas.

» **Types of Solid Waste**: Solid waste can include plastics, metals, synthetic fibers, discarded machinery, and chemical containers. Plastics, commonly found in solid waste, are particularly problematic as they break down into microplastics over time. These tiny plastic particles can be carried by wind, water, or direct soil contact and integrate into the soil matrix. Microplastics in the soil impact the physical structure, limiting water retention and affecting plant root development.

» **Environmental Persistence**: Many solid wastes, especially synthetic plastics and metals, are resistant to natural degradation processes and can persist in the environment for decades. As these materials accumulate in the soil, they can leach harmful substances, including additives in plastics and metal residues, which may enter groundwater or soil systems, impacting soil health and potentially contaminating crops. For instance, metals in electronic waste, such as lead and cadmium, can leach into soils, resulting in heavy metal toxicity that affects plant root systems and nutrient uptake.

» **Effects on Soil and Plant Health**: Accumulated solid waste disrupts soil aeration and water infiltration, which can lead to soil compaction and reduced root growth. In fruit crops, where root health is vital for nutrient uptake and fruit quality, the presence of solid waste can lead to poor root structure, limited water availability, and compromised plant stability. As a result, fruit trees and other perennial crops may experience stunted growth, reduced yield, and lower resistance to environmental stressors.

Liquid Waste

Liquid industrial waste includes wastewater, chemical effluents, and byproducts of various manufacturing processes. Liquid wastes often contain high concentrations of organic matter, heavy metals, toxic chemicals, and other pollutants, which, if released untreated, pose significant risks to agricultural soils and water sources.

» **Sources of Liquid Waste**: Liquid waste primarily comes from industries such as textile dyeing, chemical manufacturing, pharmaceuticals, oil refining, and metal finishing. Wastewater and effluents from these industries frequently contain dyes, acids, salts, and other chemicals that, when discharged, can enter nearby soils, rivers, and groundwater systems.

» **Composition and Environmental Impact**: Liquid industrial wastes are often rich in hazardous substances such as heavy metals (lead, cadmium, chromium), volatile organic compounds (VOCs), and acids that alter soil pH and salinity levels. For example, effluents from textile dyeing processes contain chemical dyes and salts that can increase soil salinity, disrupting plant growth and microbial activity in agricultural soils.

» **Soil and Crop Implications**: The presence of untreated liquid waste in agricultural soils can lead to multiple adverse effects. Heavy metals and chemicals can accumulate in the soil, making it difficult for plants to absorb water and nutrients effectively. High salt concentrations can cause osmotic stress in plants, leading to dehydration, poor growth, and leaf burn. In fruit crops, soil salinization and heavy metal contamination can result in reduced fruit size, compromised flavor, and decreased marketability. Moreover, toxic compounds in liquid waste may alter microbial diversity in the soil, reducing beneficial organisms essential for nutrient cycling.

Gaseous Emissions

Industrial processes release a range of gaseous pollutants into the atmosphere, including sulfur dioxide (SO_2), carbon monoxide (CO), nitrogen oxides (NOx), and volatile organic compounds (VOCs). These emissions contribute to environmental issues like acid rain and particulate deposition, which have significant implications for agricultural soils and plant health.

» **Sources of Gaseous Emissions**: Common sources include power plants, refineries, chemical manufacturing, cement production, and the burning of fossil fuels. These industries emit gases directly into the atmosphere, where they can travel and settle over large areas, affecting both air and soil quality.

» **Effects of Acid Rain on Soils and Crops**: Sulfur dioxide and nitrogen oxides react with moisture in the atmosphere to form acid rain, which then falls onto the soil, altering its pH balance. Acid rain increases soil acidity, which can lead to nutrient leaching, reducing the availability of essential nutrients like calcium and magnesium for plants. For fruit crops, acid rain exposure can lead to nutrient deficiencies, reduced photosynthesis, and weaker plant structures, affecting yield and quality.

» **Particulate Matter and Crop Growth**: Particulate matter from industrial emissions, including heavy metals and other pollutants, can settle on plant surfaces, obstructing sunlight absorption and photosynthesis. Dust and particulates also block stomata, the small openings on leaves responsible for gas exchange, leading to reduced photosynthetic efficiency and increased water loss. Fruit crops affected by particulate matter may exhibit slower growth, lower fruit production, and decreased resilience to pests and diseases.

Hazardous Waste

Hazardous industrial waste includes highly toxic chemicals, radioactive substances, and heavy metals, which pose severe threats to human and environmental health. These wastes are often byproducts of chemical manufacturing, metal processing, and electronics industries, and they require specialized handling and disposal to prevent contamination.

» **Toxic Chemicals and Pesticides**: Hazardous wastes contain chemicals like pesticides, solvents, and pharmaceutical byproducts that, when introduced into the environment, can persist for extended periods. These substances can disrupt soil microbial communities, alter plant metabolism, and accumulate in food crops, posing health risks to consumers. For example, pesticides from industrial runoff can affect fruit quality and increase residues on the skin and in the flesh of fruit crops, which are particularly harmful when consumed.

» **Radioactive Waste**: Industrial activities such as mining and nuclear power generation produce radioactive waste, which can contaminate soils and water sources if not managed properly. Radioactive contaminants are especially hazardous due to their long half-lives, meaning they remain active in the environment for decades or even centuries. Radioactive contamination in soils affects plant growth, cellular health, and can lead to mutations in plants, significantly impacting crop yield and quality.

» **Heavy Metal Toxicity**: Metals like lead, cadmium, and mercury are common in hazardous waste from industrial processes. These metals disrupt enzymatic activity in plants, interfere with nutrient uptake, and accumulate in edible tissues, posing health risks when the fruits are consumed. In fruit crops, heavy metal toxicity can cause chlorosis, poor root development, and stunted growth, ultimately reducing yield and compromising fruit quality. The bioaccumulation of heavy metals in edible fruit parts is a significant concern, as these metals have been linked to various human health issues, including neurological disorders and kidney damage.

2. Impact of Industrial Waste on Soil and Plant Health

The contamination of soil by industrial waste poses serious threats to both soil and plant health, affecting agricultural productivity and food safety. Industrial pollutants, including chemicals, heavy metals, and salts, alter soil chemistry, disrupt microbial communities, and reduce nutrient availability. These changes compromise soil fertility, damage plant physiology, and allow toxic elements to enter the food chain, affecting fruit crops in particular, where quality and safety are paramount.

Soil Acidification and Alkalization

Industrial waste, especially when it includes sulfur and calcium compounds, can lead to soil acidification or alkalization, each of which disrupts nutrient balance and affects plant health. Acidic or alkaline soil conditions interfere with nutrient absorption and the overall health of crops, including high-value fruit crops that require optimal soil conditions for growth and fruit quality.

- **Soil Acidification**: Acidification is a common consequence of industrial waste containing sulfur compounds, such as sulfur dioxide from chemical plants and fossil fuel combustion. When sulfur compounds are introduced into the soil, they can form sulfuric acid, lowering the soil pH. Acidic soils reduce the availability of key nutrients like calcium, magnesium, and potassium, which are essential for plant health and fruit development. In acidified soils, plants struggle to access these nutrients, leading to poor growth, reduced flowering, and lower fruit yields. Fruit crops such as apples and grapes, which require balanced pH for quality production, are particularly vulnerable to the effects of acidified soils.
- **Soil Alkalization**: On the other hand, certain industrial wastes, especially those containing calcium and carbonate compounds, can cause soil alkalization, raising the soil pH. Alkaline soils often experience deficiencies in micronutrients like iron, manganese, and zinc, as these nutrients become insoluble and unavailable to plants in high pH environments. In fruit crops, iron deficiency causes iron chlorosis, a condition where leaves turn yellow due to insufficient chlorophyll production. Alkaline soils reduce plant vigor, leading to smaller fruits, poor color, and reduced nutritional value. Crops such as citrus and blueberries, which are sensitive to pH changes, may suffer yield loss and decreased fruit quality in alkaline soils.

Heavy Metal Toxicity

Heavy metals from industrial waste, including lead, cadmium, and mercury, accumulate in soils over time, posing significant risks to plant health and human safety. These metals disrupt cellular processes, hinder enzyme function, and impair photosynthesis, ultimately impacting crop productivity and quality. The absorption of heavy metals by plants can result in bioaccumulation in edible tissues, leading to potential health risks for consumers.

» **Lead (Pb) Toxicity**: Lead is a common pollutant in soils near industrial sites, particularly those related to metal processing, mining, and battery production. Lead inhibits plant root development, reducing the plant's ability to absorb water and nutrients. In fruit crops, lead toxicity manifests as leaf chlorosis, reduced leaf size, and stunted growth, which result in lower fruit yields and compromised quality. Lead accumulates in plant tissues, especially in the leaves and fruits, posing health risks when ingested by humans. Chronic exposure to lead through consumption of contaminated fruits has been linked to neurological disorders and kidney damage.

» **Cadmium (Cd) Toxicity**: Cadmium is found in industrial waste from electroplating, paint, and plastic manufacturing. It interferes with plant photosynthesis and nutrient transport by displacing essential elements like calcium and zinc. Cadmium toxicity leads to leaf chlorosis, wilting, and poor root growth, which restricts the plant's capacity to thrive. In fruit crops, cadmium accumulation in tissues affects fruit size and flavor, making them less marketable. Cadmium's potential to accumulate in edible parts of the plant also poses health risks to consumers, as it is a carcinogenic metal linked to kidney and bone damage.

» **Mercury (Hg) Toxicity**: Mercury, primarily released from mining and industrial waste, disrupts enzymatic activity and damages plant cells. This toxic metal interferes with the plant's hormonal balance and energy production processes. In fruit crops, mercury toxicity leads to reduced flowering and fruit set, lowering yields and overall quality. Mercury contamination in fruits presents health risks, as even low levels of mercury are toxic and linked to neurological and developmental disorders.

Salinity and Soil Structure

Industrial effluents often contain high levels of salts, leading to soil salinization and degradation of soil structure. Excessive salinity alters water availability, affects soil compaction, and impacts plant physiology, particularly in fruit crops that require balanced soil conditions for optimal growth and flavor.

» **Soil Salinization**: Industrial wastewater, especially from textile, leather, and chemical industries, contains salts that can accumulate in the soil. Salinity disrupts osmotic balance, making it difficult for plants to absorb water from the soil, even when moisture is present. This condition, known as osmotic stress, results in dehydration, reduced growth, and wilting. In fruit crops, salt stress impacts flowering and fruit formation, leading to smaller, less flavorful fruits. For example, grapes grown in saline soils produce fruit with lower sugar content and diminished flavor, reducing their market value and appeal.

» **Impact on Soil Structure and Water Retention**: High salt concentrations in soil lead to soil compaction and poor water infiltration, as salts cause soil particles to clump together. Compacted soils have reduced porosity, limiting root growth and reducing water retention capacity. Fruit crops growing in compacted, saline soils experience restricted root development, affecting their ability to access water and nutrients. This results in poor plant stability, lower yields, and increased susceptibility to environmental stressors.

» **Reduced Water-Use Efficiency**: In saline conditions, plants expend more energy to manage osmotic stress, reducing their water-use efficiency. This inefficiency can result in lower yields, as plants allocate more resources toward survival than growth. For high-value fruit crops, lower water-use efficiency due to salinity stress decreases productivity and compromises fruit quality, making it difficult for growers to meet market demands for size, flavor, and appearance.

Disruption of Soil Microbial Communities

Soil microbial communities play a critical role in nutrient cycling, organic matter decomposition, and disease suppression, all of which are essential for maintaining soil health and supporting plant growth. Industrial pollutants, including toxic chemicals and heavy metals, disrupt microbial populations, reducing microbial diversity and compromising soil health.

» **Impact on Beneficial Microbes**: Soil contamination reduces the population of beneficial microbes, including nitrogen-fixing bacteria and mycorrhizal fungi. These organisms are vital for nutrient cycling and plant health. Nitrogen-fixing bacteria, for example, convert atmospheric nitrogen into a form that plants can use, while mycorrhizal fungi enhance nutrient and water uptake through symbiotic relationships with plant roots. When pollutants disrupt these microbes, nutrient availability declines, affecting plant vigor and resilience. For fruit crops, reduced access to nitrogen and phosphorus due to microbial loss results in poor growth, reduced flowering, and lower fruit production.

» **Alteration in Soil Enzyme Activity**: Soil microbes produce enzymes that are crucial for organic matter decomposition and nutrient mineralization, making nutrients available for plants. Industrial pollutants, particularly heavy metals and chemical residues, inhibit enzyme activity, slowing down nutrient cycling and reducing soil fertility. For example, enzymes involved in phosphorus mineralization become less active in contaminated soils, limiting phosphorus availability for crops. In fruit crops, phosphorus is essential for root development and fruit quality, so reduced enzyme activity can directly affect yield and marketability.

» **Increase in Pathogenic Microbes**: While beneficial microbes decline in contaminated soils, some pathogenic microbes may thrive. Certain industrial pollutants create conditions that favor the growth of pathogenic bacteria and fungi, increasing the risk of diseases in crops. For fruit crops, diseases such as root rot and fungal infections become more common in contaminated soils, as pathogen populations increase in the absence of beneficial microbes. The higher incidence of disease adds to the production challenges, as growers must invest more in pest and disease management to maintain crop health.

3. Impacts on Fruit Crop Quality and Yield

Fruit crops are particularly sensitive to soil quality and environmental conditions, and exposure to industrial waste can significantly impair their growth, quality, and marketability. Industrial pollutants like heavy metals, toxic compounds, and chemical residues affect various aspects of crop physiology, leading to changes in fruit size, nutritional content, flavor, texture, and susceptibility to pests and diseases. These changes can reduce yields and affect consumer demand, as quality and nutritional value are crucial in the competitive fruit market. Below are the key ways industrial waste affects fruit crop quality and yield.

Reduced Fruit Size and Quality

Industrial waste contaminants, especially heavy metals and toxic compounds can interfere with nutrient absorption in plants, impacting fruit size, weight, and overall quality. These effects are especially noticeable in high-value fruit crops like apples, strawberries, and grapes, where even slight variations in size or appearance can reduce consumer appeal and market value.

- **Nutrient Uptake Inhibition**: Heavy metals like lead, cadmium, and mercury bind with soil particles and plant roots, blocking the uptake of essential nutrients such as nitrogen, potassium, and calcium. This nutrient limitation disrupts plant growth, causing deficiencies that stunt fruit development. In fruit crops, smaller fruits result in lower yields and reduced marketability, as consumers often equate larger, well-formed fruits with higher quality. For example, apples grown in contaminated soils may be undersized or develop deformities, making them less appealing to consumers and reducing their commercial value.
- **Impact on High-Value Crops**: For premium fruit crops like strawberries and grapes, where size and visual quality are critical for market success, industrial waste contamination can be particularly damaging. Reduced fruit size and compromised quality affect these crops' reputation in the market, where appearance is directly linked to consumer preference and pricing. Such impacts can lead to financial losses for growers, as smaller or damaged fruits may be sold at lower prices or be unsuitable for fresh-market sale, limiting their market potential.

Altered Nutritional Composition

Heavy metals and other pollutants can displace essential nutrients in plant tissues, leading to fruits with lower levels of vitamins, minerals, and antioxidants. This reduction in nutritional quality affects the health benefits of fruits and may impact consumer demand, particularly among health-conscious consumers.

- **Nutrient Displacement and Deficiencies**: Industrial pollutants, particularly heavy metals, can replace essential minerals like iron, zinc, and magnesium within plant tissues. This displacement results in nutritional imbalances and lower nutrient density in fruits, diminishing their health benefits. For example, lead can interfere with the plant's ability to uptake and transport calcium, a mineral vital for fruit firmness and quality. Consequently, fruits grown in polluted soils may lack essential nutrients, reducing their appeal to health-conscious consumers seeking nutritious, mineral-rich produce.
- **Antioxidant and Vitamin Levels**: The nutritional quality of fruits is closely tied to their antioxidant, vitamin, and mineral content, which contributes to the health benefits they offer. However, industrial waste exposure can reduce levels of antioxidants like vitamin C, phenolic compounds, and flavonoids, which play a role in protecting cells and maintaining immune health. For fruit crops like blueberries and strawberries, known for their high antioxidant content, reductions in these compounds due to contamination undermine their health appeal, potentially impacting demand and sales in health-focused markets.

Flavor and Texture Changes

Industrial pollutants can disrupt the biochemical pathways responsible for fruit flavor and texture, leading to alterations that impact consumer satisfaction and marketability. Flavor and texture are key quality attributes for fruits, as they influence the eating experience and shelf life, both of which are important for fresh-market produce.

- **Disruption of Sugar and Acid Metabolism**: Heavy metals and toxic compounds interfere with the enzymatic processes that control sugar and acid levels in fruits. In plants, the balance of sugars and organic acids determines fruit flavor, contributing to sweetness and tartness. Industrial pollutants can skew this balance, leading to overly acidic or bland-tasting fruits. For example, high concentrations of cadmium in soil can affect carbohydrate metabolism, resulting in lower sugar levels in fruits like grapes, which impacts their sweetness and suitability for products like wine and juice. Altered sugar levels also affect the flavor complexity of fruits, making them less appealing to consumers who expect specific taste profiles.
- **Textural Changes and Consumer Appeal**: Industrial pollutants can affect cell wall structure and water content within fruits, leading to texture changes such as increased softness or reduced firmness. For fruit crops like peaches and cherries, where firmness is essential for shelf life and handling, reduced cell wall stability leads to bruising and quicker softening, reducing their shelf life and appeal. Softer fruits are more prone to damage during transportation and have a shorter storage period, increasing waste and lowering profitability. For growers and retailers, maintaining quality texture is crucial for delivering fruits in optimal condition to consumers.

Increased Susceptibility to Pests and Diseases

Industrial pollutants weaken plants' natural defenses, making them more vulnerable to pests and diseases. This increased susceptibility not only reduces crop yields but also raises production costs, as growers may need to apply additional pesticides or adopt intensive management practices to protect their crops.

- **Weakened Plant Defense Mechanisms**: Pollutants like heavy metals impair plants' natural defense pathways, affecting their ability to produce defense compounds such as phytoalexins, which help resist pathogens and pests. When plants lack these natural defenses, they become more susceptible to pest attacks and diseases. In fruit crops, this vulnerability can lead to increased pest infestation, fungal infections, and root diseases, which reduce yields and compromise fruit quality. For example, apples and strawberries exposed to contaminated soils may exhibit higher rates of powdery mildew and fungal root rot due to weakened resistance, making pest and disease management more challenging and costly.
- **Increased Pest Pressure and Crop Damage**: Industrial waste contamination can lead to imbalances in soil nutrients and weaken plants' structural integrity, making them more attractive to pests. Pests such as aphids, mites, and caterpillars often target nutrient-deficient or weakened plants, exacerbating crop damage. In fruit crops, increased pest pressure reduces fruit set, causes leaf and stem damage, and diminishes overall productivity. For growers, higher pest incidence necessitates additional pest control measures, which increase labor and material costs and may contribute to chemical residue accumulation on fruits, further affecting marketability.

- **Higher Disease Incidence and Production Costs**: With increased susceptibility to diseases in contaminated soils, fruit crop production requires additional inputs to maintain healthy crops. Growers may need to apply fungicides and bactericides more frequently to protect against diseases like blight, rust, and bacterial canker, raising production costs. For organic growers, who rely on natural pest control methods, managing diseases in polluted soils can be particularly challenging, as many natural treatments are less effective under high disease pressure. Increased chemical use also poses environmental risks, as pesticide residues can accumulate in soils and water systems, impacting soil health and biodiversity.

4. Long-Term Environmental Impacts and Ecosystem Degradation

The long-term presence of industrial waste in agricultural soils has profound environmental impacts that extend beyond the immediate effects on crop growth. Pollutants from industrial waste persist in ecosystems, often becoming more harmful over time as they accumulate in soil and water sources. These contaminants contribute to land degradation, loss of biodiversity, and ecosystem imbalances, all of which affect plant and animal life in complex ways.

Soil Degradation and Erosion

Soil degradation is one of the primary consequences of industrial waste contamination, as pollutants disrupt the physical structure and fertility of the soil. Contaminated soils often exhibit compaction and poor water infiltration, which can lead to increased erosion and nutrient loss. Over time, these effects degrade the soil, reducing its ability to support crops and contributing to food security challenges.

- **Compaction and Reduced Water Infiltration**: Industrial waste, particularly when it contains heavy metals and chemical residues, alters soil composition and structure. Pollutants cause soil particles to clump together, leading to compaction and reducing the soil's porosity. Compacted soil has limited space for water infiltration and root growth, making it harder for plants to access the moisture and nutrients they need. In fruit crop production, where healthy root systems are essential for productivity and fruit quality, compacted soils result in stunted plant growth and poor yields.
- **Increased Risk of Erosion**: Compacted and contaminated soils are highly susceptible to erosion, especially during heavy rainfall. Erosion removes the nutrient-rich topsoil layer, which is essential for plant growth, leaving behind nutrient-poor subsoil. As the topsoil is lost, soil depth decreases, and the ability of the soil to support diverse crops diminishes. For fruit crops, erosion not only impacts current yields but also limits the soil's long-term productivity, making it difficult for growers to maintain sustainable production.
- **Impact on Soil Fertility**: Erosion and soil degradation lead to nutrient depletion, as essential elements like nitrogen, phosphorus, and potassium are washed away along with the topsoil. As fertility declines, farmers must rely on increased fertilization to restore nutrients, which can be costly and environmentally unsustainable. For regions already dealing with limited resources, soil degradation and erosion due to industrial pollutants further threaten food security by reducing the land's agricultural potential.

Bioaccumulation and Biomagnification

Bioaccumulation and biomagnification refer to the processes by which pollutants, particularly heavy metals, accumulate in organisms and magnify up the food chain. Persistent contaminants like lead, cadmium, and mercury can bioaccumulate in plants and subsequently biomagnify through the food chain, affecting not only crop consumers but also higher trophic levels, such as animals and humans.

» **Bioaccumulation in Plant Tissues**: Pollutants from industrial waste are taken up by plants through their roots and accumulate in tissues over time. Heavy metals such as cadmium and lead are particularly prone to bioaccumulation, as plants cannot easily break them down. These metals interfere with metabolic functions and cellular health, impacting growth and productivity. In fruit crops, bioaccumulation affects the quality of edible parts, making them unsafe for consumption. Consuming contaminated fruits poses health risks, including heavy metal poisoning, which can lead to kidney damage, neurological disorders, and other chronic health issues in humans.

» **Biomagnification in the Food Chain**: When herbivores or omnivores consume contaminated plants, the pollutants concentrate in their bodies and move up the food chain. Predators that feed on these animals accumulate even higher levels of toxins, a process known as biomagnification. This phenomenon is particularly dangerous for wildlife and human health, as organisms at the top of the food chain, including humans, are exposed to high concentrations of toxins. For agricultural ecosystems, biomagnification poses threats to biodiversity, as species exposed to high levels of contaminants may experience reproductive issues, behavioral changes, or death.

» **Impact on Wildlife and Human Health**: Biomagnification of heavy metals and other toxic compounds affects not only agricultural areas but also surrounding wildlife habitats. Birds, mammals, and aquatic life that consume contaminated plants and water are at risk of poisoning, which can disrupt local ecosystems. For human populations, consuming fruits and other food products grown in contaminated soils increases exposure to hazardous substances, leading to potential long-term health risks.

Loss of Biodiversity

Industrial waste contamination leads to a decline in soil biodiversity, as many beneficial organisms are unable to survive in polluted environments. Soil biodiversity is crucial for maintaining ecosystem health, as it supports nutrient cycling, organic matter decomposition, and pest control. A loss in biodiversity compromises these ecosystem services, impacting crop productivity and environmental resilience.

» **Decline in Microbial Diversity**: Soil microbes, including bacteria, fungi, and actinomycetes, play essential roles in nutrient cycling and soil structure maintenance. Heavy metals and toxic chemicals in industrial waste inhibit microbial activity and reduce microbial diversity. Beneficial microbes, such as nitrogen-fixing bacteria and mycorrhizal fungi, struggle to survive in contaminated soils, leading to declines in nutrient availability and soil fertility. In fruit crop systems, reduced microbial diversity impacts plant growth, as nitrogen fixation and phosphorus solubilization become limited, resulting in poor crop development and yield.

- **Reduced Pest Control and Pollination Services**: Soil contamination and biodiversity loss affect pest control and pollination, which are vital for sustainable agriculture. Many beneficial insects, such as ground beetles and parasitoid wasps, rely on healthy soil ecosystems to thrive. However, pollution disrupts their habitats, leading to declines in their populations. Pollinators like bees, essential for fruit crops, are also affected by exposure to toxins, which can reduce pollination rates and fruit set. As pest control and pollination services decline, growers face increased challenges in maintaining crop health and productivity.
- **Impact on Ecosystem Resilience**: A diverse soil ecosystem is more resilient to environmental stressors, such as drought, pests, and disease. When biodiversity declines due to pollution, ecosystems become less adaptable to changing conditions. For agricultural systems, this loss of resilience means that crops are more vulnerable to fluctuations in weather, pest outbreaks, and disease. In the long term, the decline in biodiversity from industrial waste contamination threatens the stability of agricultural production and reduces the land's capacity to support healthy crops.

Groundwater Contamination

Industrial pollutants, particularly in liquid form, have the potential to leach into groundwater, contaminating water sources for agriculture and human use. Groundwater contamination poses significant health risks, as pollutants like heavy metals and nitrates accumulate in drinking water supplies. This contamination creates a cycle of pollution, as contaminated water exacerbates soil pollution when used for irrigation.

- **Leaching of Pollutants**: Pollutants from industrial waste, especially soluble substances like nitrates, phosphates, and heavy metals, can leach through the soil profile and enter groundwater. This process is accelerated in soils with poor structure or high permeability. In agricultural regions, the use of contaminated groundwater for irrigation further spreads pollutants to the soil, making it difficult to break the cycle of contamination.
- **Health Risks of Contaminated Water**: Groundwater contaminated with heavy metals, nitrates, and other pollutants poses severe health risks to humans. Nitrates, for example, are linked to methemoglobinemia (blue baby syndrome) in infants, while prolonged exposure to heavy metals in drinking water has been associated with kidney disease, neurological disorders, and cancer. In fruit crop production, reliance on contaminated groundwater introduces pollutants into the soil, affecting crop safety and posing risks to consumers who ingest contaminated produce.
- **Challenges in Remediation and Water Management**: Addressing groundwater contamination from industrial pollutants is challenging and costly. Traditional water treatment methods may not effectively remove all contaminants, and restoring contaminated groundwater is often a long-term process. For agriculture, limited access to clean water sources affects productivity, as reliance on polluted water restricts crop choice and requires additional remediation efforts. Sustainable water management practices, such as rainwater harvesting and regulated use of clean water for irrigation, are essential to reduce reliance on contaminated sources and maintain soil health.

Broader Environmental and Agricultural Implications

The long-term environmental impacts of industrial waste in agricultural soils go beyond immediate crop productivity, affecting the stability and sustainability of entire ecosystems. Soil degradation, bioaccumulation, loss of biodiversity, and groundwater contamination present complex challenges that reduce land productivity, threaten food security, and endanger public health. For fruit crop production, these long-term impacts translate into lower yields, reduced quality, and increased production costs, as growers must address the effects of environmental degradation on soil and water resources.

Addressing the long-term impacts of industrial waste requires a combination of sustainable waste management, pollution prevention, soil remediation, and adaptive agricultural practices. By reducing industrial waste discharge, restoring contaminated soils, and promoting biodiversity, agricultural ecosystems can regain resilience and productivity, ensuring that fruit crops continue to provide safe, high-quality food for consumers while preserving the health of the environment.

5. Management and Mitigation Strategies

Effective management and mitigation of industrial waste impacts on agriculture require a combination of preventative measures, soil remediation, and adaptive crop management practices. These strategies can help restore soil health, reduce pollution levels, and improve crop resilience.

» **Soil Remediation Techniques**: Soil remediation involves removing or neutralizing contaminants to restore soil health. Techniques such as phytoremediation (using plants to absorb contaminants), bioremediation (using microbes to break down pollutants), and soil washing are commonly used to clean up contaminated soils. In fruit orchards, phytoremediation with hyperaccumulator plants like mustard and sunflower can help remove heavy metals, while compost and biochar amendments can improve soil structure and reduce metal availability.

» **Buffer Zones and Vegetative Barriers**: Creating buffer zones with vegetation between industrial sites and agricultural fields can help trap pollutants before they reach crop-growing areas. Trees and grasses act as physical barriers, capturing airborne particles and absorbing pollutants from runoff. Buffer zones also reduce wind erosion, preventing the spread of contaminants to nearby agricultural soils. For fruit crops, buffer zones help protect sensitive plants from direct exposure to pollutants, reducing contamination risk.

» **Improved Waste Management Practices**: Industries can adopt more responsible waste management practices to reduce pollutant release into the environment. Treating wastewater, recycling solid waste, and using closed-loop systems to prevent waste discharge are essential for minimizing pollution. Regulatory measures that enforce waste treatment standards and encourage industries to adopt sustainable practices play a vital role in protecting agricultural lands.

» **Adaptive Crop Management**: In areas with mild contamination, growers can adopt crop management practices that minimize pollutant uptake and mitigate stress. Techniques such as crop rotation, organic amendments, and careful nutrient management can help reduce the impact of contaminants on crops. For example, applying organic matter like compost can improve soil structure, increase microbial activity, and immobilize certain pollutants, making them less available to plants. Additionally, selecting crops or varieties with greater tolerance to heavy metals or salinity can improve productivity in contaminated soils.

6. Policy and Research Initiatives for Sustainable Agriculture

Addressing the impacts of industrial waste on agriculture requires a coordinated effort between policymakers, researchers, and industry stakeholders. Policies and research initiatives that support sustainable agriculture and pollution prevention can help mitigate industrial waste impacts and promote environmental health.

» **Regulatory Standards and Enforcement**: Governments can implement stricter regulations on industrial waste disposal, requiring industries to treat waste before discharge. Policies that set limits on pollutant concentrations in air, water, and soil help reduce the environmental burden of industrial activities. Enforcement mechanisms, such as regular inspections and penalties for non-compliance, ensure industries adhere to these standards, protecting agricultural lands from contamination.

» **Incentives for Waste Reduction and Recycling**: Providing incentives for industries to reduce waste generation and adopt recycling practices can encourage sustainable waste management. Tax breaks, grants, and subsidies for companies that implement eco-friendly technologies promote waste reduction and recycling. For agriculture, reduced industrial waste pollution leads to cleaner soils and safer water sources, supporting sustainable crop production.

» **Research and Development in Soil Health and Pollution Remediation**: Investment in research to understand the impacts of industrial waste on soil health and develop innovative remediation techniques is critical. Research on pollution-tolerant crop varieties, soil amendments, and biological remediation methods can improve soil resilience and enhance the productivity of contaminated lands. Collaborations between universities, research institutions, and the agricultural industry can accelerate the development of solutions for industrial waste management and agricultural sustainability.

An overview of the impact of various types of stress on different fruit crops

S. No	Fruit Crop	Type of Stress	Impact on Crop Production
1	Apple	Drought Stress	Reduces fruit size and overall yield due to limited water availability.
			Leads to early fruit drop and reduced sugar content, impacting flavor and marketability.
			Affects fruit firmness and increases susceptibility to disorders like bitter pit due to impaired nutrient uptake.
		Heat Stress	Causes sunburn on fruit surface, reducing visual quality.
			Leads to poor fruit set and smaller fruit size in high temperatures.
			Alters sugar-acid balance, affecting taste.

		Salinity Stress	Impairs nutrient uptake, leading to potassium and calcium deficiencies, causing disorders like leaf burn and smaller fruit size.
			Reduces overall growth and yield in saline conditions.
		Pest Attack	Aphid and codling moth infestations lead to fruit damage and reduced market value.
			Fungal diseases like apple scab and powdery mildew reduce fruit quality and storage potential.
2	Grape	Drought Stress	Reduces berry size, cluster weight, and overall yield.
			Increases sugar concentration, affecting flavor and wine quality in high-stress conditions.
		Heat Stress	Sunburn on berries leads to color loss and cracking, affecting marketability and wine quality.
			Increases respiration rate, leading to faster ripening and shorter harvest window.
		Salinity Stress	Causes leaf burn, reduces photosynthesis, and affects vine vigor.
			Leads to lower yield and smaller, less flavorful berries.
		Pest Attack	Downy and powdery mildew reduce leaf function, limiting photosynthesis and berry quality.
			Infestation by grapevine moths and mealybugs damages berries, reducing marketable yield.
3	Citrus	Drought Stress	Leads to smaller fruit size, reduced yield, and increased fruit drop.
			Alters juice content and acidity, impacting flavor.
		Heat Stress	Causes sunburn, which leads to scalding and lower visual quality.
			Affects fruit set, reducing yield.
		Salinity Stress	Accumulation of sodium and chloride ions affects leaf and root function, leading to leaf burn and poor growth.
			Reduces yield and quality of fruits in saline soils.
		Pest Attack	Aphids, mites, and citrus leaf miners cause leaf damage, reducing photosynthesis.
			Citrus canker and greening disease decrease fruit quality and yield, with significant economic impact.

4	Peach	Drought Stress	Decreases fruit size and sugar content, impacting taste and marketability.
			Causes early fruit drop and lowers overall yield.
		Heat Stress	Leads to sunburn and skin damage, reducing fruit quality and shelf life.
			Reduces fruit set and may cause poor ripening in high temperatures.
		Salinity Stress	Impairs root function and nutrient uptake, causing leaf burn and chlorosis.
			Decreases yield and leads to poor fruit quality in saline soils.
		Pest Attack	Peach leaf curl and bacterial spot reduce leaf health, affecting photosynthesis and growth.
			Oriental fruit moth and plum curculio damage fruits, leading to yield loss.
5	Strawberry	Drought Stress	Reduces berry size, yield, and sugar content, affecting flavor and market value.
			Causes wilting and early plant senescence under prolonged drought.
		Heat Stress	Leads to flower and fruit abortion, reducing yield significantly.
			Causes berry softening, reducing shelf life and increasing susceptibility to rot.
		Salinity Stress	Leads to leaf burn, chlorosis, and reduced growth.
			Lowers fruit quality by affecting flavor and berry size.
		Pest Attack	Spider mites, aphids, and thrips cause leaf damage, reducing photosynthesis.
			Botrytis fruit rot and anthracnose reduce berry quality and storage potential.
6	Banana	Drought Stress	Reduces leaf area, delaying fruit development and lowering yield.
			Reduces bunch weight, affecting profitability.
		Heat Stress	Causes leaf burn and reduces photosynthetic activity.
			Impairs fruit quality, leading to smaller fruits and reduced shelf life.
		Salinity Stress	Impairs root function, causing stunted growth and leaf chlorosis.
			Reduces fruit yield and quality in saline soils.
		Pest Attack	Nematodes and banana weevils damage roots, limiting water and nutrient uptake.
			Fusarium wilt and Black Sigatoka disease reduce yield and plant lifespan, with severe economic impact.

7	Mango	Drought Stress	Reduces fruit size and yield, and increases the risk of fruit drop.
			Reduces sugar content, impacting flavor quality.
		Heat Stress	Causes sunburn on fruit, leading to scalding and skin damage.
			Reduces flowering and fruit set, lowering yield.
		Salinity Stress	Affects nutrient uptake, leading to leaf burn and reduced growth.
			Lowers fruit quality in saline environments, impacting market value.
		Pest Attack	Mango hoppers and mealybugs damage leaves and fruit, reducing yield.
			Anthracnose and powdery mildew reduce fruit quality and increase post-harvest losses.
8	Pineapple	Drought Stress	Limits leaf development, reducing photosynthesis and fruit size.
			Delays fruit maturation and reduces yield.
		Heat Stress	Causes leaf scorching and reduces growth rate.
			Leads to poor fruit development and reduced fruit weight.
		Salinity Stress	Reduces growth rate and causes nutrient imbalances.
			Affects fruit flavor and texture, impacting quality
		Pest Attack	Mealybugs and scales damage plant tissues, affecting growth and yield.
			Phytophthora root rot and heart rot reduce plant health and productivity.
9	Avocado	Drought Stress	Reduces fruit size, oil content, and yield.
			Increases fruit drop, impacting marketable yield.
		Heat Stress	Causes leaf scorch and premature fruit ripening, reducing shelf life.
			Alters oil content, affecting flavor quality.
		Salinity Stress	Causes leaf burn and reduces root function, impacting growth and yield.
		Pest Attack	Thrips and mites cause leaf and fruit damage, lowering quality.
			Phytophthora root rot reduces root function and can lead to plant death.

III.Stress Indices in Fruit Crop Production

Stress indices are quantitative tools used to assess and monitor the degree of stress affecting plants, especially in response to abiotic factors such as water scarcity, temperature extremes, and salinity. These indices help researchers, farmers, and agronomists understand how environmental stressors impact crop growth, physiology, and yield, enabling more informed decisions about crop management and intervention strategies. In fruit crop production, stress indices are particularly valuable for evaluating drought tolerance, assessing temperature stress, and gauging the effects of salinity, as these stresses directly influence yield, quality, and marketability.

1. Types of Stress Indices

Stress indices are categorized based on the type of stress being measured and often involve a combination of physiological, biochemical, and morphological assessments. Some of the key stress indices used in fruit crop production include:

Drought Stress Indices

Drought stress indices are essential tools in agriculture, especially for fruit crop production, as they measure a plant's response to water scarcity. By assessing various physiological, morphological, and biochemical parameters, these indices provide insights into a plant's capacity to endure limited water availability, guiding the selection of drought-tolerant varieties and enabling efficient water management strategies. Below are some commonly used drought stress indices in fruit crops:

Relative Water Content (RWC)

Relative Water Content (RWC) is a widely used index that measures the water status in plant tissues, indicating the plant's ability to retain moisture under drought conditions. RWC reflects cellular hydration levels, which are critical for sustaining physiological functions, including photosynthesis, nutrient transport, and growth.

- » **Calculation of RWC**: RWC is calculated as the percentage of water retained in plant tissues relative to their fully hydrated state. The formula for RWC is:
- » Here, "Fresh Weight" represents the initial weight of the tissue, "Dry Weight" is the weight after drying, and "Turgid Weight" is the weight when the tissue is fully hydrated. Higher RWC values indicate better drought tolerance, as plants that maintain higher water content can sustain cellular functions and growth even during water stress.
- » **Importance in Drought Tolerance**: High RWC suggests that a plant has efficient mechanisms for water retention and cellular hydration, which are essential for enduring drought stress. For fruit crops, maintaining adequate hydration under drought conditions ensures that photosynthetic and metabolic processes remain active, supporting continued growth and productivity. In crops like grapes, apples, and citrus, high RWC levels help preserve fruit quality and yield during dry spells.
- » **Application in Crop Selection**: RWC measurements aid in identifying drought-tolerant cultivars. In fruit crops, selecting varieties with higher RWC is beneficial for regions where water scarcity is common, as these plants can sustain productivity despite limited water availability. High RWC values also support fruit set and size, contributing to better marketability and economic returns.

Drought Stress Tolerance Index (DSTI)

The Drought Stress Tolerance Index (DSTI) is a comprehensive index that evaluates a crop's ability to sustain productivity under drought conditions by integrating yield and growth parameters under both normal and drought conditions. DSTI is valuable for assessing long-term drought resilience and selecting cultivars that can perform well under water-limited environments.

- **Calculation of DSTI**: DSTI is typically calculated by comparing crop yield or growth measurements under drought and non-drought conditions. A simplified DSTI formula could be:
- Where, Ys and Yn are the genotype yields (or biomass) under stress and non-stress respectively and Ms and Mn are the mean yields (or biomass) over all genotypes in the given test under stress and non-stress respectively.
- Higher DSTI values indicate greater drought resilience, as plants are able to maintain a larger proportion of their yield despite reduced water availability.
- **Importance in Crop Breeding and Selection**: DSTI provides valuable information for breeding programs focused on developing drought-resistant fruit crop cultivars. By comparing DSTI values, breeders can select varieties that show higher productivity and growth stability under drought stress, enabling them to develop cultivars suited for arid and semi-arid regions.
- **Practical Applications in Orchard Management**: For orchard management, DSTI helps identify drought-tolerant varieties that require fewer inputs and perform well under water-limited conditions. DSTI is particularly useful for long-term planning, as it aids growers in choosing cultivars that can adapt to changing water availability, reducing reliance on irrigation and increasing resilience to drought-related challenges.

Advantages of Using Drought Stress Indices in Fruit Crop Production

The application of drought stress indices offers several advantages in managing and improving fruit crop production under water-scarce conditions:

1. **Efficient Resource Management**: By identifying drought-tolerant cultivars through indices like DSTI and RWC, growers can make better use of limited water resources. This efficiency is essential in areas facing water scarcity or high irrigation costs, as drought-tolerant plants can maintain productivity with less water.
2. **Early Detection of Stress**: Remote-sensing indices like NDVI allow for early detection of drought stress, enabling timely interventions. Early response to water stress prevents crop damage and yield loss, supporting stable and high-quality fruit production.
3. **Targeted Crop Breeding**: Drought stress indices support targeted breeding programs by identifying genetic traits linked to water-use efficiency and drought tolerance. This is particularly valuable for developing new fruit crop varieties that can thrive in arid or drought-prone regions.
4. **Improved Orchard Resilience**: Selecting cultivars with lower CTD and higher RWC improves the resilience of orchards, reducing the need for frequent irrigation and protecting

crops from the impact of prolonged droughts. Resilient orchards contribute to sustainable fruit production and long-term profitability.

5. **Support for Precision Agriculture**: Drought stress indices integrate well with precision agriculture practices, allowing for data-driven management decisions. Indices like NDVI and CTD can be monitored at a large scale through drone or satellite imagery, providing real-time insights into crop health and guiding site-specific irrigation management.

2. Temperature Stress Indices

Temperature stress indices are tools used to evaluate a plant's response to extreme temperatures, both high and low, which can impair critical physiological processes, including growth, photosynthesis, and fruit development. These indices help assess a plant's tolerance to temperature extremes, aiding in the selection of resilient cultivars and the implementation of protective measures to maintain productivity. In fruit crop production, temperature stress indices are particularly valuable, as temperature fluctuations directly impact flowering, fruit set, ripening, and quality. The most commonly used temperature stress indices include:

Heat Tolerance Index (HTI)

The Heat Tolerance Index (HTI) measures a plant's ability to endure high temperatures without significant losses in growth or productivity. HTI is particularly useful for assessing how well different cultivars can maintain yields under heat stress, which is essential for fruit crops grown in hot climates.

» **Calculation of HTI**: HTI is calculated by comparing yield or growth performance under heat stress conditions with that under optimal (non-stress) conditions. The HTI formula may look like this:

» Higher HTI values indicate greater heat resilience, as these plants sustain more of their yield and growth potential despite high temperatures.

» **Significance in Fruit Crop Production**: High temperatures can lead to reduced fruit size, poor coloration, and increased susceptibility to disorders like sunburn in fruit crops. By using HTI, growers and researchers can identify cultivars that retain yield and quality even under heat stress. For instance, grape, apple, and peach varieties with high HTI values are better suited to endure heat without compromising fruit set, ripening, or flavor.

» **Application in Hot Climates**: HTI is valuable for managing orchards in regions prone to high temperatures. By selecting fruit crop varieties with high HTI, growers can improve orchard resilience to heatwaves and high summer temperatures, ensuring consistent yields and reducing the risk of heat-related crop losses. HTI-based selection is also useful in breeding programs focused on developing heat-tolerant cultivars that meet quality and productivity standards.

Chlorophyll Fluorescence (CF)

Chlorophyll fluorescence (CF) is an index that indicates photosynthetic efficiency, which is highly sensitive to temperature stress. CF is a measure of the light energy emitted by chlorophyll molecules during photosynthesis, reflecting the health and activity of photosystem II (PSII), the component responsible for the initial steps of photosynthesis. Under high or low temperatures, PSII activity can decline, affecting the plant's energy production and growth.

» **Measurement of CF**: CF is measured using fluorometers, which assess the fluorescence emitted by chlorophyll molecules. One common parameter derived from CF measurements is the maximum quantum efficiency of PSII, often represented as Fv/Fm. Lower CF or Fv/Fm values indicate that PSII is stressed, usually due to high or low temperatures, affecting photosynthetic capacity.

» **Role in Detecting Temperature Stress**: Photosynthesis is particularly sensitive to temperature fluctuations, and CF serves as an early indicator of stress before visible symptoms appear. Under high temperatures, CF values may decrease as heat disrupts the photosynthetic apparatus, limiting the plant's ability to convert light energy into chemical energy. Low temperatures can also affect CF by slowing down photosynthesis and causing photoinhibition. In fruit crops, declining CF values may suggest a need for shading, misting, or other cooling strategies to protect plants from further heat damage.

» **Guiding Protective Measures**: CF measurements enable growers to monitor the impact of temperature stress on photosynthesis in real-time, guiding them in applying protective measures. In hot climates, shading nets or misting systems can help reduce canopy temperature, preserving CF values and maintaining photosynthetic efficiency. For fruit crops like citrus and mango, monitoring CF can support the maintenance of fruit quality and size even under extreme temperature conditions.

Cold Tolerance Index (CTI)

The Cold Tolerance Index (CTI) evaluates a plant's ability to survive and recover from low temperatures, including chilling and freezing conditions. CTI is crucial for assessing fruit crop varieties in regions with frost risk, as cold stress can cause cellular damage, impair growth, and reduce yield. The CTI measures parameters like electrolyte leakage, which occurs when cell membranes are damaged by freezing temperatures.

» **Measurement of CTI**: CTI is assessed by comparing the crop's physiological or growth response under freezing or chilling conditions with its response under optimal temperatures. Electrolyte leakage is commonly measured to estimate cell membrane stability under cold stress, as damaged cells release more electrolytes into surrounding tissues. Higher CTI values indicate better cold tolerance, as plants with minimal electrolyte leakage have more resilient cell membranes and are less affected by frost.

» **Importance in Frost-Prone Areas**: Cold temperatures cause water within cells to freeze, leading to the formation of ice crystals that can puncture cell membranes and cause dehydration. High CTI values are particularly important for fruit crops grown in temperate regions, where late spring frosts or early fall chills are common. Fruit crops like apples, cherries, and blueberries are sensitive to frost, which can damage flower buds, reduce fruit set, and impact overall yield.

» **Application for Cold-Climate Fruit Production**: Monitoring CTI helps identify cultivars that are better suited to withstand frost events. High CTI values are beneficial for choosing varieties that can maintain productivity and quality in colder climates. For example, apple and cherry cultivars with high CTI are less likely to suffer from frost damage during critical growth stages, reducing crop loss and ensuring a more stable harvest.

Advantages of Temperature Stress Indices in Fruit Crop Management

Using temperature stress indices like HTI, CF, and CTI offers several advantages in managing fruit crop production:

1. **Enhanced Selection of Resilient Cultivars**: By evaluating HTI, CF, and CTI, growers can select fruit crop cultivars that are more resilient to temperature extremes, reducing the risk of yield loss and ensuring quality even in challenging climates. This selection process is especially valuable for breeding programs focused on developing heat- and cold-tolerant cultivars.
2. **Early Detection and Intervention**: Temperature stress indices, particularly CF, allow for early detection of temperature-related stress before visible symptoms appear. This proactive approach enables growers to implement timely protective measures, such as shading or misting in high temperatures, reducing the risk of long-term damage.
3. **Targeted Crop Management Strategies**: HTI, CF, and CTI provide insights into the specific temperature tolerance needs of each crop, supporting targeted management practices. For instance, cultivars with low HTI may benefit from cooling systems or shaded environments, while those with high CTI can be planted in frost-prone areas.
4. **Support for Climate Adaptation**: As climate change increases the frequency of temperature extremes, temperature stress indices offer a scientific basis for adapting fruit crop production to these changes. By selecting and managing temperature-resilient cultivars, growers can reduce their vulnerability to climate-related risks and improve long-term productivity.
5. **Informed Irrigation and Water Management**: Temperature stress often affects water usage and transpiration rates. Understanding a crop's HTI or CTI can help optimize irrigation, as heat-tolerant plants may require more water, while cold-tolerant plants might need less during certain stages. In this way, temperature stress indices aid in water conservation and resource efficiency.

3. Salinity Stress Indices

Salinity stress indices evaluate a plant's tolerance to high salt concentrations in the soil, which disrupt water absorption, nutrient uptake, and overall growth. Salinity is a common problem in arid and semi-arid regions, where irrigation water often contains dissolved salts, leading to salt buildup in the soil over time. High salt concentrations impose osmotic stress on plants, affecting cellular hydration and interfering with nutrient balance, especially in fruit crops, which are sensitive to soil quality. The following salinity stress indices are commonly used to assess salinity tolerance in fruit crops:

Sodium (Na^+) and Potassium (K^+) Ratio

The sodium (Na^+) and potassium (K^+) ratio in plant tissues is a critical indicator of salinity tolerance, as excessive sodium in the soil can interfere with potassium uptake. Potassium is essential for many cellular processes, including enzyme activation, stomatal regulation, and photosynthesis. Under salinity stress, plants with a low Na^+/K^+ ratio are generally more tolerant, as they effectively exclude sodium and retain higher potassium levels.

- **Importance of Na^+/K^+ Balance**: Sodium and potassium are both absorbed through root systems, but high sodium levels in saline soils create competition, displacing potassium. This imbalance disrupts cellular function and metabolic activities, leading to physiological stress and growth inhibition. A lower Na^+/K^+ ratio in plant tissues indicates that the plant can maintain potassium levels despite the presence of sodium, ensuring proper physiological function.
- **Measurement and Interpretation**: The Na^+/K^+ ratio is determined by analyzing the concentration of sodium and potassium in plant tissues, particularly in leaves, where nutrient imbalances first manifest. Lower ratios are desirable, as they indicate better salt exclusion and potassium retention. For fruit crops like citrus and grapes, maintaining a low Na^+/K^+ ratio is essential for salt tolerance, as high potassium supports key physiological functions that directly impact yield and fruit quality.
- **Application in Cultivar Selection**: Monitoring the Na^+/K^+ ratio helps identify salt-tolerant cultivars, as those with lower ratios are more effective in managing ion balance under saline conditions. Fruit crops with a low Na^+/K^+ ratio are better able to maintain growth, flowering, and fruit set in saline environments, supporting stable yields even under salinity stress.

Osmotic Adjustment Capacity (OAC)

Osmotic Adjustment Capacity (OAC) refers to a plant's ability to accumulate compatible solutes, such as proline, glycine betaine, and soluble sugars, in response to salt stress. These solutes help maintain cellular turgor, allowing plants to cope with the osmotic pressure caused by high salt concentrations in the soil. High OAC values indicate that a plant has a stronger capacity to adapt to saline conditions, sustaining cellular hydration and growth.

- **Role of Compatible Solutes**: When exposed to high salinity, plants produce compatible solutes that do not interfere with cellular functions but help maintain osmotic balance. These compounds allow cells to retain water and sustain turgor, which is essential for nutrient transport, photosynthesis, and growth. In fruit crops, osmotic adjustment is critical for preventing dehydration, as water loss can lead to reduced fruit size, yield, and quality.
- **Measurement of OAC**: OAC is measured by assessing the concentration of osmolytes, such as proline and sugars, in plant tissues. Higher concentrations of these solutes indicate a greater capacity for osmotic adjustment. For example, fruit crops like pomegranate and olive exhibit high OAC under salinity stress, allowing them to adapt better to arid and semi-arid regions where saline conditions are common.
- **Impact on Crop Yield and Quality**: High OAC in fruit crops prevents osmotic stress and dehydration, supporting continued growth and development. In crops like tomatoes and grapes, which are highly sensitive to water availability, osmotic adjustment helps maintain fruit size and flavor even under saline conditions. Cultivars with high OAC values are thus ideal for saline-prone regions, as they are less susceptible to yield and quality reductions due to salt stress.

Electrolyte Leakage (EL)

Electrolyte Leakage (EL) is an index that measures the extent of cell membrane damage due to salt-induced osmotic stress. High salinity levels can lead to cellular dehydration and ion toxicity, causing damage to cell membranes and resulting in the release (leakage) of ions and other cell contents. Higher EL values indicate greater membrane damage and a lower tolerance to salt stress.

» **Mechanism of Membrane Damage**: Under salinity stress, excessive sodium and chloride ions can accumulate within cells, leading to osmotic imbalances that cause cell membranes to become permeable and lose structural integrity. This damage results in electrolyte leakage, as essential ions such as potassium and calcium are lost, impacting metabolic functions and reducing cell viability.

» **Measurement of EL**: EL is typically measured by placing plant tissues in distilled water and analyzing the ion concentration in the surrounding solution. The degree of leakage reflects the extent of membrane damage. High EL values indicate more severe salt stress, as the cell membranes are less stable and more prone to injury. For fruit crops, maintaining low EL levels is crucial to support growth and protect tissue integrity.

» **Significance in Salinity Tolerance**: Low EL values indicate stronger cell membrane stability, as salt-tolerant plants can better regulate ion transport and exclude toxic sodium ions. Fruit crops with lower EL values are more resilient in saline soils, maintaining productivity and quality despite high salt concentrations. This makes EL a valuable index for identifying salt-tolerant fruit crop cultivars, as those with stable membranes are less affected by osmotic stress and maintain higher yields.

Benefits of Salinity Stress Indices in Fruit Crop Management

Using salinity stress indices like Na^+/K^+ ratio, OAC, and EL in fruit crop management offers several advantages for understanding plant responses to salinity, selecting tolerant cultivars, and implementing effective management practices:

1. **Selection of Salt-Tolerant Cultivars**: Salinity stress indices help identify cultivars that perform well under high salinity, enabling growers to select varieties suited to saline soils. For instance, fruit crops with lower Na^+/K^+ ratios and higher OAC values are better able to adapt to saline conditions, supporting sustainable production even in challenging environments.
2. **Early Detection of Salinity Stress**: Measuring indices like EL provides early insights into the impact of salinity on cellular health. By identifying early signs of membrane damage, growers can take preventive measures, such as adjusting irrigation practices or applying soil amendments, to minimize further stress and protect crop yield and quality.
3. **Guidance for Soil and Water Management**: Salinity indices guide management practices by indicating the plant's specific salinity tolerance requirements. For instance, crops with low Na^+/K^+ ratios may benefit from irrigation practices that minimize sodium uptake, while those with high OAC can better withstand water deficits, enabling more efficient water use.
4. **Support for Saline Soil Remediation**: Understanding salinity stress indices aids in the development of soil remediation practices. For example, adding gypsum to soil reduces sodium

levels, supporting crops with a high Na^+/K^+ sensitivity. Additionally, identifying salt-tolerant cultivars with high OAC can reduce the impact of salinity on crop productivity, making it feasible to cultivate fruit crops in saline-affected areas.

5. **Improved Yield and Quality under Salinity Stress**: By selecting fruit crop varieties with favorable salinity stress indices, growers can enhance resilience to salt stress, ensuring consistent yields and maintaining fruit quality. For high-value crops like citrus, grapes, and berries, where quality and flavor are crucial for marketability, salinity stress indices support the maintenance of premium fruit standards.

4. Canopy Temperature Depression (CTD)

Canopy Temperature Depression (CTD) is an index used to measure a plant's cooling ability through transpiration by assessing the difference between the temperature of the surrounding air and the temperature of the leaf surface. This metric provides insights into a plant's water-use efficiency and tolerance to drought. Under drought conditions, plants often close their stomata to conserve water, resulting in reduced transpiration and, consequently, higher leaf surface temperatures. CTD is particularly valuable for managing fruit crops, as efficient cooling helps maintain photosynthesis and productivity, even in water-limited conditions.

Measurement of CTD

CTD is determined by calculating the difference between air temperature and leaf surface temperature, which provides an estimate of the plant's cooling capacity.

- » **Infrared Thermometers and Thermal Imaging**: The leaf temperature can be measured using infrared thermometers or more advanced thermal imaging technology. Infrared thermometers offer a quick, point-based measurement of leaf temperature, while thermal imaging provides a comprehensive view, capturing temperature variations across the entire plant canopy.
- » **Calculation**: CTD is calculated using the formula:
- » Higher canopy temperatures indicate reduced cooling and transpiration, while lower canopy temperatures indicate efficient transpiration and water use. Lower CTD values reflect better cooling and transpiration, indicative of a plant's drought resilience and water-use efficiency.
- » **Practicality for Field Monitoring**: CTD measurement is quick and non-invasive, allowing growers to monitor crop health in real-time. In orchards, portable infrared thermometers and drones equipped with thermal imaging cameras are used to assess CTD, offering valuable information on plant stress and guiding irrigation strategies.

Significance in Water-Use Efficiency

CTD provides critical insights into a plant's ability to use water efficiently, which is essential in drought-prone environments. Plants with lower CTD values are likely to have effective water-use mechanisms that allow them to maintain transpiration and cooling despite limited water availability.

- » **Transpiration and Photosynthesis**: Transpiration is the process through which water evaporates from the leaf surface, cooling the plant. This process not only regulates leaf temperature but also aids in the uptake of carbon dioxide (CO_2), which is necessary for photosynthesis. In fruit crops, efficient transpiration directly supports photosynthesis, ensuring carbohydrate production and energy availability for growth and fruit development. For example, apples and peaches with lower CTD values under drought conditions maintain higher photosynthetic rates and resist yield losses.

» **Stomatal Behavior and Cooling Efficiency**: Plants with efficient water-use strategies are better able to regulate their stomata, the tiny openings on leaves that control gas exchange and water vapor release. By modulating stomatal closure under drought conditions, these plants can optimize water retention without completely inhibiting transpiration. As a result, such plants exhibit lower CTD values, reflecting their capacity to balance water conservation with cooling. This is particularly beneficial in fruit crops, where even slight reductions in cooling can affect fruit set, size, and quality.

» **Implications for Fruit Quality**: Maintaining lower CTD values contributes to fruit quality, as effective cooling and photosynthesis support nutrient transport and accumulation. In crops like peaches, nectarines, and citrus, efficient water use under stress is associated with better fruit color, sugar content, and firmness. Thus, monitoring CTD is integral to managing both yield and quality in drought-prone fruit orchards.

Application in Drought-Prone Orchards

Monitoring CTD is a practical approach for identifying fruit crop varieties that are more efficient in water use, which is essential for orchards located in drought-prone regions. Selecting cultivars with lower CTD values can help improve orchard resilience, reduce the need for frequent irrigation, and ultimately conserve water resources.

» **Variety Selection for Drought Resilience**: CTD measurements are highly useful in breeding programs and variety selection, as they help identify drought-tolerant cultivars that can perform well in regions with limited water availability. By choosing fruit cultivars with low CTD values, growers can ensure that their orchards remain productive and sustainable even under prolonged dry conditions. For instance, grape, apple, and peach cultivars with lower CTD values are better suited to arid climates, as they maintain cooling and photosynthesis without excessive water loss.

» **Irrigation Management**: In drought-prone orchards, CTD is an effective tool for optimizing irrigation schedules. By monitoring CTD values, growers can determine when plants begin to experience water stress and need supplemental irrigation. Precision irrigation based on CTD can help reduce water wastage and prevent over-irrigation, leading to cost savings and more efficient water use. Additionally, adjusting irrigation based on CTD data minimizes water-related diseases, as excessive irrigation is avoided.

» **Integrating CTD with Precision Agriculture**: With advancements in technology, CTD monitoring can be integrated into precision agriculture systems. Drones and thermal sensors provide real-time, field-wide CTD data, helping growers detect spatial variations in plant stress across the orchard. This information enables site-specific irrigation, targeting only the areas in need, which further conserves water and maximizes productivity.

Benefits of Using CTD in Fruit Crop Production

1. **Enhanced Selection of Drought-Resilient Varieties**: By measuring CTD, growers and researchers can identify fruit crop varieties that maintain better cooling and transpiration under drought stress. Selecting cultivars with lower CTD values contributes to more resilient orchards, supporting stable yields and high-quality produce in water-limited regions.
2. **Optimized Water Management**: CTD data provides insights into water needs, helping growers optimize irrigation timing and frequency. Instead of uniform irrigation, CTD-based schedules focus on actual plant stress, allowing for more targeted water use, reducing waste, and conserving resources.
3. **Improved Fruit Quality and Yield**: Lower CTD values are associated with efficient photosynthesis and cooling, which are critical for maintaining fruit quality under drought stress. For high-value crops like grapes, apples, and peaches, CTD monitoring ensures that fruit size, flavor, and appearance are not compromised due to water scarcity.
4. **Adaptation to Climate Change**: As climate change increases the frequency of droughts and heatwaves, CTD provides a scientific basis for adapting fruit crop production to these challenges. By focusing on cultivars with lower CTD values and implementing efficient irrigation practices, growers can increase orchard resilience to climate variability, ensuring consistent production and profitability.
5. **Support for Sustainable Practices**: CTD monitoring aligns with sustainable agriculture practices by promoting water conservation and reducing dependency on water resources. Precision irrigation based on CTD data minimizes water inputs, lowering operational costs and supporting environmental sustainability in agricultural systems.

5. Normalized Difference Vegetation Index (NDVI)

The Normalized Difference Vegetation Index (NDVI) is a widely used remote-sensing index that provides insights into plant health, vigor, and stress levels based on light reflection data. NDVI is calculated by analyzing the difference between near-infrared (NIR) and red-light reflectance from plant leaves, which is closely associated with chlorophyll content and photosynthetic activity. In fruit crop production, NDVI is an effective tool for assessing crop health, identifying drought stress, and optimizing irrigation management.

Calculation of NDVI

NDVI is calculated using the following formula:

$$\text{NDVI} = \frac{\text{NIR-Red}}{\text{NIR+Red}}$$

where:

- **NIR** represents the near-infrared reflectance, which is strongly reflected by healthy vegetation.
- **Red** represents the red-light reflectance, which is absorbed by chlorophyll during photosynthesis.

NDVI values range from -1 to +1:

- **High NDVI values** (0.6 to 0.9) indicate healthy, vigorously growing vegetation with high chlorophyll content, typically reflecting good plant health and adequate water availability.
- **Low NDVI values** (below 0.3) suggest stressed or sparse vegetation, which may be due to drought, disease, nutrient deficiencies, or senescence.
- **Negative NDVI values** indicate non-vegetated surfaces, such as soil or water.

The NDVI calculation allows for consistent, large-scale assessment of vegetation health, making it a valuable tool in agriculture, forestry, and environmental monitoring.

Significance of NDVI in Monitoring Plant Health and Drought Stress

NDVI is highly sensitive to changes in chlorophyll content and photosynthetic activity, both of which are impacted by drought stress, nutrient deficiencies, and other environmental factors. In fruit crops, NDVI monitoring helps growers detect early signs of stress, ensuring timely interventions to prevent yield losses.

- **Chlorophyll Content and Photosynthetic Activity**: NDVI provides an indirect measure of chlorophyll content, as healthy leaves with high chlorophyll absorb more red light and reflect more NIR. When plants experience drought stress, they often close their stomata, limiting CO_2 uptake and reducing photosynthesis. This reduction leads to chlorophyll degradation, which in turn lowers NDVI values, signaling a decline in plant health.
- **Drought Detection and Stress Monitoring**: NDVI is particularly useful for assessing drought stress. In water-limited conditions, plants may exhibit reduced leaf area, lower chlorophyll content, and less dense canopies, all of which reduce NIR reflectance and increase red-light reflectance. Monitoring NDVI values over time enables growers to identify areas in need of irrigation, adjust water schedules, and minimize drought impact.
- **Early Stress Detection**: NDVI allows for early detection of stress before visible symptoms appear, enabling preventive measures. For example, a decrease in NDVI may signal water stress or nutrient deficiencies, prompting timely interventions such as targeted irrigation or fertilization to preserve crop health and prevent yield losses.

Applications of NDVI in Fruit Crop Production

NDVI is an integral tool in fruit crop production, supporting decision-making in irrigation, nutrient management, and crop monitoring. By providing real-time data on crop health, NDVI enables precision agriculture practices, optimizing resource use and improving yield and quality.

Irrigation Management

Efficient water management is essential for maintaining crop health, particularly in drought-prone regions. NDVI helps monitor the spatial variability of water stress in orchards, guiding targeted irrigation practices.

- **Precision Irrigation**: NDVI data from satellites, drones, or ground-based sensors identify areas of an orchard experiencing water stress, enabling site-specific irrigation. This approach conserves water by directing it to areas in need, reducing water usage and minimizing costs. In drought-prone regions, NDVI-guided irrigation improves water-use efficiency, supporting crop resilience and sustainability.

» **Adjusting Irrigation Schedules**: By tracking NDVI values over time, growers can adjust irrigation schedules based on crop needs. For example, a consistent decline in NDVI may indicate increasing water stress, prompting adjustments to irrigation frequency or intensity to restore plant health. In fruit crops like citrus, apple, and grape, maintaining stable NDVI values throughout the growing season helps ensure consistent fruit development and quality.

Nutrient Management

NDVI also provides insights into nutrient status, as nutrient deficiencies often manifest as reduced chlorophyll content, which lowers NDVI values. This information helps growers optimize nutrient applications for healthy crop growth.

» **Detecting Nutrient Deficiencies**: Nutrient deficiencies, particularly nitrogen, reduce chlorophyll content and photosynthetic activity, which can lead to lower NDVI values. By identifying areas with declining NDVI, growers can apply fertilizers precisely where they are needed, improving nutrient efficiency and reducing environmental impact.

» **Optimizing Fertilizer Application**: NDVI-guided nutrient management ensures that fertilizer is only applied to areas in need, minimizing waste and preventing over-application. This approach is beneficial for fruit crops, where balanced nutrition is essential for fruit size, flavor, and marketability. For example, NDVI monitoring in vineyards can guide nitrogen application, helping maintain optimal leaf health and grape quality.

Yield Prediction and Quality Assessment

NDVI data can provide insights into crop yield potential and quality by reflecting overall crop vigor and health. This is particularly useful for high-value fruit crops, where yield and quality directly impact profitability.

» **Estimating Yield Potential**: NDVI values are often correlated with crop yield, as healthy, vigorously growing plants produce more biomass and, consequently, higher yields. In fruit crops, high NDVI values during critical growth stages, such as flowering and fruit set, indicate favorable conditions for yield potential. Growers can use this information to forecast harvest volumes and make informed marketing decisions.

» **Assessing Fruit Quality**: NDVI can also help estimate fruit quality by monitoring canopy health, which influences fruit size, sugar content, and color. For example, high NDVI values during the maturation phase in grapevines are associated with better grape quality, which is critical for winemaking. In apples and other fruit crops, NDVI-guided quality assessment helps growers manage crop health for premium market standards.

Advantages of Using NDVI in Fruit Crop Management

1. **Large-Scale Monitoring**: NDVI allows for large-scale monitoring of crop health across entire orchards, making it a cost-effective tool for growers managing extensive fruit crop

areas. Drones and satellites provide high-resolution NDVI data, enabling efficient monitoring of crop conditions.

2. **Early Detection of Stress**: NDVI offers early detection of drought stress, nutrient deficiencies, and disease, allowing for timely interventions. Early detection is critical in fruit crops, where stress can affect yield and quality, impacting profitability.
3. **Support for Precision Agriculture**: NDVI data enables precision agriculture by guiding irrigation, nutrient application, and pest management practices. This approach optimizes resource use, reduces costs, and enhances environmental sustainability, supporting long-term productivity.
4. **Enhanced Yield and Quality Management**: By maintaining optimal NDVI values, growers can ensure that crops receive adequate water, nutrients, and care, maximizing yield and maintaining fruit quality. This is especially beneficial for high-value crops, where quality and yield directly influence market value.
5. **Climate Adaptation and Resilience**: NDVI monitoring supports climate adaptation efforts by identifying stress patterns linked to changing weather conditions. By analyzing NDVI trends over time, growers can adjust practices to increase crop resilience to drought and other climate-related stresses.

2. Applications of Stress Indices in Fruit Crop Management

Stress indices play a vital role in the management of fruit crops, providing growers and researchers with valuable data to assess stress levels, identify resilient cultivars, and optimize cultural practices. These indices allow for precise interventions to mitigate the effects of drought, temperature extremes, salinity, and other environmental stressors, ultimately minimizing yield losses and maintaining fruit quality.

Selection of Stress-Tolerant Cultivars

One of the primary applications of stress indices is in the selection and breeding of fruit crop varieties that exhibit resilience to specific environmental stresses. By measuring stress tolerance indices, researchers can identify cultivars that perform well under drought, heat, salinity, or other adverse conditions, enabling the development of varieties suited for specific climates or challenging environments. The following sections highlight applications in selecting cultivars tolerant to various stressors.

Drought-Tolerant Varieties

Drought stress indices, such as the Drought Stress Tolerance Index (DSTI) and Relative Water Content (RWC), are commonly used to identify cultivars that can sustain growth and productivity even under limited water availability. These indices help assess a plant's water-use efficiency, cellular hydration, and drought adaptability, making them essential tools in breeding programs for regions prone to drought.

» **DSTI and RWC as Selection Tools**: DSTI measures the relative yield of a cultivar under drought and non-drought conditions, providing a direct comparison of performance under water stress. RWC, on the other hand, evaluates a plant's ability to retain water in its tissues, offering insights into cellular hydration and turgor maintenance under drought. Cultivars with high DSTI and RWC values are better equipped to survive in drought-prone regions, as they retain productivity and growth.

- **Applications in Apple and Grape Production**: In crops like apples and grapes, where drought stress can reduce fruit size, sweetness, and overall quality, drought-tolerant varieties are highly valuable. Selecting varieties with high DSTI and RWC ensures that these crops maintain consistent yields and high-quality fruit despite water scarcity. Additionally, drought-resistant apple and grape varieties help reduce irrigation needs, lowering water costs and conserving resources.
- **Impact on Sustainable Agriculture**: Selecting drought-tolerant varieties also contributes to sustainable agriculture by reducing dependence on irrigation. As water scarcity becomes a growing concern in agriculture, drought-resistant cultivars are essential for ensuring food security and supporting environmentally friendly practices.

Temperature-Resilient Varieties

Temperature stress indices, such as the Heat Tolerance Index (HTI), are instrumental in identifying cultivars that can tolerate high temperatures without experiencing yield or quality losses. Temperature resilience is especially important for crops grown in regions with high summer temperatures or heatwave events.

- **HTI for Heat Tolerance**: HTI measures yield performance under heat stress relative to normal conditions, identifying cultivars that can maintain productivity even in hot climates. Plants with high HTI values have a greater capacity for heat tolerance, enabling them to survive and perform under high temperatures without significant physiological disruptions.
- **Applications in Peaches and Nectarines**: In crops like peaches and nectarines, high temperatures can disrupt fruit development, reduce firmness, and affect flavor. Selecting cultivars with high HTI values allows growers to produce quality fruits despite high summer temperatures. In regions prone to heatwaves, these heat-tolerant varieties are crucial for preventing crop loss and maintaining market standards.
- **Long-Term Benefits of Temperature-Resilient Cultivars**: Heat-tolerant varieties are particularly valuable as climate change increases the frequency of extreme temperatures. Selecting cultivars with high HTI can help growers adapt to shifting temperature patterns and maintain stable yields, reducing the impact of climate-induced stress on fruit crop production.

Salt-Tolerant Varieties

Salinity stress indices, including the Sodium (Na^+) and Potassium (K^+) Ratio and Osmotic Adjustment Capacity (OAC), are essential for selecting cultivars adapted to saline environments. In areas where irrigation water contains high salt concentrations, salinity-tolerant varieties can help sustain crop health and productivity.

- **Na^+/K^+ Ratio and OAC as Indicators**: The Na^+/K^+ ratio assesses a plant's ability to exclude sodium and retain potassium, which is critical for cellular function under salt stress. Low Na^+/K^+ ratios indicate better salt tolerance. OAC measures the plant's ability to accumulate solutes that maintain cellular turgor under saline conditions. High OAC values reflect a plant's adaptability to saline environments, as it prevents dehydration and maintains growth.

- **Applications in Salinity-Prone Areas**: In crops like citrus and tomatoes, salinity stress can impair growth, reduce fruit size, and alter taste. Selecting varieties with low Na^+/K^+ ratios and high OAC ensures that these crops can thrive in regions with saline soils or saline irrigation water. This approach is particularly useful for regions where salinity poses a continuous challenge to crop production.
- **Supporting Crop Sustainability in Saline Soils**: By selecting salt-tolerant cultivars, growers can maintain productivity in regions with naturally saline soils or limited access to low-salinity water sources. Salt-tolerant varieties also reduce the need for soil amendments and frequent leaching, making agriculture more sustainable in salinity-affected regions.

Benefits of Using Stress Indices in Fruit Crop Management

Using stress indices for selecting and managing stress-tolerant varieties provides numerous benefits, helping growers ensure high yields, maintain fruit quality, and reduce resource dependency. Below are some key advantages of applying stress indices in fruit crop management:

1. **Enhanced Crop Resilience**: Stress indices enable growers to select resilient cultivars that can withstand specific environmental stresses, ensuring consistent yields even under challenging conditions. This resilience is particularly valuable for high-value fruit crops, where yield consistency and quality are crucial for profitability.
2. **Reduced Resource Use**: Drought- and salinity-tolerant varieties often require less irrigation and fewer soil amendments, reducing input costs and conserving natural resources. By selecting varieties that perform well under low-resource conditions, growers support more sustainable agricultural practices.
3. **Improved Yield and Quality**: By choosing cultivars that are resilient to drought, heat, or salinity, growers can maintain higher fruit quality and yields, even under stress conditions. Stress-tolerant varieties tend to have better fruit set, size, and flavor, contributing to marketability and consumer satisfaction.
4. **Climate Adaptation**: Stress indices provide valuable data for adapting to climate variability. Selecting stress-tolerant cultivars helps growers mitigate the impact of climate-related stress, including drought, heatwaves, and saline water use, ensuring long-term viability of fruit production in changing climates.
5. **Support for Breeding Programs**: Stress indices aid breeding programs focused on developing stress-resilient cultivars. By using indices to assess stress tolerance in potential varieties, breeders can accelerate the development of cultivars adapted to specific environments, contributing to more resilient agricultural systems.

Monitoring Stress Levels for Timely Interventions

Stress indices are also useful for monitoring current stress levels in orchards and guiding timely interventions, such as irrigation adjustments, shade installation, or nutrient application. For example:

- **Irrigation Scheduling**: Drought stress indices, such as NDVI and CTD, provide early warning signs of water stress. By monitoring these indices, growers can optimize irrigation schedules, ensuring that crops receive water when they need it most. In water-scarce regions, this approach reduces water wastage while maintaining crop health.
- **Temperature Management**: In regions prone to heatwaves, CF measurements can help assess how well crops are coping with temperature stress. If CF levels drop significantly, growers can implement cooling measures, such as misting systems or shade nets, to reduce leaf temperature and prevent heat-induced damage.
- **Salinity Management**: Salinity stress indices help identify fields where salt buildup is causing stress to crops. Monitoring electrolyte leakage and Na^+/K^+ ratios can guide the use of salt leaching techniques, gypsum applications, or salt-tolerant rootstocks to minimize damage.

Precision Agriculture and Remote Sensing

Many stress indices, especially those involving vegetation indices like NDVI, are used in precision agriculture to monitor crop health across large fields. Remote sensing technology enables real-time assessment of stress levels, providing actionable data that can improve decision-making in orchard management.

- **Remote Sensing for Large-Scale Monitoring**: Satellite and drone-based remote sensing technologies can detect variations in NDVI across large orchards, identifying areas under stress due to drought or salinity. This information helps growers target interventions where they are needed most, optimizing resource use and reducing crop losses.
- **Precision Input Application**: By using stress indices, farmers can apply inputs, such as water and nutrients, in a more precise manner. For example, identifying areas of an orchard with high CTD values (indicating heat stress) enables targeted cooling or watering, reducing input costs and maximizing effectiveness.

3. Future Developments in Stress Indices

The future of stress indices in fruit crop production is being shaped by rapid advancements in technology, such as machine learning, genetic research, and IoT-enabled systems. These innovations promise to make stress indices more accurate, real-time, and integrated with on-farm decision-making, allowing for dynamic responses to changing environmental conditions. By combining genetic markers, IoT technology, and predictive models, stress indices are evolving into powerful tools that support proactive, precision agriculture.

Molecular Markers for Stress Tolerance

Molecular markers, or genetic markers, are specific DNA sequences associated with desirable traits such as stress tolerance. Through molecular biology and genomic research, scientists can identify markers linked to genes responsible for drought, heat, and salinity tolerance, allowing breeders to more effectively select for stress-resistant traits.

- **Marker-Assisted Selection (MAS)**: Marker-assisted selection involves using specific molecular markers to identify plants with desired genetic traits early in the breeding process.

For instance, markers linked to genes that enhance drought tolerance enable breeders to select for plants that retain water, maintain higher photosynthetic rates, or adjust osmotic balance effectively under drought conditions. This approach accelerates the development of drought-resistant varieties.

» **Improving Heat and Salinity Tolerance**: For fruit crops exposed to high temperatures or salinity, markers associated with heat shock proteins or sodium-exclusion mechanisms help breeders develop cultivars that better withstand these stresses. For example, in crops like tomatoes and grapes, molecular markers for heat tolerance allow for the selection of varieties that continue to flower and set fruit even in extreme heat, while markers for salinity tolerance enable better potassium and sodium regulation in saline soils.

» **Benefits for Sustainable Agriculture**: Molecular markers streamline the breeding process, reducing the time and resources required to develop new cultivars. By increasing the availability of stress-resistant fruit varieties, these advancements support sustainable agriculture, as growers can rely less on chemical inputs and more on the inherent resilience of their crops.

Integration with Internet of Things (IoT) Technology

IoT technology has the potential to transform stress index monitoring by providing real-time, continuous data on crop conditions. IoT-enabled sensors, such as soil moisture sensors, temperature and humidity monitors, and chlorophyll meters, can track stress indicators in real-time, enabling growers to respond dynamically to changes in crop health.

» **Real-Time Monitoring of Stress Indices**: IoT sensors can continuously track metrics like Relative Water Content (RWC), Canopy Temperature Depression (CTD), and soil salinity levels, providing immediate insights into crop conditions. This real-time monitoring allows for more precise adjustments in irrigation, nutrient applications, and temperature control based on current stress levels, reducing water and input waste.

» **Automated Irrigation and Nutrient Management**: IoT-based systems can be linked with automated irrigation and fertilization systems, allowing growers to make data-driven adjustments. For example, if sensors detect a rise in canopy temperature (indicating drought stress), an IoT-based system could automatically increase irrigation to maintain optimal crop hydration. In salinity-affected areas, soil sensors could detect changes in sodium concentration, triggering targeted applications of calcium to reduce sodium uptake.

» **Data Collection and Integration**: IoT devices collect large amounts of data over time, creating comprehensive records of crop health and stress levels. By integrating these data with farm management systems, growers can track crop responses to stress, optimize management practices, and improve long-term productivity. For example, in orchards, historical data on temperature stress can guide decisions on protective measures, such as installing shade nets or misting systems.

» **Sustainability and Cost Savings**: IoT technology improves resource efficiency by delivering inputs like water and nutrients only when they are needed. By reducing over-application of resources, IoT-enabled stress indices help lower input costs and support sustainable agricultural practices, benefiting both the environment and farm profitability.

Machine Learning for Predictive Modeling

Machine learning (ML) has the potential to enhance stress indices by analyzing complex data sets from multiple sources to predict crop responses to stress. Machine learning algorithms can process environmental variables, crop data, and historical stress index readings to create predictive models that help growers make proactive management decisions.

- » **Data Analysis and Pattern Recognition**: ML algorithms can analyze data from multiple stress indices, such as NDVI, CTD, and EL, alongside environmental variables like temperature, humidity, and soil moisture. By recognizing patterns and relationships in this data, ML models can identify early warning signs of stress and predict how crops will respond under various conditions.
- » **Predicting Crop Responses to Stress**: ML models can forecast crop performance under predicted environmental conditions, enabling growers to implement preventative measures before stress becomes severe. For example, an ML model might predict that an upcoming heatwave will increase canopy temperature to levels that could impair fruit set in apples. In response, growers could take preventive actions such as adjusting irrigation schedules or setting up temporary shading.
- » **Optimizing Resource Use**: Predictive modeling helps optimize resource allocation, as ML models can determine the precise amount of water, nutrients, or cooling needed based on crop-specific stress tolerance levels. For instance, an ML model could recommend the exact amount of irrigation required to mitigate drought stress without causing waterlogging, maximizing water efficiency.
- » **Integrating with Remote Sensing Data**: Machine learning can enhance remote-sensing applications, such as drone-based NDVI monitoring. By integrating NDVI data with environmental and crop-specific information, ML algorithms can produce more accurate stress maps, enabling site-specific management. In fruit orchards, these maps help growers apply resources only where needed, saving inputs and protecting yield.
- » **Supporting Long-Term Planning**: Machine learning models that continuously learn from historical data can also assist in long-term planning. For example, by analyzing seasonal temperature and drought trends, ML models can help growers plan for expected climate patterns, select appropriate cultivars, and adjust planting schedules to optimize crop resilience.

Potential Benefits of Future Developments in Stress Indices

The incorporation of molecular markers, IoT technology, and machine learning into stress index development offers a range of benefits that promise to transform fruit crop management:

1. **Accelerated Breeding and Selection**: With molecular markers, breeders can more quickly identify and select for traits associated with stress resilience, reducing the time and cost of breeding programs and enabling growers to adapt to changing environmental conditions.
2. **Real-Time, Proactive Management**: IoT-enabled sensors provide continuous, real-time monitoring of crop stress levels, allowing for immediate interventions. This proactive approach helps minimize stress impacts, ensuring better yield stability and fruit quality, particularly under extreme environmental conditions.

3. **Increased Predictive Accuracy**: Machine learning models analyze vast amounts of data to predict crop responses, helping growers implement stress-mitigation strategies before damage occurs. These predictive capabilities support more accurate and efficient use of water, fertilizers, and other inputs.
4. **Enhanced Precision Agriculture**: By combining IoT monitoring and machine learning analysis, growers can implement precision agriculture practices that optimize resource allocation and reduce waste. Targeted interventions based on real-time data help maximize crop health, yield, and quality, all while minimizing environmental impact.
5. **Adaptation to Climate Change**: As climate variability continues to pose challenges for agriculture, future stress indices will be invaluable in helping growers adapt to fluctuating weather patterns, such as droughts, heatwaves, and changes in rainfall. Integrating these indices into adaptive management practices will allow for greater resilience to climate-related stressors.
6. **Improved Sustainability and Economic Viability**: Future stress indices will make agriculture more sustainable by improving resource use efficiency and reducing dependence on chemical inputs. This not only reduces production costs but also minimizes the environmental footprint of fruit crop production, supporting long-term profitability.

Table 7: Physiological and Biochemical factors associated with various types of environmental stresses

S. No.	Stress Type	Physiological and Biochemical Factors	Effects on Crop Health and Productivity
1	Drought Stress	Reduced Relative Water Content (RWC)	Decreased growth and yield
		Stomatal Closure	Reduced CO_2 uptake, leading to lower photosynthesis
		Accumulation of Osmolytes (e.g., proline, glycine betaine)	Maintains cellular turgor but is energy-intensive
		Increased Abscisic Acid (ABA) levels	Stomatal regulation to conserve water; may reduce growth
		Decline in Chlorophyll Content	Reduced photosynthetic efficiency and leaf senescence
2	Heat Stress	Enzyme Denaturation and Reduced Enzyme Activity	Impairs metabolism and reduces growth
		Production of Heat Shock Proteins (HSPs)	Provides temporary protection to proteins, but energy-intensive
		Increased Ethylene Production	Premature fruit ripening and senescence
		Membrane Fluidity Changes	Increased ion leakage and dehydration
		Accumulation of Reactive Oxygen Species (ROS)	Oxidative stress damages cellular components

3	Cold Stress	Reduced Enzyme Activity	Slowed metabolism and stunted growth
		Production of Cold-Responsive Proteins (e.g., COR proteins)	Enhances cellular stability under low temperatures
		Increased Abscisic Acid (ABA)	Stomatal closure to reduce water loss, but limits CO_2 uptake
		Accumulation of Cryoprotectants (e.g., proline, sugars)	Prevents ice crystal formation within cells, but may limit growth due to resource diversion
		Membrane Rigidity	Decreased fluidity affects nutrient and water transport
4	Salinity Stress	High Na^+/K^+ Ratio	Interferes with potassium-dependent processes, affecting enzyme activity and nutrient transport
		Accumulation of Compatible Solutes (e.g., proline, glycine betaine)	Maintains cellular turgor but diverts energy from growth
		Increased Reactive Oxygen Species (ROS)	Causes oxidative damage to lipids, proteins, and DNA
		Membrane Permeability Changes	Ion leakage disrupts cellular homeostasis
		Reduced Chlorophyll Content	Lower photosynthesis reduces yield and quality
5	Flooding (Waterlogging)	Oxygen Deficiency (Hypoxia)	Reduced root respiration and energy deficits
		Accumulation of Ethylene	Promotes root and leaf senescence, and reduces fruit set
		Formation of Aerenchyma in Roots	Improves oxygen diffusion within roots but slows growth
		Increased Accumulation of Toxic Compounds (e.g., ethanol, lactic acid)	Cellular toxicity due to anaerobic respiration
		Reduced Nutrient Uptake	Nutrient deficiencies impair growth and yield

6	Nutrient Deficiency	Reduced Chlorophyll Content (especially with nitrogen deficiency)	Decreased photosynthesis and chlorosis
		Altered Hormone Levels (e.g., gibberellins decrease with phosphorus deficiency)	Stunted growth and poor flowering
		Reduced Protein Synthesis	Low biomass production and poor fruit quality
		Enhanced Root Growth (in response to phosphorus and nitrogen deficiencies)	Increased energy expenditure to access nutrients
		Accumulation of Stress Hormones (e.g., abscisic acid in potassium deficiency)	Reduced stomatal conductance, affecting photosynthesis
7	Heavy Metal Toxicity	Increased Reactive Oxygen Species (ROS)	Oxidative stress damages cellular structures
		Chelation and Compartmentalization of Metals (e.g., sequestration in vacuoles)	Limits metal toxicity but diverts energy from growth
		Reduced Photosynthetic Pigments	Lower chlorophyll content and decreased photosynthetic efficiency
		Disruption of Enzyme Activity (due to metal binding)	Inhibited metabolic pathways and reduced growth
		Altered Nutrient Uptake	Imbalance of essential nutrients, affecting yield
8	Oxidative Stress	Increased Production of Antioxidant Enzymes (e.g., superoxide dismutase, catalase)	Detoxifies ROS but requires significant energy
		Accumulation of Non-Enzymatic Antioxidants (e.g., ascorbate, glutathione)	Protects cells from oxidative damage
		Lipid Peroxidation	Membrane damage and ion leakage
		Reduced Photosynthetic Efficiency	Low energy production and impaired growth
		DNA and Protein Damage	Mutations and dysfunction in cellular machinery

9	Radiation Stress	Increased Production of UV-Absorbing Compounds (e.g., flavonoids)	Protects against UV damage but may reduce resources for growth
		Accumulation of Reactive Oxygen Species (ROS)	Oxidative damage to proteins, DNA, and lipids
		Reduced Photosynthetic Efficiency (photoinhibition)	Lower chlorophyll function and energy production
		Membrane Damage	Increased permeability affects nutrient and water transport
		Altered Growth Patterns	Stunted growth and leaf damage
10	Wind Stress	Increased Ethylene Production	Promotes leaf drop, senescence, and fruit abscission
		Mechanical Damage to Tissues	Physical breakage and reduced photosynthetic area
		Increased Transpiration Rates	Water loss and potential dehydration
		Altered Root-to-Shoot Ratio	Resource allocation to stabilize root systems
		Reduced Leaf Area	Lower photosynthesis due to decreased leaf area

Outlines of physiological and biochemical factors associated with various biotic stresses in fruit crops, along with their effects on crop health and productivity:

Table 8: Biotic Stresses trigger physiological and biochemical responses in fruit crops

S. No.	**Biotic Stress Type**	**Physiological and Biochemical Factors**	**Effects on Crop Health and Productivity**
1	Insect Pest Attack	Increased Production of Defensive Compounds (e.g., alkaloids, phenolics)	Deters feeding but diverts resources from growth and fruiting
		Induced Production of Jasmonic Acid (JA)	Activates pest defense pathways, but may reduce growth and reproductive output
		Cell Wall Thickening	Provides physical barrier against pests but slows overall plant growth
		Production of Volatile Organic Compounds (VOCs)	Attracts natural enemies of pests, enhancing biological control
		Reduced Photosynthetic Rate	Leaf damage limits CO_2 uptake, reducing energy for growth and yield

2	Fungal Infection	Production of Pathogenesis-Related Proteins (PR proteins)	Inhibits fungal growth but requires significant energy, impacting growth
		Accumulation of Phytoalexins (e.g., resveratrol in grapes)	Toxic to fungi and reduces infection spread, but diverts energy from other processes
		Increased Salicylic Acid (SA) Levels	Activates systemic acquired resistance but can slow growth in high concentrations
		Cell Wall Lignification	Strengthens tissues against fungal penetration but may limit expansion
		Increased Reactive Oxygen Species (ROS)	Promotes cell death in infected areas to limit spread but causes oxidative stress
3	Bacterial Infection	Accumulation of Antimicrobial Compounds (e.g., phenolics, flavonoids)	Limits bacterial proliferation but diverts resources from growth
		Increased Production of Hydrogen Peroxide (H_2O_2)	Creates oxidative burst that kills bacteria but can damage host cells
		Strengthening of Cell Walls	Forms a physical barrier against bacterial entry, reducing infection risk
		Stomatal Closure	Prevents bacterial entry through stomata but reduces photosynthesis due to limited CO_2 uptake
		Systemic Acquired Resistance (SAR) Activation	Enhances long-term resistance but is energy-intensive
4	Viral Infection	Suppression of Host RNA Silencing Mechanisms	Viral proteins interfere with host defense, facilitating infection
		Altered Hormone Levels (e.g., increased abscisic acid, reduced gibberellins)	Abnormal growth patterns, stunted growth, and leaf curling
		Degradation of Chloroplasts	Reduces photosynthetic efficiency, leading to chlorosis (leaf yellowing)
		Production of Antiviral Proteins	Limits viral replication but diverts energy from other metabolic processes
		Increased Reactive Oxygen Species (ROS)	Oxidative damage to infected tissues but may also harm healthy cells

5	Nematode Infestation	Formation of Root Galls	Reduces root efficiency, impairing water and nutrient uptake
		Production of Phenolic Compounds	Provides localized defense but reduces energy for growth
		Increased Salicylic Acid (SA) and Jasmonic Acid (JA) Pathways Activation	Activates resistance but may reduce overall plant vigor
		Altered Root Architecture	Impaired root system impacts nutrient absorption and water uptake
		Reduced Photosynthetic Rate	Lower nutrient and water uptake limits leaf development and chlorophyll production
6	Herbivory (Large Pests)	Increased Production of Physical Defenses (e.g., thorns, tougher leaves)	Physical barriers reduce herbivory but divert resources from fruit development
		Production of Secondary Metabolites (e.g., tannins, terpenoids)	Deters herbivores but affects fruit palatability if compounds accumulate in edible parts
		Altered Hormone Levels (e.g., increased ethylene and jasmonic acid)	Activates defense mechanisms but may lead to premature leaf senescence and fruit drop
		Production of Signaling Compounds (e.g., VOCs)	Attracts predators of herbivores, supporting natural pest control
		Reduced Leaf Area	Less photosynthetic surface affects energy production and growth
7	Parasitic Plants	Nutrient Diversion to Parasite	Reduced nutrient availability for host plant growth
		Altered Water Dynamics	Water diverted to parasite, causing host dehydration
		Activation of Defense Genes	Limits parasitic growth but consumes energy, reducing crop growth and yield
		Production of Phytoalexins	Toxic to parasitic plants but diverts resources
		Stunted Growth and Reduced Yield	Reduced photosynthesis and nutrient access limit productivity

8	Insect Vectors of Disease	Production of Volatile Organic Compounds (VOCs)	Attracts natural enemies of insect vectors, aiding in pest control
		Increased Jasmonic Acid (JA) and Salicylic Acid (SA) Pathways Activation	Activates resistance to vector-borne diseases but may reduce growth
		Cell Wall Fortification	Reduces vector feeding but limits cell expansion
		Increased Production of Defensive Enzymes	Limits vector entry and disease spread but diverts resources
		Reduced Photosynthetic Efficiency	Leaf damage from vectors limits photosynthesis, reducing growth
9	Fungal-Like Pathogens	Production of Callose and Lignin in Cell Walls	Strengthens cell walls to inhibit pathogen entry but slows plant growth
		Synthesis of Pathogenesis-Related (PR) Proteins	Inhibits pathogen growth but requires substantial energy
		Increased Levels of Salicylic Acid (SA)	Activates long-term resistance but may inhibit growth
		Accumulation of Reactive Oxygen Species (ROS)	Triggers cell death to contain infection but can damage surrounding cells
		Stunted Growth and Chlorosis	Reduced nutrient uptake and photosynthesis limit productivity

IV. Fruit Crops suitable for different stress situations

Fruit crops often face various environmental stresses that impact their growth, yield, and quality. However, certain fruit crops have natural adaptations or can be selectively bred to better withstand specific types of stress.

1. Drought Stress

Drought stress is one of the most challenging conditions for fruit crops, as it limits water availability, affecting crop growth, photosynthesis, and fruit development. However, several fruit crops have evolved to withstand water scarcity through physiological adaptations and efficient water management.

Key Crops and Their Adaptations

» **Olive**: Olives are highly adapted to drought-prone Mediterranean climates, with deep root systems that allow access to subsurface moisture. Their small, waxy leaves reduce water loss through transpiration. Olive trees use an osmotic adjustment mechanism that retains cellular hydration during dry periods, making them suitable for areas with erratic rainfall. Notable varieties like 'Arbequina' and 'Koroneiki' maintain productivity in dry conditions, making olives an essential crop in arid regions.

- **Pomegranate**: Known for its resilience in semi-arid regions, pomegranate has high drought tolerance. It uses a combination of deep roots, water-storing tissue, and osmotic regulation to adapt to water scarcity. Varieties like 'Wonderful' are commonly grown in water-limited environments. Pomegranate can shift its metabolic activity to maintain productivity under drought, making it a sustainable choice in regions with irregular precipitation.
- **Date Palm**: The date palm is an iconic desert plant. It has a specialized root system capable of extracting deep groundwater, and it minimizes water loss by regulating stomatal closure during extreme heat and dryness. Varieties like 'Medjool' and 'Deglet Noor' are widely cultivated in arid regions. Date palms not only survive but thrive under minimal water conditions, providing valuable nutrition in harsh climates.
- **Fig**: With a root system that can access deep soil moisture and a high tolerance for dry conditions, figs are naturally suited to drought-prone regions. Their leaves have a thick cuticle that limits water loss, and they can survive long dry spells with minimal irrigation. Figs are cultivated in regions with hot, dry summers, where they produce high-quality fruit even under limited water availability.
- **Grape (Wine Varieties)**: Many wine grape varieties, such as 'Grenache' and 'Syrah', are drought-tolerant. Grapes manage water stress by developing smaller leaves, reducing transpiration, and concentrating sugars in the fruit under water stress, enhancing wine quality. This makes drought-tolerant grape varieties suitable for viticulture in arid and semi-arid climates.
- **Almond**: Almond trees are deep-rooted and efficient in their water use. They can survive and produce under reduced water availability, making them ideal for Mediterranean and dry regions. Almond varieties like 'Nonpareil' are resilient under drought, supporting stable yields with minimal irrigation after establishment.
- **Prickly Pear (Opuntia)**: This cactus fruit is highly drought-resistant due to its ability to store water in its fleshy pads. Opuntia can thrive with little water, as it has CAM (Crassulacean Acid Metabolism) photosynthesis, which allows it to open stomata at night to minimize water loss.

Management Practices for Drought-Resilient Crops

To maximize productivity under drought conditions, growers often use methods like mulching to conserve soil moisture, and drip irrigation to supply precise amounts of water. Selection of drought-tolerant varieties, combined with soil amendments that improve water retention, can support sustained production in water-scarce environments.

2. Heat Stress

Heat stress can affect flowering, pollination, and fruit development in fruit crops. High temperatures accelerate respiration, reduce chlorophyll content, and disrupt enzyme functions, ultimately impacting yield. However, some fruit crops are inherently adapted to high temperatures.

Key Crops and Their Adaptations

- **Mango**: Mango trees are naturally suited to warm tropical climates. They exhibit mechanisms such as stomatal control to limit water loss and avoid overheating. Varieties like 'Haden' and 'Tommy Atkins' are known for their resilience to high temperatures, enabling consistent flowering and fruit set even in intense heat.
- **Banana**: Bananas are adapted to hot, humid climates, where they perform well with ample moisture. Their broad leaves are structured to facilitate efficient cooling through transpiration. Bananas can grow rapidly in high temperatures, though adequate irrigation is critical to maintain fruit quality.
- **Papaya**: Papaya thrives in tropical and sub-tropical climates, handling temperatures above 30°C well. The plant's ability to open its stomata during cooler parts of the day helps manage heat stress. Varieties like 'Red Lady' are known to be heat-tolerant, maintaining growth and fruit quality in warm regions.
- **Pineapple**: Pineapples can handle high temperatures, thanks to their ability to store water in thick leaves and reduce transpiration. Varieties like 'MD-2' produce quality fruit in tropical regions. The CAM photosynthetic pathway in pineapple also supports its survival in hot, arid areas by reducing water loss.
- **Guava**: Guava is heat-tolerant, with varieties like 'Allahabad Safeda' performing well in warm climates. Guavas continue to produce fruit despite high temperatures, making them suitable for tropical and semi-tropical areas with variable weather.
- **Watermelon**: This crop is adapted to high temperatures and grows well in summer. Watermelons thrive in sunny, hot conditions and have a strong root system that supports their water requirements. Varieties like 'Sugar Baby' and 'Charleston Gray' are commonly grown in warm climates.

Management Practices for Heat-Resilient Crops

Heat stress management includes using shade nets to reduce direct sun exposure, optimizing irrigation to maintain soil moisture, and applying mulches to cool the root zone. Selecting heat-tolerant varieties and providing adequate nutrition can support yield stability under high temperatures.

3. Cold Stress

Cold and frost damage can limit crop production in temperate regions. Chilling and frost tolerance in some fruit crops allow them to survive and produce in cold climates.

Key Crops and Their Adaptations

- **Apple**: Many apple varieties require cold temperatures for bud dormancy and chilling accumulation. Varieties such as 'Honeycrisp' and 'Granny Smith' have high tolerance to frost, making apples suitable for temperate climates with chilling requirements for flowering.
- **Pear**: Pear varieties like 'Bartlett' and 'Bosc' thrive in cooler regions due to their chilling requirements. They can tolerate moderate frost, which is essential for areas with seasonal temperature fluctuations.

- **Plum**: European plums, such as 'Stanley', are frost-resistant and adapted to cold environments. Plums have high chilling requirements, making them productive in temperate zones with long winters.
- **Cherry**: Cherry varieties like 'Bing' and 'Stella' require winter chilling for optimal fruit set. These varieties tolerate frost and perform well in cold climates, where they develop firm, high-quality fruit.
- **Raspberry**: Raspberries, such as 'Heritage' and 'Latham', are cold-hardy and grow well in temperate climates. They have mechanisms to survive winter dormancy and can withstand frost.
- **Blueberry**: Northern highbush blueberries, like 'Bluecrop' and 'Patriot', tolerate cold conditions, with high chilling requirements for proper flowering. Blueberries also benefit from cold-induced dormancy, which enhances fruiting.
- **Cranberry**: Cranberries are adapted to cold bog conditions and can handle winter frost. They are typically grown in northern regions where winter chilling benefits fruiting.

Management Practices for Cold-Resilient Crops

Cold stress management includes site selection in frost-free areas, utilizing windbreaks, and frost protection measures such as row covers or irrigation during frost events. Proper variety selection based on regional climate conditions ensures better survival and yield in cold areas.

4. Salinity Stress

High soil salinity interferes with nutrient uptake, water availability, and cellular functions in plants. However, some fruit crops can tolerate saline conditions by regulating ion balance and osmosis.

Key Crops and Their Adaptations

- **Date Palm**: Date palms are highly salt-tolerant and can grow in saline soils. They effectively exclude sodium from roots and maintain cellular ion balance, allowing them to thrive in arid and saline environments.
- **Pomegranate**: Pomegranates are moderately salt-tolerant and can grow in soils with mild to moderate salinity. The crop maintains cellular hydration through osmotic adjustment, making it suitable for semi-arid regions with saline irrigation.
- **Fig**: Figs tolerate moderate salinity due to their ability to manage osmotic stress and exclude excess sodium. They grow well in semi-arid regions with saline soils, where they produce high-quality fruit.
- **Grape (Salt-Tolerant Rootstocks)**: Grapevines grafted onto salt-tolerant rootstocks can thrive in saline conditions. Salt-tolerant rootstocks exclude sodium, allowing the scion variety to maintain productivity and quality under salinity stress.
- **Guava**: Guavas have moderate salt tolerance and can grow in areas with mild salinity. The crop regulates ion concentration to maintain nutrient uptake and growth.
- **Olive**: Olive trees are salt-tolerant, excluding excess sodium and chloride ions, which allows

them to survive in saline environments. They are commonly grown in coastal and arid areas with saline soils.

Management Practices for Saline Soils

Management of salinity stress includes using salt-tolerant rootstocks, applying gypsum to mitigate sodium buildup, and using drip irrigation to flush salts from the root zone. Selecting salt-tolerant varieties and maintaining proper drainage improves productivity in saline environments.

5. Flooding (Waterlogging) Tolerance

Flooding can lead to oxygen deficiency in the root zone, affecting root respiration and nutrient uptake. Some fruit crops can survive short periods of waterlogging due to adaptations like aerenchyma formation.

Key Crops and Their Adaptations

» **Banana**: Bananas tolerate short-term waterlogging as they have a fibrous root system and the ability to form aerenchyma. This allows them to survive heavy rains in tropical climates.

» **Pineapple**: Pineapple can handle waterlogged conditions temporarily. The crop has shallow roots, which tolerate brief flooding.

» **Papaya**: Papaya trees can survive occasional waterlogging in tropical environments due to their tolerance to root oxygen deficiency. However, drainage is essential to avoid prolonged water stress.

» **Coconut**: Naturally adapted to coastal regions, coconuts tolerate saline and waterlogged conditions, making them resilient in areas prone to flooding.

» **Mango**: Mango can withstand occasional flooding if grown on resistant rootstocks. The tree adjusts its root function to maintain growth under temporary waterlogged conditions.

Management Practices for Flood-Resilient Crops

Flood management includes proper site drainage, raised beds, and selecting rootstocks resistant to waterlogging. Temporary flooding tolerance is beneficial in monsoon-prone regions, supporting crop survival during heavy rainfall.

Table 9: Various fruit crops suited to different types of environmental stress

Stress Type	Suitable Fruit Crops	Adaptations/Reasons
Drought Stress	Olive	Deep root systems; high water-use efficiency and tolerance to low soil moisture
	Pomegranate	Can survive in arid conditions; has drought-tolerant cultivars; efficient water-use
	Date Palm	Tolerant to extreme drought and heat; adapted to desert environments
	Fig	Thrives in dry climates with minimal water needs

	Grape (especially wine grapes)	Some varieties are drought-tolerant and can concentrate sugars in low-water conditions, enhancing flavor
	Almond	Deep root systems and efficient water management; requires minimal water once established
	Prickly Pear (Opuntia)	High tolerance to drought; stores water in its fleshy pads
Heat Stress	Mango	Grows well in tropical and subtropical regions; tolerates high summer temperatures
	Banana	Adapted to hot and humid climates; some cultivars bred for heat tolerance
	Papaya	Thrives in warm conditions; tolerates high temperatures with minimal yield loss
	Pineapple	Tolerant to heat; performs well in tropical climates
	Guava	Heat-tolerant, especially in tropical and semi-tropical regions
	Watermelon	Has high heat tolerance; commonly grown in hot climates
Cold Stress	Apple (certain varieties)	Many varieties are cold-hardy and can withstand frost; chilling requirement for flowering
	Pear	Cold-hardy species adapted to temperate regions; chilling requirement for bud formation
	Plum	Cold-hardy cultivars available, particularly European plums
	Cherry	Requires chilling hours; many varieties are frost-tolerant
	Raspberry	Cold-tolerant, especially varieties bred for temperate climates
	Blueberry	Northern varieties adapted to cold climates
	Cranberry	Naturally adapted to cold, bog-like environments
Salinity Stress	Date Palm	High tolerance to saline soils and water
	Pomegranate	Moderate salt tolerance; performs better than many fruit crops in saline conditions
	Fig	Tolerates moderate salinity; can grow in semi-arid saline soils
	Grape (salt-tolerant rootstocks)	Certain rootstocks are selected for salt tolerance, enabling growth in moderately saline conditions
	Guava	Shows moderate salt tolerance, making it suitable for areas with mild salinity
	Olive	Salt-tolerant; can thrive in saline soils

Flooding (Waterlogging)	Banana	Tolerant to short-term waterlogging; used to high rainfall in natural habitat
	Pineapple	Tolerates waterlogging due to its shallow root system, especially in tropical climates
	Papaya	Can withstand temporary waterlogging; fast-draining soils help manage excess moisture
	Coconut	Naturally grows in coastal areas with periodic flooding
	Mango	Can tolerate occasional flooding; certain rootstocks can improve flood tolerance
Heavy Metal Stress	Coconut	Tolerant to certain heavy metals and capable of growing in soils affected by mining residues
	Grape (some rootstocks)	Select rootstocks can tolerate heavy metal presence in the soil, enabling grape cultivation in contaminated areas
	Olive	Shows moderate tolerance to heavy metals, especially in soils with cadmium and lead contamination
	Jujube	Tolerant to heavy metals like zinc and lead, often used in reclamation of contaminated lands
	Pomegranate	Moderate tolerance to heavy metals; can grow in contaminated soils
Pest and Disease Resistance	Avocado (certain varieties)	Certain varieties are resistant to common pests like thrips and have natural defenses against Phytophthora root rot
	Citrus (disease-resistant rootstocks)	Resistant rootstocks available for citrus greening and root rot resistance
	Apple (disease-resistant varieties)	Varieties such as Enterprise and Liberty are resistant to apple scab and fire blight
	Pear	Varieties like Harrow Sweet show resistance to fire blight
	Grape (disease-resistant hybrids)	Hybrids like Concord and Norton show resistance to fungal diseases such as downy and powdery mildew

Chapter - 3

Assessment, Physiology and Performance of Stress on Fruit Crops

1. Crop Modeling for Stress Situations

Crop modeling has emerged as an invaluable tool in modern agriculture, allowing for predictive insights and management strategies tailored to various environmental stressors, including drought, salinity, heat, and nutrient deficiencies. In fruit crop production, these models enable growers, agronomists, and researchers to make informed decisions, especially when confronting climate variability and resource constraints. Crop models like DSSAT (Decision Support System for Agrotechnology Transfer), Aqua Crop, and APSIM (Agricultural Production Systems Stimulator) have been at the forefront of this effort, helping to simulate crop growth, yield outcomes, and stress responses.

a. Crop Models Used for Stress Prediction and Management

Crop models have become essential tools for predicting and managing the impacts of environmental stresses on agricultural productivity. For fruit crops, which are highly sensitive to variations in water availability, temperature, and soil conditions, crop models offer insights that aid in maintaining yield and quality under stress conditions. Among the most prominent models for this purpose are DSSAT, AquaCrop, and APSIM. Each model has specific capabilities and focuses on distinct environmental factors, making them valuable for targeted stress management in fruit production.

1. DSSAT Model

The Decision Support System for Agrotechnology Transfer (DSSAT) is recognized as one of the most advanced and versatile crop models, supporting over 40 crop species, including various fruit crops like grapes, citrus, and apples. Developed through collaboration among several institutions, DSSAT is highly adaptable and integrates an array of environmental and physiological parameters. This flexibility makes it particularly effective for predicting and managing stress conditions in fruit crops.

DSSAT's strengths lie in its ability to simulate complex interactions between plants and environmental factors. It incorporates variables such as soil moisture, temperature, nutrient availability, and crop growth stages, allowing it to accurately represent how different stress factors, like drought, salinity, and temperature fluctuations, affect crop performance. For example, DSSAT can model the impact of soil moisture levels on the flowering and fruiting stages, providing growers with actionable insights for scheduling irrigation, fertilization, and harvesting to mitigate the adverse effects of water

DOI: 10.1201/9781003717409-3

scarcity. In grape production, this is particularly valuable because drought stress can influence berry size, sugar concentration, and overall fruit quality, all of which are critical for wine production.

DSSAT is also valuable in nutrient management, as it can simulate nutrient uptake and predict deficiencies under different environmental conditions. This capability is particularly beneficial in fruit crops, where specific nutrient balances are crucial for maintaining quality attributes such as color, flavor, and firmness. By simulating nutrient dynamics, DSSAT can assist growers in planning fertilization schedules that align with crop growth stages and stress responses, minimizing waste and optimizing crop health.Another key feature of DSSAT is its ability to simulate the effects of temperature stress. High temperatures can lead to physiological damage in fruit crops, affecting fruit set and ripening. DSSAT's temperature modules allow users to assess how heat stress affects different growth stages, supporting the development of heat mitigation strategies. For instance, in citrus orchards, high temperatures during fruit set can lead to reduced yield. By using DSSAT, growers can evaluate the effectiveness of interventions such as shade netting or adjusted planting dates, ensuring that crops maintain their productivity even under fluctuating temperature conditions.Overall, DSSAT provides a comprehensive platform for assessing how multiple stressors interact to influence crop performance. Its versatility and detailed parameterization make it a valuable tool for optimizing resource use and enhancing stress resilience in fruit production systems.

2. Aqua Crop Model

Aqua Crop, developed by the Food and Agriculture Organization (FAO), is a crop model specifically designed to evaluate the impacts of water-related stresses, such as drought and salinity, on crop yield. Unlike other models that focus broadly on multiple environmental factors, Aqua Crop is tailored for scenarios where water availability is the primary constraint, making it particularly well-suited for arid and semi-arid regions.

One of Aqua Crop's primary strengths is its focus on water-use efficiency, an essential factor for fruit crops grown in water-limited environments. The model is structured to simulate soil moisture dynamics, allowing it to predict how different irrigation regimes affect crop growth and productivity. Aqua Crop calculates the soil water balance, taking into account variables such as soil type, root depth, and evapotranspiration. This enables it to accurately simulate how fruit crops respond to variations in soil moisture, providing growers with data to make informed irrigation decisions. For instance, pomegranate and olive trees, which are moderately drought-tolerant, benefit significantly from Aqua Crop's irrigation management insights, allowing growers to maintain optimal soil moisture levels without overusing water resources.

In addition to drought stress, Aqua Crop is effective in assessing the impacts of salinity on crop yield. Salinity stress can inhibit water uptake and disrupt nutrient absorption, which is particularly detrimental in fruit crops that require balanced nutrition for high-quality yields. AquaCrop simulates how salinity in the root zone affects water availability to plants, predicting the extent to which salt accumulation can impair growth. This is crucial for regions that rely on saline water sources for irrigation, as it helps growers develop strategies to manage salt buildup, such as applying gypsum amendments or leaching salts through controlled irrigation. For fruit crops like grapes and citrus, Aqua Crop's ability to predict salinity effects enables growers to mitigate potential quality losses due to salt stress.

Another valuable application of Aqua Crop is its capacity to simulate yield outcomes under different climate scenarios. By adjusting parameters related to precipitation, temperature, and evapotranspiration, Aqua Crop can project how future climate conditions may impact water availability and crop productivity. This predictive capability supports long-term planning in fruit orchards, where investment and planning decisions often span several years. AquaCrop's simplicity and emphasis on water-related parameters make it accessible and practical for growers who need targeted water management solutions to support sustainable production.

3. APSIM Model

The Agricultural Production Systems Simulator (APSIM) is a powerful tool designed to simulate the biological and environmental interactions that influence crop productivity. It is widely used to assess the impacts of heat and nutrient stresses on crops, making it particularly valuable in fruit production in regions with high temperatures or soils prone to nutrient deficiencies. APSIM was developed by a consortium of researchers and integrates multiple environmental and physiological processes, offering a holistic view of crop-environment interactions.

APSIM incorporates variables such as weather data, soil properties, and crop growth stages to simulate how different stress factors affect crop performance. One of APSIM's key features is its ability to model heat stress, a significant concern for fruit crops like strawberries and grapes, which are sensitive to high temperatures. APSIM uses temperature thresholds to determine the effects of heat stress on processes like flowering, pollination, and fruit set. By predicting how extreme temperatures may influence these stages, APSIM enables growers to adjust planting dates, implement shading techniques, or apply cooling measures to protect crops from heat-induced yield losses

Nutrient stress management is another area where APSIM excels. The model simulates soil nutrient dynamics and plant uptake, providing insights into how nutrient availability fluctuates under different environmental conditions. This capability is especially important for fruit crops, where nutrient balance affects fruit size, taste, and nutritional content. APSIM enables growers to create nutrient management plans that are responsive to crop needs and environmental stressors. For example, in strawberry production, nutrient deficiencies can reduce fruit sweetness and firmness, impacting marketability. APSIM's nutrient modules allow growers to determine optimal fertilizer applications that support healthy fruit development, ensuring high-quality produce.

APSIM also has applications in long-term resource management and planning. The model can simulate multi-year scenarios, accounting for cumulative stress impacts and helping growers assess the sustainability of different management practices. This is particularly useful for perennial fruit crops, where the effects of environmental stressors accumulate over time. APSIM's ability to predict the long-term impacts of soil degradation, nutrient depletion, and climate change on fruit production supports the development of resilient cropping systems that can maintain productivity under changing conditions. Through its detailed simulations of heat and nutrient stress responses, APSIM provides critical data to support precision agriculture practices in fruit production. By optimizing planting dates, nutrient application schedules, and soil management practices, APSIM helps growers enhance resilience and productivity in challenging environments.

b. How do Crop Models Incorporate Variables for Simulating Stress Impacts?

Crop models provide critical insights into how environmental variables influence crop growth, productivity, and resilience under stress. In fruit production, where stress factors such as drought, heat, and nutrient deficiencies significantly impact quality and yield, crop models integrate specific environmental and physiological data to predict stress responses.

Soil Moisture and Drought Stress

Drought stress is among the most significant factors affecting the productivity of fruit crops, as water scarcity directly influences photosynthesis, nutrient uptake, and fruit development. Crop models, including DSSAT, Aqua Crop, and APSIM, incorporate soil moisture as a core variable to simulate how drought conditions impact fruit crops. By continuously monitoring changes in soil moisture, these models can predict water availability throughout the growing season and simulate the plant's physiological response to water stress.

In Aqua Crop, for instance, soil moisture balance is the main driver. The model considers variables such as soil type, depth, and root distribution to assess how well plants access water. This soil moisture monitoring helps Aqua Crop accurately simulate water uptake and predict drought impacts, guiding irrigation management practices. DSSAT similarly uses soil moisture data to model water stress effects, allowing for precise irrigation scheduling. For example, in vineyards, where grape quality and sugar concentration are directly influenced by water availability, DSSAT can help determine irrigation intervals that minimize drought stress while conserving water resources.

These models also consider how different soil types impact moisture retention, helping growers decide whether additional soil amendments or mulching might be necessary to enhance water-holding capacity. Through soil moisture-based predictions, crop models support efficient irrigation strategies tailored to the specific water needs of each fruit crop. This capability is crucial for conserving water and ensuring that crops remain productive even during prolonged dry spells.

Temperature and Heat Stress

High temperatures induce various physiological responses in fruit crops, from altered growth rates to sunburn and reduced fruit quality. Temperature data, such as daily highs, lows, and duration of extreme heat, are critical inputs in crop models that simulate the impact of heat stress on fruit production. APSIM, for example, is particularly adept at evaluating heat stress effects by integrating temperature thresholds that align with different stages of crop growth.

APSIM's temperature-driven modules allow growers to predict how high temperatures may impact flowering, fruit set, and ripening, all of which are sensitive stages for fruit crops. By understanding how extreme temperatures affect these processes, APSIM can help implement heat mitigation strategies. For example, high temperatures during the flowering stage in crops like apples can reduce pollen viability, lowering fruit set and ultimately affecting yield. Using APSIM, growers can anticipate periods of high heat and take measures such as installing shade nets or applying cooling sprays to mitigate the effects of heat stress.

DSSAT also uses temperature data to model potential heat stress impacts. In citrus production, where heat can cause rind damage and affect juice quality, DSSAT can help schedule harvests to avoid periods of intense heat that may compromise fruit quality. Additionally, these models assess cumulative heat stress over the growing season, helping growers evaluate whether to adjust planting dates or change varieties to those more tolerant to high temperatures. By simulating temperature effects on fruit development, crop models enable more informed management practices that reduce yield losses and maintain quality even in hot climates.

Crop Growth Stages and Stress Sensitivity

Fruit crops have varying sensitivities to environmental stress at different growth stages. The stages of flowering, fruit set, and fruit ripening are especially vulnerable to stresses like drought, heat, and nutrient deficiency. Crop models such as DSSAT and Aqua Crop incorporate growth stage data to simulate how stresses specifically affect each phase, allowing growers to take timely and targeted actions to minimize stress impacts.For example, during the flowering stage, water stress can lead to poor pollination and reduced fruit set in crops like grapes and peaches. Aqua Crop models these growth stages with sensitivity parameters that reflect how water availability at critical stages impacts fruit yield and quality. This level of detail helps growers decide when supplemental irrigation is necessary to support flowering and fruit set, particularly in regions where water is limited.

Similarly, nutrient stress during fruit development can affect fruit size, color, and taste. DSSAT models the nutrient needs of crops across growth stages, enabling growers to plan nutrient applications that align with peak demand periods. In apple orchards, for instance, nitrogen and potassium are critical during fruit expansion. DSSAT's growth stage simulations can guide optimal nutrient timing, ensuring that fruit crops receive adequate nutrition to produce high-quality fruit. By understanding growth stage-specific stress sensitivities, growers can tailor their management practices to address the unique needs of each crop phase, ultimately preserving both yield and fruit quality.

Yield Data and Stress Response

Yield prediction is a primary function of crop models, and it incorporates the cumulative effects of various stresses throughout the growing season. Models like DSSAT, AquaCrop, and APSIM consider multiple environmental variables—including soil fertility, water availability, temperature, and light exposure—to simulate how stress impacts expected yield and quality in fruit crops.

DSSAT, for instance, calculates yield potential based on growth data across the season, adjusting for any stressors encountered at each stage. By predicting the expected yield under different stress scenarios, DSSAT helps growers allocate resources effectively. For example, if a model predicts reduced yield due to anticipated drought stress, growers can choose to invest in additional irrigation or modify pruning techniques to alleviate stress. This level of yield simulation is particularly valuable for high-value crops like wine grapes, where quality and yield both determine market value.

AquaCrop's yield simulation focuses on water-driven productivity, making it useful for estimating yield outcomes under various irrigation regimes. In water-scarce regions, AquaCrop can model yield response to limited irrigation, helping growers optimize their water allocation without sacrificing productivity. For crops like olives and pomegranates, which can tolerate some drought, AquaCrop's yield predictions provide a basis for adjusting water use to balance economic and environmental considerations.APSIM integrates environmental stress data with phenological stages to predict cumulative yield and crop performance. By accounting for the combined impact of heat, drought, and nutrient stress, APSIM helps growers identify the most yield-effective management practices. For fruit crops, where consistent quality and quantity are essential for market success, the ability to anticipate and respond to stress-induced yield fluctuations is critical. Crop models thus offer data-driven insights that support both resource efficiency and yield stability, ensuring that fruit production remains viable under a range of environmental conditions.

c. Practical Applications in Decision-Making, Resource Management, and Adaptation Strategies in Fruit Production

Crop models have become instrumental in guiding decision-making, resource allocation, and adaptation strategies in fruit production. These models provide insights into managing water, nutrients, and other essential resources, allowing growers to maintain productivity and quality in the face of environmental challenges.

Decision-Making in Irrigation and Water Management

Efficient water management is critical in fruit production, particularly in regions prone to drought or where water resources are limited. Crop models such as Aqua Crop play a crucial role in optimizing irrigation by predicting soil moisture needs, simulating drought scenarios, and offering data-driven guidance on when and how much to irrigate. Aqua Crop, developed specifically to address water-related challenges, incorporates soil moisture, climate data, and crop growth stages to simulate how water stress impacts yield and quality.

In pomegranate and olive orchards, where crops are moderately drought-tolerant, AquaCrop assists in creating irrigation schedules that conserve water while ensuring sufficient hydration to sustain growth. For instance, AquaCrop can predict periods of peak water demand, allowing growers to provide supplemental irrigation only when necessary. This precision prevents over-irrigation, which can lead to soil degradation, nutrient leaching, and increased operational costs. For regions facing recurrent droughts, AquaCrop's ability to forecast soil moisture levels helps in balancing the limited water supply, especially in crops like grapes, where precise water control is crucial to achieving desired sugar concentrations and flavor profiles.

In apple orchards, AquaCrop can be used to determine the exact amount of water required during fruit set and ripening stages, as these phases are highly sensitive to water stress. By integrating soil and climate data, AquaCrop aids in adjusting irrigation schedules to coincide with these critical growth stages, thereby preserving fruit quality and ensuring a steady supply. For high-value crops such as apples and grapes, where quality dictates market price, AquaCrop's irrigation management tools enable growers to protect their investment while supporting sustainable water practices.

Resource Allocation for Nutrient Management

Nutrient management is a cornerstone of quality fruit production, as nutrient imbalances can significantly affect yield, fruit size, flavor, and shelf life. Crop models like APSIM are particularly useful for nutrient management, as they assess real-time soil nutrient levels, predict deficiencies or toxicities, and recommend targeted fertilization strategies. APSIM's ability to monitor nutrient dynamics and forecast crop responses to nutrient inputs makes it invaluable in managing the delicate nutrient balance required for premium fruit crops.

In fruit crops such as citrus and strawberries, where nutrient needs fluctuate with growth stages, APSIM provides insights into optimal fertilization timing and quantities. The model accounts for environmental factors that influence nutrient availability, such as soil pH and temperature, as well as crop-specific nutrient requirements. For example, nitrogen is essential during the early growth stages of strawberries, while potassium is critical during fruiting to enhance sugar content and firmness. APSIM can simulate how the soil's nitrogen levels respond to different application rates, helping growers avoid both deficiency and excess.

In regions where soil nutrient levels vary significantly, APSIM allows for site-specific nutrient management by integrating spatial data from soil testing and remote sensing. This is particularly useful for fruit orchards with diverse soil types, as it ensures each section of the orchard receives the nutrients it needs without over-application. This targeted approach helps growers minimize fertilizer costs and reduce the risk of environmental pollution from nutrient runoff, supporting sustainable production.By tailoring nutrient applications to meet the precise needs of each crop and stage, APSIM enables growers to enhance fruit quality while reducing waste and environmental impact. For crops like mango and papaya, which require balanced nutrient inputs for optimal fruit flavor and shelf life, APSIM provides data-driven recommendations that support market-ready produce.

Adaptation Strategies for Climate Change Resilience

With climate change increasing the frequency and intensity of environmental stresses, crop models have become essential for developing adaptation strategies in fruit production. By simulating future climate scenarios, models such as DSSAT and APSIM provide a roadmap for growers to proactively adjust their practices and maintain productivity under changing conditions.

In grapevines, for instance, heat and drought are becoming more common, affecting berry size, acidity, and sugar concentration. DSSAT can simulate how these climatic variables impact grape quality, guiding decisions such as selecting drought-resistant varieties, adjusting planting dates, or modifying vineyard orientation to minimize heat exposure. These adjustments help maintain grape quality and yield despite rising temperatures, ensuring that growers continue to meet market demands for high-quality wine grapes.

Similarly, APSIM's climate modules allow growers to assess the long-term impacts of temperature and rainfall changes on fruit crops. For crops like apples and pears, which have specific chilling requirements for bud break and fruit set, warming temperatures may reduce the number of chilling hours available each winter. APSIM can simulate how reduced chill hours will affect yield, prompting growers to consider alternative varieties with lower chilling requirements or adapt orchard management practices, such as the use of dormancy-breaking sprays to compensate for insufficient chilling.

Crop models also support water management strategies that are resilient to climate variability. By simulating rainfall patterns, models like AquaCrop help growers prepare for irregular precipitation, allowing for the storage and efficient use of water resources. For example, in regions where rainfall is expected to become more erratic, AquaCrop can guide the installation of water storage systems that capture and store rainwater during wet periods for use during dry spells. In mango orchards, where water stress affects flowering and fruit set, such storage systems help ensure consistent yield despite climate-induced water shortages.Through their ability to simulate the effects of environmental changes, crop models enable growers to implement adaptation strategies that protect both yield and quality in the face of climate uncertainty. These models are invaluable in fostering resilience in fruit production systems, helping growers maintain stable productivity, profitability, and marketability over time.

2. Cropping Systems and Stress Mitigation

Cropping systems play a crucial role in managing environmental stresses in fruit production. Different systems—such as monocropping, intercropping, and agroforestry—each offer unique advantages that can help mitigate stress factors, such as drought, nutrient deficiencies, temperature extremes, and pest pressure.

a. Cropping Systems

Monocropping

Monocropping, the practice of growing a single crop variety over a large area, is commonly used in commercial fruit production due to its streamlined management and operational efficiency. While monocropping is often associated with increased susceptibility to specific stress factors like pests and soil degradation, it can be effective when combined with strategic stress mitigation practices.

Stress Mitigation in Monocropping:

1. **Optimized Water and Nutrient Management**: In monocropping systems, the uniformity of crop requirements allows for precise irrigation and fertilization schedules. For instance, in vineyards, where the uniform growth stage of grapevines enables precise timing for irrigation, growers can use models like AquaCrop to implement drought-responsive watering practices that conserve resources and optimize grape quality.
2. **Specialized Pest and Disease Control**: Monocropping often requires enhanced pest and disease management, as a single crop system can lead to pest population build-up. However, by adopting integrated pest management (IPM) strategies, such as pheromone traps and biological controls, growers can minimize pest damage. For instance, apple orchards in monocropping systems can benefit from targeted pest monitoring and species-specific biocontrol methods, helping to reduce pesticide reliance and maintain ecological balance.
3. **Adaptation to Climate Variability**: Monocropping systems in fruit production can also mitigate climate-induced stresses by selecting crop varieties that are resilient to specific stresses, such as heat-tolerant or drought-resistant cultivars. In tropical climates, for example, heat-tolerant banana cultivars are cultivated to reduce the risk of fruit damage and ensure yield stability under rising temperatures.

While monocropping requires intensive management and carries inherent risks, adopting targeted stress mitigation practices within a monoculture framework can help sustain productivity, especially in large-scale commercial fruit production.

Intercropping

Intercropping, or growing two or more crops in close proximity, is an effective strategy for enhancing resource use efficiency and reducing environmental stress impacts. In fruit production, intercropping is often employed to balance nutrient uptake, manage pests naturally, and optimize water use, especially in systems where resilience against climate variability is critical.

Stress Mitigation in Intercropping:

1. **Nutrient Balance and Soil Fertility**: In intercropping systems, complementary crops can improve nutrient cycling, reducing the risk of nutrient stress. For example, intercropping citrus with leguminous cover crops, like clover, improves nitrogen availability in the soil through biological nitrogen fixation. This practice not only supports the nutrient demands of the fruit crop but also enhances soil fertility and reduces the need for synthetic fertilizers.
2. **Enhanced Water-Use Efficiency**: Intercropping with crops that have varying root depths helps optimize soil moisture by utilizing water from different soil layers. In mango orchards, intercropping with shallow-rooted crops like beans allows deeper-rooted mango trees to access subsoil moisture, improving drought resilience. This stratification of water use helps maintain soil moisture, reducing the risk of drought stress and increasing overall water-use efficiency.
3. **Natural Pest and Disease Control**: Intercropping also promotes biodiversity, which reduces pest pressure by interrupting pest life cycles and providing habitats for natural predators. In strawberry fields, intercropping with marigold (which has pest-repelling properties) can reduce nematode populations and limit disease spread. This integration of pest-resistant plants within the cropping system enhances ecological resilience and reduces pesticide dependency.

Through nutrient optimization, water-use efficiency, and pest control, intercropping provides a natural buffer against environmental stresses. This approach is particularly beneficial for smallholder fruit farmers who rely on diversified systems for income stability and resilience.

Agroforestry

Agroforestry, which integrates trees with crops and livestock in a symbiotic environment, is a sustainable cropping system that offers numerous stress mitigation benefits, especially in fruit production. Agroforestry systems are known for their ability to improve microclimate conditions, enhance biodiversity, and reduce soil erosion, making them highly resilient in the face of climate and environmental stresses.

Stress Mitigation in Agroforestry:

1. **Microclimate Modification**: Trees in agroforestry systems create shaded areas that help regulate temperature and humidity levels, protecting fruit crops from extreme heat and reducing the risk of sunburn on sensitive fruits like coffee berries and cocoa pods. Shade-providing trees also reduce soil temperature fluctuations, supporting root health and promoting moisture retention, which is crucial in drought-prone areas.
2. **Enhanced Soil Structure and Fertility**: Trees in agroforestry systems contribute organic matter through leaf litter, which decomposes to enrich soil fertility. This organic matter improves soil structure, increases water-holding capacity, and supports beneficial microbial activity, mitigating nutrient stress. For example, integrating nitrogen-fixing trees like Acacia in avocado orchards improves nitrogen availability, supporting avocado growth without additional fertilizer inputs.
3. **Water Conservation and Erosion Control**: Agroforestry systems improve water infiltration and reduce runoff, which minimizes soil erosion and conserves water. In hilly areas prone

to heavy rainfall, terraced agroforestry systems, where fruit trees like apple or peach are intercropped with shrubs, grasses, or other trees, help retain soil and prevent erosion. This system enhances soil stability and water retention, supporting crop resilience during both wet and dry seasons.

4. **Biodiversity and Pest Management**: Agroforestry increases biodiversity by providing habitats for a wide range of organisms, which helps control pest populations naturally. Birds, beneficial insects, and other organisms within agroforestry ecosystems contribute to a balanced pest-predator dynamic, reducing the need for chemical interventions. In cacao production, agroforestry with shade trees has shown to reduce pest pressure and improve cocoa yield and quality.

Agroforestry provides a holistic approach to stress mitigation, enhancing ecosystem services while supporting high-quality fruit production. This system is particularly advantageous for organic fruit farmers and for cultivation in marginal lands, where resilience and biodiversity are critical for sustainable production.

b. Examination of How Diverse Cropping Systems Help Reduce Stress Impacts in Fruit Crop Production

Cropping systems are the foundation of sustainable agriculture, playing an essential role in managing and reducing environmental stress impacts. By selecting the appropriate cropping system, growers can address specific stresses—such as drought, heat, nutrient limitations, and pest and disease pressures—in a manner that promotes healthy plant growth, increases yield, and improves resilience in fruit crop production. Each cropping system—monocropping, intercropping, agroforestry, and polyculture—offers unique mechanisms for stress reduction, from enhancing soil moisture retention and nutrient cycling to creating a favorable microclimate that reduces heat and supports plant health.

1. Monocropping and Its Targeted Stress Management Approaches

Monocropping, the practice of cultivating a single crop over a large area, is widely used in commercial fruit production for its operational efficiency and ease of management. While monocropping may increase vulnerability to certain stressors, it offers the potential for targeted interventions to reduce specific stresses through precision agriculture, irrigation management, and focused nutrient strategies.

Soil Moisture Conservation in Monocropping Systems

In monocropping systems, where uniform crop characteristics allow for precise management, growers can optimize irrigation based on the crop's specific water needs. Soil moisture conservation in monocropping is achieved by using precision irrigation methods, such as drip irrigation, which delivers water directly to the root zone and minimizes water loss. In high-value crops like grapevines and apples, drip irrigation allows for effective soil moisture retention by reducing evaporation, which is especially beneficial in arid regions. Additionally, mulching is commonly used in monocropped fruit orchards, where a layer of organic or inorganic material is spread on the soil surface to reduce evaporation and improve soil moisture levels.

Temperature Regulation through Soil and Canopy Management

Monocropping allows for focused canopy and soil management practices that help mitigate temperature extremes. Canopy management in vineyards, for instance, involves training and pruning techniques that control leaf exposure to sunlight, reducing the risk of sunburn on grape clusters. The uniformity in a monocropped system also facilitates the use of shading nets during high-temperature periods, which helps prevent heat stress. Moreover, soil cover crops, such as grasses or legumes, can be planted between rows in monocropped orchards to provide ground cover, which helps reduce soil temperature and moisture loss.

Nutrient Management for Improved Availability

In monocropping systems, nutrient availability is managed with precision fertilization, which aligns with the uniform nutrient demands of the crop. The homogeneity of the crop allows for targeted nutrient application based on growth stages, reducing nutrient competition and ensuring that essential nutrients are available at critical periods. For example, in monocropped apple orchards, potassium and nitrogen are applied in precise amounts to support both vegetative growth and fruit development. Soil amendments, such as organic matter and compost, are also frequently used in monocropping to improve soil structure and nutrient retention, which supports root health and nutrient uptake.

2. Intercropping as a Strategy for Multi-Layered Stress Mitigation

Intercropping, which involves growing two or more crops in proximity, is a valuable strategy in fruit production for managing stressors related to water, nutrients, and pest pressures. Intercropping systems exploit the complementary characteristics of different crops to create a balanced environment, thereby enhancing resilience and reducing competition for resources.

Enhanced Soil Moisture Retention and Conservation

Intercropping helps conserve soil moisture by maximizing soil coverage, which reduces evaporation and improves water retention. Crops with different root structures are selected to access water from various soil depths, minimizing competition and ensuring efficient water use. In mango orchards, for instance, shallow-rooted legumes are intercropped with the deeper-rooted mango trees, which prevents water loss from the topsoil and helps retain soil moisture for longer periods. Additionally, legumes fix nitrogen in the soil, which benefits the mango trees by enhancing soil fertility without requiring extra nitrogen inputs.

Temperature Regulation through Canopy Structuring

The layering effect created in intercropping systems also contributes to a cooler microclimate by providing shade and reducing soil temperature. In arid regions, intercropping mango or citrus trees with smaller understory plants, like herbs or shade-loving vegetables, helps mitigate the impact of high temperatures on the soil. The shade created by the taller fruit trees reduces sun exposure for the understory crops, while the understory plants act as a ground cover that further insulates the soil, creating a more stable temperature environment. This structure reduces heat stress, enabling both crops to thrive under high temperatures.

Improved Nutrient Cycling and Reduced Competition

Intercropping fosters a symbiotic relationship where nutrient uptake is optimized for all plants involved. Nutrient cycling is enhanced as plants with different nutrient needs extract nutrients from different soil layers, reducing competition and enhancing nutrient availability. For instance, when intercropping citrus trees with nitrogen-fixing legumes, the legumes contribute to soil fertility by enriching nitrogen levels, which supports the nutrient needs of the citrus trees. In this way, intercropping improves nutrient availability while reducing the need for synthetic fertilizers, thereby supporting sustainable fruit production.

Natural Pest and Disease Management

Intercropping also serves as a natural pest deterrent by disrupting the monoculture landscape that often favors pest build-up. The diversity of plant species creates a habitat for beneficial insects and natural predators, which help control pest populations and reduce the need for chemical pesticides. For example, in strawberry fields intercropped with marigold, the marigolds repel nematodes and pests that typically affect strawberries. This polycultural system reduces the impact of biotic stress and supports healthier crops without the environmental burden of pesticides.

3. Agroforestry Systems for Integrated Stress Mitigation

Agroforestry combines trees with crops in a mutually beneficial system that enhances biodiversity and ecosystem resilience. In fruit production, agroforestry provides an array of stress mitigation benefits, such as improving microclimate conditions, enhancing soil health, and supporting pest control through natural habitat diversity.

Soil Moisture Conservation and Improved Water Use Efficiency

Trees in agroforestry systems play a critical role in maintaining soil moisture by reducing wind speed and direct sunlight exposure. This reduction in evapotranspiration helps conserve soil moisture, which is particularly beneficial for fruit crops grown in semi-arid and arid environments. In tropical regions, coconut trees are often intercropped with bananas, with the taller coconut trees providing shade and reducing water loss from the banana plants' soil. Furthermore, the deep root systems of trees help in water infiltration, improving soil structure and water availability for fruit crops during dry periods.

Temperature Regulation and Microclimate Improvement

Agroforestry significantly improves the microclimate by providing shade and reducing temperature extremes. The tree canopy acts as a natural shield against high temperatures, protecting understory crops from direct sun exposure. In coffee and cacao production, for example, agroforestry systems with shade-providing trees like Acacia and Inga reduce the risk of sunburn on fruits and improve humidity levels, which is ideal for crops that thrive in moderate temperatures. By maintaining a cooler environment, agroforestry systems prevent heat-induced stress and support optimal growth conditions.

Enhanced Soil Fertility and Nutrient Recycling

The integration of trees in agroforestry systems supports soil fertility by contributing organic matter through leaf litter, which decomposes and enriches the soil with essential nutrients. This process enhances nutrient availability, reducing the need for synthetic fertilizers and supporting sustainable fruit production. For example, nitrogen-fixing trees such as Leucaena in mango agroforestry systems improve soil nitrogen levels, supporting mango growth and fruit quality. Additionally, the root systems of trees bring up nutrients from deeper soil layers, making them accessible to shallower-rooted fruit crops.

Biodiversity and Natural Pest Management

Agroforestry increases biodiversity by creating a complex habitat that supports beneficial insects, birds, and other wildlife, which naturally control pest populations. This diversity reduces the risk of pest infestations and enhances resilience to biotic stresses. For instance, in cocoa agroforestry systems, shade trees provide habitat for predatory ants, which reduce pest populations on cocoa plants. This integrated pest management approach reduces dependency on chemical pesticides and promotes a balanced ecosystem that supports long-term productivity.

4. Polyculture Systems and Their Role in Stress Mitigation

Polyculture systems, which involve growing multiple crops within the same area, enhance resilience by promoting biodiversity and ecological interactions that support stress mitigation. In fruit production, polyculture offers advantages such as improved soil health, enhanced nutrient cycling, and natural pest control.

Soil Moisture Retention and Conservation

Polyculture systems maintain soil moisture by maximizing ground cover, which reduces water evaporation and enhances soil structure. The varied root architectures of different crops in a polyculture help capture water from multiple soil depths, ensuring efficient water use. For example, in apple orchards interplanted with shallow-rooted vegetables and deep-rooted legumes, each crop layer contributes to overall water retention and reduces the need for frequent irrigation. This efficient water use system is particularly beneficial in areas where water availability is limited.

Temperature Moderation through Crop Diversity

Polyculture systems contribute to temperature regulation by creating a multi-layered structure that mitigates the effects of extreme weather. In polycultures that combine tall fruit trees, medium-height shrubs, and ground-cover plants, each layer provides shade and reduces soil temperature. This stratified structure is beneficial in regions with high daytime temperatures, as it prevents heat stress in sensitive crops. Polyculture is commonly applied in tropical regions, where crops like banana, coffee, and pineapple are grown together to create a favorable microclimate that supports diverse plant growth.

Enhanced Nutrient Availability and Soil Health

Polyculture improves soil fertility by enhancing nutrient availability through complementary plant interactions. Different plants have distinct nutrient needs, reducing competition and allowing for more efficient nutrient uptake. For instance, intercropping citrus trees with nitrogen-fixing cover crops like clover provides a continuous supply of nitrogen to the soil, improving citrus growth and reducing the need for synthetic fertilizers. This nutrient recycling benefits long-term soil health, as diverse plants contribute various organic matter types, supporting a rich microbial ecosystem that facilitates nutrient cycling.

Natural Pest and Disease Management

Polyculture's biodiversity reduces pest pressure by attracting a variety of beneficial insects and natural predators. This ecological balance minimizes pest infestations, enhancing plant health and reducing the need for chemical interventions. For example, in blueberry fields intercropped with pest-repellent herbs, the herbs discourage harmful insects from attacking the blueberries, resulting in healthier crops with fewer disease outbreaks. The presence of diverse plant species disrupts pest cycles and supports a balanced ecosystem, reducing the reliance on pesticides and supporting sustainable fruit production.

c. Case Studies of Specific Fruit Cropping Systems for Stress Mitigation

Fruit cropping systems such as mango-based agroforestry and grape intercropping provide compelling examples of how tailored systems can reduce environmental stress impacts. These systems leverage strategic crop combinations, natural ecosystem processes, and optimal land use to improve resilience, conserve resources, and enhance productivity.

1. Mango-Based Agroforestry in Semi-Arid Regions

Mango agroforestry systems are prevalent in regions where water availability, soil fertility, and high temperatures present challenges to sustainable fruit production. By integrating mango trees with other compatible plants, such as legumes, timber trees, and medicinal plants, agroforestry systems enhance resource use efficiency, reduce stress impacts, and improve overall orchard productivity.

Location: Maharashtra, India

Overview and Objectives: In Maharashtra, a semi-arid state in India, mango-based agroforestry has been adopted as a strategy to address water scarcity and soil degradation. Farmers grow mango trees in combination with legumes, native timber species, and medicinal plants to create a multi-functional system that maximizes land productivity and supports diverse outputs.

Stress Mitigation Mechanisms:

» **Soil Moisture Conservation:** In mango agroforestry systems, soil moisture conservation is achieved through multiple layers of vegetation that reduce soil evaporation rates. The shade provided by the mango canopy cools the soil and reduces direct sunlight exposure, helping maintain moisture levels. Additionally, the intercropping of leguminous plants enhances ground cover, creating a layer of mulch that further prevents moisture loss and promotes better water retention in the root zone.

» **Temperature Regulation and Microclimate Modification:** Agroforestry systems create a favorable microclimate that reduces heat stress on the mango trees and intercrops. Shade trees like Acacia and Albizia are commonly planted alongside mango trees, providing a buffer against high temperatures. These shade trees lower ambient temperatures, which helps to protect mango blossoms and developing fruits from sunburn and heat-induced damage. This cooler microenvironment promotes optimal growth and development for mangoes, leading to higher yields and improved fruit quality.

» **Nutrient Cycling and Soil Fertility Improvement:** Leguminous intercrops like pigeon peas and chickpeas contribute to nitrogen fixation, which enhances soil fertility and supports the nutrient needs of mango trees. These legumes release organic matter into the soil upon decomposition, enriching the soil with nitrogen and other essential nutrients. Additionally, the use of deep-rooted native timber trees allows for nutrient recycling from deeper soil layers, providing mango trees with a steady supply of minerals that improve overall orchard health and resilience.

Outcomes and Benefits: The mango-based agroforestry system in Maharashtra has proven effective in increasing soil moisture retention, reducing water requirements, and improving the soil nutrient profile. By enhancing resilience to drought and heat stress, this system allows farmers to achieve more consistent yields and higher-quality fruits even under challenging conditions. Furthermore, the diversified outputs from legumes, timber, and medicinal plants contribute to economic stability, providing additional income sources for farmers and reducing reliance on a single crop.

2. Grape Intercropping in Mediterranean Climates

Grape intercropping has been widely implemented in Mediterranean regions, where hot, dry summers and limited water availability pose significant challenges. By integrating grapevines with cover crops or compatible plants, growers can enhance water-use efficiency, improve soil structure, and reduce weed pressure, which collectively mitigates environmental stresses.

Location: Tuscany, Italy

Overview and Objectives: In Tuscany, Italy, grape growers have adopted intercropping as a sustainable practice to manage soil health and water availability, particularly in vineyards where irrigation resources are scarce. Grapevines are intercropped with cover crops such as clover, vetch, and chicory to optimize soil moisture, enhance soil fertility, and support biodiversity in the vineyard ecosystem.

Stress Mitigation Mechanisms:

» **Enhanced Water-Use Efficiency and Soil Moisture Conservation:** In the intercropped vineyards, cover crops reduce soil evaporation by providing ground cover, which helps retain moisture in the soil. The selected cover crops have different root depths, allowing them to access water from various soil layers without competing with the grapevines for water. Clover and vetch, for instance, have shallow roots that utilize water from the topsoil, leaving subsoil moisture for the grapevines' deeper root systems. This stratification of water use contributes to overall water efficiency and reduces the need for supplemental irrigation.

» **Soil Structure Improvement and Nutrient Availability:** The cover crops used in grape intercropping systems improve soil structure by preventing compaction and enhancing aeration. As the roots of these cover crops penetrate the soil, they create channels that increase water infiltration and support root expansion. This enhanced soil structure facilitates better access to nutrients for grapevines, reducing nutrient stress. Additionally, nitrogen-fixing cover crops like clover contribute to soil fertility by naturally increasing nitrogen levels, supporting vine health and reducing the need for chemical fertilizers.

» **Temperature Moderation and Microclimate Enhancement:** In Mediterranean climates, the soil can reach high temperatures during the summer, which can stress grapevine roots and reduce overall grape quality. The presence of cover crops provides shade to the soil, lowering surface temperatures and reducing the risk of heat stress on the grapevines. This cooling effect creates a more stable microclimate that enhances vine resilience during peak summer heat, supporting consistent grape development and quality.

Outcomes and Benefits: The intercropped vineyards in Tuscany demonstrate improved soil moisture retention, enhanced nutrient cycling, and greater resilience to heat stress. This cropping system allows grape growers to reduce dependency on external inputs, such as water and fertilizers, and promotes healthier vineyard ecosystems through increased biodiversity. The use of intercropping in Tuscany's vineyards not only improves grape quality by supporting optimal growing conditions but also aligns with sustainable agriculture practices, helping growers reduce environmental impact and preserve the local ecosystem.

3. Citrus-Guava Intercropping System for Pest Management and Resource Optimization

Intercropping citrus with guava is practiced in regions prone to pest pressures and nutrient deficiencies, as the combination of these two fruit crops provides mutual benefits that enhance resilience and productivity. This intercropping system is particularly valuable in tropical and subtropical climates, where high humidity creates favorable conditions for pests.

Location: Punjab, Pakistan

Overview and Objectives: In Punjab, Pakistan, citrus-guava intercropping is used as a strategy to reduce pest populations, improve soil fertility, and optimize resource use. Farmers integrate guava trees with citrus trees to create a diverse ecosystem that supports natural pest control, reduces soil degradation, and enhances yield stability.

Stress Mitigation Mechanisms:

» **Natural Pest Control through Plant Diversity:** Guava is known to repel certain pests that affect citrus trees, particularly fruit flies. By intercropping guava with citrus, growers leverage the natural pest-repellent properties of guava, which reduces the need for chemical pesticides. This approach creates a more resilient cropping system that maintains pest control through natural means, preserving fruit quality and reducing the environmental impact associated with pesticide use.

» **Soil Fertility and Nutrient Optimization:** In the citrus-guava intercropping system, organic matter from guava leaf litter enriches the soil, enhancing nutrient availability for both crops. Additionally, the root systems of guava and citrus trees complement each other, with guava's root structure improving soil aeration and nutrient absorption. This compatibility allows both trees to access a balanced supply of nutrients, reducing nutrient competition and supporting healthy growth.

» **Enhanced Water-Use Efficiency:** The intercropping of citrus with guava contributes to improved water-use efficiency by maximizing canopy cover, which reduces soil evaporation. Guava trees provide partial shade to the citrus trees, which helps retain soil moisture, especially during dry periods. This mutual shading effect conserves water, enabling both crops to thrive with reduced irrigation requirements, particularly in water-scarce regions.

Outcomes and Benefits: The citrus-guava intercropping system in Punjab has proven effective in reducing pest pressures, improving soil fertility, and conserving water. By minimizing pesticide use and optimizing resource allocation, this system enhances resilience to environmental stresses, ensuring stable yields and high-quality fruits. The dual-crop output also provides farmers with diversified income sources, contributing to economic resilience in the region.

4. Apple-Alley Cropping with Nitrogen-Fixing Cover Crops for Nutrient Management

Alley cropping involves growing rows of fruit trees with spaces in between that are cultivated with nitrogen-fixing cover crops. This system is widely used in apple orchards to manage soil fertility, conserve water, and prevent soil erosion.

Location: Washington State, USA

Overview and Objectives: In Washington State, apple growers have adopted alley cropping with nitrogen-fixing legumes, such as clover and alfalfa, planted between rows of apple trees. This system addresses the need for enhanced soil fertility, water conservation, and erosion control in apple orchards, where nutrient demands are high and soil degradation is a potential risk.

Stress Mitigation Mechanisms:

» **Nutrient Enrichment and Soil Fertility Improvement:** The nitrogen-fixing cover crops in alley cropping systems enrich the soil with bioavailable nitrogen, reducing the need for synthetic fertilizers. As legumes fix nitrogen in the soil, they provide a continuous supply of nutrients that supports apple tree growth, particularly during high-demand periods like flowering and fruit set. The decaying cover crop residue also adds organic matter to the soil, enhancing its nutrient-holding capacity and promoting a balanced soil ecosystem.

» **Soil Moisture Conservation and Reduced Erosion:** The cover crops in alley cropping systems act as a living mulch, protecting the soil surface from direct sunlight and reducing moisture loss through evaporation. This ground cover also prevents soil erosion, which is critical on sloped terrains where apple orchards are often located. By retaining moisture and stabilizing the soil, alley cropping helps apple trees access water and nutrients more effectively, even during dry periods.

» **Temperature Moderation and Microclimate Regulation:** Alley cropping with cover crops creates a cooler microenvironment around apple trees, reducing heat stress on the soil and roots. The shading effect of cover crops minimizes soil temperature fluctuations, which supports root health and enhances apple tree resilience during hot summer months.

Outcomes and Benefits: The apple-alley cropping system in Washington State effectively manages nutrient demands, conserves water, and controls soil erosion. By integrating nitrogen-fixing crops, apple growers benefit from improved soil fertility and reduced fertilizer costs. This system also supports sustainable practices by reducing environmental impact, enhancing apple quality, and supporting resilient orchard ecosystems.

3. Assessing Stress Through Remote Sensing

Remote sensing technologies have become integral to modern agricultural practices, providing advanced tools to monitor and assess plant health and environmental stressors across large areas efficiently. Technologies like satellite imagery, drones, and multispectral/hyperspectral sensors allow growers to assess crop stress indicators by capturing data that may not be visible to the naked eye. These tools provide valuable insights into vegetation health, water stress, nutrient deficiencies, and disease impacts, enabling proactive management in fruit crop production. Here's a detailed look at how each remote sensing technology works and its applications in assessing stress in fruit crops.

a. Explanation of Remote Sensing Technologies for Stress Assessment

Remote sensing in agriculture relies on capturing data across various spectral bands to interpret plant health indicators. Key technologies used include satellite imagery, drones, and multispectral/hyperspectral sensors, each with specific applications that provide actionable insights into crop stress. Vegetation indices such as the Normalized Difference Vegetation Index (NDVI) and the Enhanced Vegetation Index (EVI) are widely used to quantify plant vigor and detect early signs of stress.

1. Satellite Imagery

Satellite imagery provides a broad and consistent view of agricultural landscapes, allowing for large-scale monitoring of crop health and stress indicators. Satellites like Landsat, Sentinel-2, and MODIS capture high-resolution images across various spectral bands, making them invaluable for tracking changes in vegetation over time.

Applications in Stress Assessment:

- **Monitoring Vegetation Health**: Satellite imagery enables the calculation of vegetation indices like NDVI and EVI, which provide insights into plant health based on light reflection data. NDVI, calculated using near-infrared (NIR) and red-light reflectance, measures chlorophyll content and photosynthetic activity. High NDVI values indicate healthy, photosynthetically active plants, while lower values may signal stress from drought, nutrient deficiencies, or disease.
- **Tracking Drought Stress**: Satellite imagery is particularly useful for identifying drought stress in fruit orchards. By comparing historical and current NDVI values, growers can detect areas with reduced vegetation cover or weakened photosynthetic activity, indicative of water stress. For instance, in apple and grape vineyards, a decline in NDVI values over time can highlight areas that need immediate irrigation intervention.
- **Assessing Soil Moisture and Crop Water Requirements**: Certain satellite platforms, such as Sentinel-1 and SMAP (Soil Moisture Active Passive), can estimate soil moisture content, providing indirect measurements of drought stress. This information helps growers plan water use more efficiently, particularly in water-scarce regions where precise irrigation management is crucial for maintaining fruit quality.

Advantages and Limitations: Satellite imagery covers extensive areas and provides consistent data over time, making it ideal for long-term monitoring and regional analysis. However, cloud cover can interfere with data acquisition, and the resolution may not capture detailed variations within smaller orchards, limiting its use for site-specific management.

2. Drones (Unmanned Aerial Vehicles)

Drones offer greater flexibility and higher spatial resolution than satellites, making them ideal for targeted monitoring of smaller or more varied agricultural areas. Equipped with multispectral, thermal, and visible light cameras, drones can capture real-time data across various spectral bands, enabling highly detailed stress assessments.

Applications in Stress Assessment:

» **High-Resolution Vegetation Indices**: Drones equipped with multispectral sensors provide high-resolution NDVI and EVI maps that reveal precise variations in plant health across individual rows and plots. In vineyards, for example, drones can identify specific grapevines under stress, allowing for targeted irrigation, pruning, or nutrient applications. This precision helps maintain uniform crop health, improving both yield and quality.

» **Detecting Water Stress and Canopy Temperature**: Thermal cameras on drones detect temperature variations across the canopy, identifying areas where plants may be experiencing water stress. Higher canopy temperatures suggest reduced transpiration rates, often due to stomatal closure in response to drought. By analyzing these temperature maps, growers can identify areas that require water intervention, conserving resources by applying water only where it's needed.

» **Pest and Disease Detection**: Drones can identify signs of pest infestations or disease by detecting subtle changes in vegetation indices. Early detection is critical in managing diseases like powdery mildew in grapevines or fungal infections in strawberries, where rapid intervention can prevent extensive crop loss. By capturing high-resolution images, drones allow for quick identification of stressed areas, enabling timely treatment that minimizes damage.

Advantages and Limitations: Drones provide detailed, on-demand data, allowing growers to conduct frequent monitoring and obtain high-resolution images that highlight specific stress areas. However, drones are limited by flight range, battery life, and regulatory restrictions, which may make frequent or large-scale monitoring challenging.

3. Multispectral and Hyperspectral Sensors

Multispectral and hyperspectral sensors are essential for capturing detailed spectral data across multiple wavelengths, offering a fine-tuned approach to detecting specific types of plant stress. Multispectral sensors capture a few specific bands (e.g., red, green, blue, and NIR), while hyperspectral sensors measure hundreds of narrow bands across the electromagnetic spectrum. These sensors can be mounted on drones, satellites, or even handheld devices, making them versatile tools in stress assessment.

Applications in Stress Assessment:

» **Detailed Nutrient and Water Deficiency Mapping**: Hyperspectral sensors are particularly useful for detecting subtle changes in plant physiology caused by nutrient deficiencies or water stress. By analyzing spectral signatures across multiple wavelengths, hyperspectral sensors can identify nutrient stress symptoms like chlorosis (yellowing due to nitrogen deficiency) or anthocyanin accumulation (a common response to phosphorus deficiency). In fruit crops like citrus and blueberries, early detection of nutrient stress through hyperspectral imaging enables growers to apply corrective measures before fruit quality is compromised.

» **Identification of Disease and Pest Stress**: Hyperspectral data can detect biochemical changes associated with pest or disease attacks, as infected plants reflect light differently across certain wavelengths. For instance, in apple orchards, hyperspectral sensors can identify early signs of apple scab by detecting biochemical stress responses, allowing for timely intervention. This level of precision is especially beneficial for organic orchards, where pest and disease control relies on early detection to reduce the need for chemical treatments.

» **Enhanced Vegetation Indices**: Besides NDVI and EVI, hyperspectral sensors enable the calculation of advanced indices that provide specific insights into plant stress. For example, the Photochemical Reflectance Index (PRI) detects changes in photosynthetic activity related to stress, while the Water Index (WI) helps assess water content in plant tissues. These indices are particularly valuable for crops sensitive to water stress, such as avocados and mangoes, where maintaining optimal water levels is critical to fruit quality.

Advantages and Limitations: Multispectral and hyperspectral sensors offer detailed information about plant physiology, enabling the detection of subtle stress indicators. However, hyperspectral sensors are costly and generate large amounts of data that require advanced processing, which may limit accessibility for some growers.

Key Vegetation Indices for Stress Detection

Vegetation indices derived from remote sensing data provide simplified metrics that indicate plant health and stress. NDVI and EVI are two widely used indices, but other indices like the PRI and WI also contribute valuable insights for monitoring crop stress.

» **Normalized Difference Vegetation Index (NDVI)**: NDVI is a widely used index that measures the difference between NIR and red light reflectance to estimate plant vigor. Healthy, unstressed plants reflect more NIR light and less red light, resulting in higher NDVI values. Low NDVI values indicate stress, such as drought or nutrient deficiency, making NDVI an effective early warning indicator for growers.

» **Enhanced Vegetation Index (EVI)**: EVI improves on NDVI by incorporating additional corrections for atmospheric effects and canopy background signals, making it particularly useful in areas with dense canopies. EVI is beneficial in monitoring stress in crops with thick foliage, such as citrus orchards, where accurate vegetation assessment is essential for managing nutrient and water needs.

» **Photochemical Reflectance Index (PRI)**: PRI detects changes in the xanthophyll pigment cycle, which is related to photosynthetic efficiency and stress. Lower PRI values indicate reduced photosynthetic activity, often caused by heat or drought stress. In fruit crops like apples and grapes, monitoring PRI helps assess the immediate physiological effects of environmental stressors, enabling more responsive management.

» **Water Index (WI)**: WI uses NIR reflectance to estimate water content in plant tissues, providing direct insights into water stress levels. This index is especially valuable in assessing drought stress, as lower WI values suggest reduced water content, prompting targeted irrigation. For crops like mango and avocado, where water availability significantly impacts fruit quality, WI offers a precise metric for water management.

b. Description of How Remote Sensing Supports Early Detection of Drought, Pest, and Disease Stress

Remote sensing has revolutionized agricultural management by enabling early detection of stress factors such as drought, pest infestations, and diseases. By analyzing physiological markers like canopy temperature, chlorophyll content, and other key indicators, remote sensing provides growers with real-time data that allows for timely intervention, which is critical in managing stress in fruit crops.

1. Early Detection of Drought Stress

Drought stress significantly impacts fruit crop productivity by reducing water availability and affecting plant physiological functions such as photosynthesis, transpiration, and nutrient uptake. Remote sensing technologies allow for early detection of drought stress by monitoring indicators like canopy temperature and water content, providing actionable insights for efficient water management.

Canopy Temperature as an Indicator of Drought Stress

One of the primary indicators of drought stress detectable via remote sensing is canopy temperature. When fruit crops are water-stressed, their stomata tend to close to conserve water, reducing transpiration. This leads to an increase in canopy temperature because transpiration is a cooling process that lowers leaf temperature.

- **Thermal Imaging with Drones and Satellites**: Drones and satellites equipped with thermal sensors can measure the surface temperature of crop canopies. By mapping temperature variations across an orchard, thermal imaging helps identify areas experiencing higher temperatures, which likely correspond to zones under drought stress. High-resolution thermal imagery from drones provides detailed insights, enabling growers to address water deficiencies in specific areas rather than irrigating the entire field.
- **Canopy Temperature Depression (CTD)**: CTD is a drought stress index derived from canopy temperature data, measuring the difference between air temperature and canopy temperature. Lower CTD values indicate a plant's ability to cool itself through transpiration, signaling adequate water availability. Conversely, a decrease in CTD values signifies reduced transpiration and thus potential drought stress, allowing for targeted irrigation in response to stress signals.

Water Content Measurement and Vegetation Indices

Remote sensing also detects water stress by estimating the water content within plant tissues using specific vegetation indices and wavelengths.

- **Normalized Difference Water Index (NDWI)** and **Water Index (WI)**: NDWI and WI are calculated using NIR and shortwave infrared (SWIR) wavelengths to assess the relative water content in plant tissues. Lower NDWI and WI values indicate reduced water content, signaling early stages of drought stress. Monitoring these indices over time allows growers to detect water deficiencies before visible symptoms appear, enabling them to implement corrective measures, such as supplemental irrigation or mulching, to conserve soil moisture.

- **NDVI (Normalized Difference Vegetation Index)**: NDVI, although primarily a vegetation health index, also reflects drought stress indirectly. Drought-stressed plants generally have lower NDVI values due to reduced chlorophyll and photosynthetic activity, allowing growers to assess overall canopy health and address stress at an early stage.

By identifying changes in canopy temperature and water indices, remote sensing allows for efficient water management, particularly valuable in water-scarce regions where drought stress is a common challenge in fruit production.

2. Early Detection of Pest Stress

Pests pose a significant threat to fruit crops, as infestations can rapidly spread and damage large portions of an orchard if left unmanaged. Remote sensing technologies enable early pest detection by capturing subtle physiological and biochemical changes in plants, often before visible signs of damage appear.

Vegetation Indices for Pest Detection

When plants experience pest stress, they exhibit physiological changes that affect their light absorption and reflectance patterns, which remote sensing can capture.

- **Chlorophyll Fluorescence (CF)**: Chlorophyll fluorescence is a sensitive indicator of photosynthetic efficiency, which can decline under pest stress. Remote sensing technologies measure CF to detect reductions in chlorophyll activity associated with pest infestations. For instance, in apple orchards, pests like aphids and leaf miners affect photosynthesis, causing reductions in CF. Early detection of declining CF values allows growers to investigate and address pest presence before extensive damage occurs.
- **Photochemical Reflectance Index (PRI)**: PRI detects changes in photosynthetic pigments linked to stress responses. Pest-infested plants often produce stress-induced pigments like anthocyanins, which change leaf reflectance properties. Monitoring PRI helps detect areas with lower reflectance, indicating possible pest stress. This index is particularly useful in detecting early stages of pest damage in crops like strawberries and grapes, where prompt intervention is crucial to prevent yield losses.

High-Resolution Imaging for Precise Detection

Drones equipped with high-resolution multispectral or hyperspectral cameras provide detailed imagery that captures early signs of pest stress across specific sections of an orchard.

- **Multispectral and Hyperspectral Imaging**: Pests like spider mites or thrips cause leaf discoloration and chlorosis that are detectable through multispectral or hyperspectral sensors. These sensors capture narrow spectral bands, allowing for the identification of changes in leaf pigment and structure indicative of pest damage. By analyzing specific wavelengths, growers can detect infestations in their initial stages, facilitating immediate response to limit pest spread.
- **Thermal Imaging to Detect Pest-Induced Temperature Changes**: Some pests cause

changes in plant transpiration rates and canopy temperature. For example, a pest infestation that damages plant tissues can reduce transpiration, leading to increased leaf temperatures. Thermal imaging can detect these temperature differences, enabling growers to locate areas with unusual heat patterns indicative of pest activity.

Through vegetation indices and high-resolution imaging, remote sensing supports proactive pest management, allowing for targeted pesticide application, biological control, or mechanical removal of pests in specific affected zones.

3. Early Detection of Disease Stress

Diseases are another major stress factor in fruit crops, as pathogens can spread rapidly and severely affect yield and fruit quality. Remote sensing facilitates early disease detection by capturing physiological changes such as chlorophyll reduction, leaf discoloration, and structural alterations in plant tissues.

Chlorophyll Content as an Indicator of Disease Stress

Diseased plants often exhibit reduced chlorophyll content, leading to lower photosynthetic rates. Remote sensing detects these changes through specific indices that reflect variations in chlorophyll and other pigments.

» **Normalized Difference Vegetation Index (NDVI)**: NDVI is highly effective for detecting early-stage disease stress. When a disease affects chlorophyll content, NDVI values decline, reflecting reduced photosynthetic activity. For instance, in citrus orchards affected by bacterial infections or fungal pathogens, a decrease in NDVI may indicate disease presence even before symptoms become visible. This enables early intervention, such as pruning infected branches or applying targeted treatments, to contain disease spread.

» **Disease-Specific Spectral Indices**: Certain diseases produce unique spectral signatures that can be detected through hyperspectral imaging. Hyperspectral sensors capture hundreds of spectral bands, allowing for the identification of biochemical markers specific to certain pathogens. In vineyards, hyperspectral imaging can detect early signs of powdery mildew by identifying spectral patterns associated with fungal infections, helping growers respond promptly and prevent extensive crop loss.

Structural and Thermal Changes Associated with Disease Stress

Remote sensing technologies detect structural changes and thermal variations in plants that are associated with disease progression.

» **Thermal Imaging for Disease Detection**: Some diseases, such as root infections, affect water uptake and lead to reduced transpiration. This results in canopy temperature increases, detectable through thermal imaging. In avocado orchards, thermal imaging can identify root rot by highlighting areas with elevated leaf temperatures due to impaired water transport. Early detection of such heat patterns allows for immediate action, including soil treatment or the application of fungicides, to minimize the disease's impact.

» **High-Resolution Imagery for Symptom Identification**: Visible symptoms like leaf necrosis, spotting, and wilting, which are indicative of disease stress, can be detected early using high-resolution imaging from drones. By flying at low altitudes, drones equipped with RGB or multispectral cameras capture detailed images that reveal subtle symptoms before they spread extensively. For example, in apple orchards affected by fire blight, drones can detect initial signs of leaf and twig wilting, prompting timely pruning to control the disease.

Physiological Markers and Reflectance Changes

Diseased plants often undergo biochemical changes that affect reflectance in certain wavelengths, which remote sensing can detect.

» **Carotenoid Reflectance Index (CRI)**: The CRI detects changes in carotenoid pigments, which are often produced in higher concentrations in diseased plants as part of the stress response. This index is useful for identifying bacterial and fungal infections in crops like grapes and strawberries, where early detection helps prevent yield loss and minimizes the spread of pathogens within the crop.

» **Leaf Area Index (LAI)**: Diseases that affect leaf health, such as leaf spot or blight, reduce the overall leaf area available for photosynthesis. LAI measures the leaf area and is often reduced in diseased plants. Monitoring LAI through remote sensing provides early warnings of disease impact, enabling growers to adopt control measures such as fungicide applications in affected zones to preserve crop yield and quality.

c. Data Integration from Remote Sensing with Crop Models for Real-Time Stress Monitoring

The integration of remote sensing data with crop models is transforming agricultural management by enabling real-time stress monitoring, timely intervention, and improved yield forecasting. This synergy allows growers to monitor environmental conditions and plant responses dynamically, helping them make informed decisions to mitigate stress factors that could impact crop productivity. By combining the detailed spatial and temporal insights from remote sensing with the predictive capabilities of crop models, this integrated approach enhances the precision of interventions and optimizes resource use in fruit crop production.

The Role of Remote Sensing Data in Crop Models

Remote sensing technologies—satellite imagery, drone surveillance, and multispectral/hyperspectral sensors—capture real-time data across various spatial and spectral dimensions. This data provides information on key indicators of crop health and environmental stress, including soil moisture, canopy temperature, chlorophyll content, and vegetation indices like NDVI and EVI. By feeding these insights into crop models, growers gain a comprehensive view of how current environmental conditions affect crop growth stages, nutrient needs, and potential yield outcomes.

1. Data Types and Their Relevance in Crop Modeling

» **Soil Moisture Data**: Satellite platforms, such as Sentinel-1 and SMAP (Soil Moisture Active Passive), provide accurate soil moisture readings that inform crop models on the availability of water in the soil. Crop models like AquaCrop and DSSAT incorporate this data to simulate drought stress effects, guiding irrigation schedules and water conservation strategies.

» **Temperature Data**: Thermal sensors on satellites or drones provide canopy temperature data, which helps in assessing plant transpiration and detecting water stress. When integrated into crop models like APSIM, temperature data enables simulations of heat stress impacts on various growth stages, predicting potential yield losses or shifts in fruit quality.

» **Vegetation Indices (NDVI, EVI)**: Vegetation indices derived from multispectral data reflect plant health and vigor. NDVI, for example, indicates chlorophyll content and photosynthetic activity, which are essential for predicting growth rates. Crop models use NDVI values to estimate biomass accumulation and yield potential, allowing for continuous adjustment of growth predictions based on real-time vegetation health.

How Remote Sensing and Crop Models Work Together for Stress Monitoring?

Integrating remote sensing with crop models enhances the accuracy and efficiency of stress monitoring in fruit crops. Real-time data from remote sensing provides crop models with current conditions, enabling these models to reflect real-world dynamics and produce accurate forecasts.

1. Real-Time Stress Detection and Monitoring

One of the main advantages of integrating remote sensing with crop models is the ability to detect and monitor stress factors like drought, nutrient deficiencies, and heat stress in real-time.

» **Drought Stress Monitoring**: Real-time soil moisture and temperature data from remote sensing, when incorporated into models like AquaCrop, allow for precise drought stress simulation. By continuously tracking changes in soil moisture, these models adjust water requirements and irrigation recommendations, ensuring that crops receive the optimal amount of water based on current conditions. In drought-prone regions, this integration supports efficient water management by signaling when to implement water-saving strategies, such as deficit irrigation.

» **Nutrient Deficiency Detection**: Multispectral and hyperspectral sensors detect chlorophyll content and nutrient deficiencies early on. When this data is integrated into crop models, the models can simulate how nutrient stress impacts growth and yield. For example, if NDVI data indicates a nitrogen deficiency, the crop model can adjust growth predictions and suggest supplemental fertilization at critical growth stages, minimizing nutrient stress impacts.

» **Heat Stress Prediction**: Heat stress data from canopy temperature measurements can be fed into models like DSSAT, which simulate the effects of high temperatures on physiological processes like flowering and fruit set. By modeling the impact of projected heat waves on yield, growers can adjust practices, such as increasing irrigation or using shading nets, to mitigate heat stress.

2. Dynamic Adjustment of Growth Projections and Interventions

Integrated remote sensing and crop modeling allow for dynamic adjustments to crop growth projections, enabling growers to respond rapidly to changing conditions.

» **Optimized Irrigation Scheduling**: Soil moisture data from remote sensing, combined with model simulations, allows for real-time irrigation scheduling. For example, in regions with limited water availability, AquaCrop uses soil moisture inputs to suggest irrigation volumes and frequencies that reduce waste and prevent water stress. This precise irrigation management is particularly beneficial in water-intensive fruit crops like citrus and grapes, where over- or under-watering can directly impact yield and quality.

» **Targeted Nutrient Applications**: Remote sensing detects areas with nutrient deficiencies by monitoring chlorophyll content and vegetation indices, which crop models use to simulate nutrient needs and allocate resources. Models like APSIM integrate this data to adjust nutrient application recommendations, guiding growers to apply fertilizers in precise amounts and locations. This reduces costs and environmental impact while ensuring that plants receive essential nutrients for optimal growth.

» **Pest and Disease Forecasting**: Hyperspectral data integrated into crop models can help forecast pest and disease outbreaks by identifying stress markers indicative of biotic threats. For instance, changes in NDVI or PRI (Photochemical Reflectance Index) values may suggest pest activity or early disease symptoms. Crop models can then simulate the potential spread of infestations and recommend intervention strategies, such as targeted pesticide application or biological control methods, to contain outbreaks effectively.

Enhancing Yield Forecasting Through Data Integration

Yield forecasting is one of the most valuable applications of remote sensing and crop model integration, allowing growers to predict harvest quantities and quality accurately.

1. Biomass and Yield Estimation Using Vegetation Indices

Vegetation indices like NDVI and EVI provide estimates of biomass, which is closely related to yield potential in fruit crops. By continuously monitoring these indices throughout the growing season and integrating the data into crop models, yield forecasts become more accurate and responsive to environmental changes.

» **Seasonal Yield Projections**: In crops like apples and peaches, integrating NDVI data with DSSAT enables early yield projections based on current plant health. If NDVI values are lower than expected, the model can adjust growth and yield predictions, allowing growers to plan for potential yield reductions and adjust market expectations accordingly.

» **Biomass Accumulation for Fruit Quality Assessment**: In fruit crops like grapes, biomass is an indicator of sugar content and fruit size, which directly affect quality and market value. By integrating NDVI data into models like APSIM, growers can monitor biomass accumulation and estimate fruit quality attributes, enabling them to make timely decisions regarding irrigation, fertilization, and harvest timing.

2. Incorporating Environmental Projections for Long-Term Planning

Remote sensing data and crop models also contribute to long-term yield forecasting by incorporating environmental projections and simulating future stress impacts.

» **Climate Change Adaptation**: By integrating temperature, precipitation, and soil moisture projections, crop models can simulate potential yield outcomes under future climate scenarios. In grape production, where temperature directly affects flavor profiles, growers can use this data to evaluate different planting dates, select heat-tolerant varieties, or alter canopy management practices to maintain quality under warming conditions.

» **Risk Assessment and Financial Planning**: Integrated yield forecasting allows growers to assess the risks of environmental stress on profitability. For example, by simulating drought impacts on avocado or mango yield, growers can determine the potential revenue loss and consider investments in water-saving technology, insurance, or alternative revenue sources.

Benefits and Challenges of Data Integration for Fruit Crop Production

Integrating remote sensing data with crop models enhances the ability of fruit growers to manage environmental stresses and optimize crop health. However, it also presents challenges that must be managed for effective implementation.

Benefits of Data Integration

» **Precision Agriculture and Resource Efficiency**: The integration of real-time data allows for site-specific management, reducing waste and improving the efficient use of water, fertilizers, and pesticides. Precision agriculture practices improve crop health while reducing input costs and environmental impacts.

» **Increased Resilience and Yield Stability**: By enabling early stress detection and timely interventions, integrated systems enhance crop resilience to unpredictable weather and stress events. This increases yield stability and minimizes the risk of significant yield losses due to extreme conditions.

» **Informed Decision-Making for Sustainable Practices**: Integrated remote sensing and crop models provide a comprehensive view of plant health and environmental conditions, allowing growers to make informed decisions that support long-term sustainability. Adjustments to practices like water management, soil health, and nutrient cycling contribute to a more resilient and sustainable fruit production system.

Challenges and Limitations

» **Data Processing and Interpretation**: Remote sensing data generates large volumes of information, requiring advanced software and expertise to process and interpret. Farmers may need access to specialized tools and training to utilize these integrated systems effectively.

» **Cost and Accessibility**: The cost of acquiring high-resolution imagery, hyperspectral sensors, and crop modeling software can be prohibitive for small-scale farmers, limiting the accessibility of integrated solutions.

» **Environmental and Sensor Limitations**: Satellite-based data collection can be impacted by cloud cover, while drone flights may be restricted by weather conditions or regulatory constraints. These limitations can disrupt the regular monitoring necessary for real-time management.

Chapter - 4

Adaptive Features of Fruit Crops for Survival Under Stress

Abiotic stress represents one of the most significant challenges in agricultural production, particularly for fruit crops that are highly sensitive to environmental fluctuations. Unlike annual crops, perennial fruit species have prolonged exposure to environmental conditions, making their resilience to abiotic stresses crucial for sustained productivity. Abiotic stresses, including drought, salinity, extreme temperatures, and nutrient deficiencies, affect multiple physiological and biochemical pathways, often triggering a cascade of responses that aim to preserve survival over productivity. In regions where climate variability is pronounced, these stresses increasingly impact both the yield and quality of fruit crops, underscoring the need for adaptive strategies in fruit science.

In horticultural science, abiotic stress adaptation refers to the mechanisms by which plants adjust to adverse environmental conditions without permanent damage. Unlike acclimatization, which is a short-term, reversible adjustment, adaptation encompasses a complex interplay of genetic, biochemical, and physiological traits that have evolved to allow plants to survive in challenging habitats. The diversity of adaptive features in fruit crops reflects both their genetic makeup and their evolutionary responses to specific environmental pressures. Understanding these adaptive traits provides valuable insights for the development of resilient crop varieties capable of maintaining productivity under stress conditions.

The physiological and morphological adaptations in fruit crops under abiotic stress serve as the first line of defence, enabling plants to cope with water, nutrient, and energy limitations. For instance, drought stress typically induces morphological changes like reduced leaf area and enhanced root development, allowing crops to minimize water loss and access deeper soil moisture. Similarly, high salinity in the soil environment drives ion regulation mechanisms, such as selective ion absorption and compartmentalization, which help maintain ionic balance within cells. These responses are often complemented by biochemical mechanisms, including the accumulation of osmoprotectants like proline and soluble sugars, which stabilize cellular structures under osmotic stress. The combination of these morphological and biochemical traits enables fruit crops to sustain growth and physiological function despite fluctuations in water availability and soil salinity.

Temperature extremes, whether high or low, impose additional stress by disrupting enzymatic activities, destabilizing cell membranes, and altering metabolic pathways. For example, high temperatures lead to the synthesis of heat shock proteins (HSPs), which protect cellular structures by refolding denatured proteins and preventing aggregation. Low temperatures, on the other hand, trigger

DOI: 10.1201/9781003717409-4

cryoprotective mechanisms, including increased membrane fluidity and the synthesis of specific proteins that protect against ice formation within cells. These temperature-induced adaptations are essential for fruit crops grown in regions with distinct seasonal variations, as they help mitigate the impact of temperature fluctuations on fruit set and development. In crops like grapes and apples, for instance, these adaptive responses allow flowering, fruiting, and maturation to proceed even under variable thermal conditions, safeguarding both yield and quality.

A critical aspect of abiotic stress adaptation is the interaction between multiple stresses, as these rarely occur in isolation. Drought and heat stress often coincide, creating a synergistic effect that intensifies the plant's response. Combined drought and heat stress typically results in accelerated water loss, heightened oxidative stress, and impaired photosynthesis, all of which can reduce fruit set and lead to premature fruit drop. Similarly, drought conditions can exacerbate salinity stress by increasing soil salt concentration through reduced leaching, particularly in arid and semi-arid regions. These compounded stressors prompt a more robust adaptive response, as plants activate multiple signaling pathways to counteract the cumulative effects. Such interactions highlight the complexity of abiotic stress adaptation, where crops must balance between competing physiological demands to survive and reproduce.

In fruit crops, the adaptive response to abiotic stress has profound implications for growth, yield, and fruit quality. At the cellular level, stress responses often involve an increase in reactive oxygen species (ROS), which, if unchecked, can lead to cellular damage. To mitigate oxidative stress, fruit crops rely on an intricate antioxidant defence system, comprising enzymatic antioxidants like superoxide dismutase (SOD) and catalase (CAT), as well as non-enzymatic compounds such as ascorbate and glutathione. This antioxidant system not only protects cellular structures but also maintains the integrity of photosynthetic processes, crucial for biomass accumulation and fruit development. However, the activation of these defence mechanisms often requires a trade-off, where resources that would normally support growth and reproduction are redirected toward survival processes. As a result, stress conditions frequently lead to reduced growth rates, smaller fruit size, and altered chemical composition, impacting marketability and consumer acceptance.

The impact of abiotic stress on fruit quality is particularly significant in crops where flavor, texture, and nutrient content are essential for market value. Drought, for instance, may enhance sugar concentration in fruits like grapes by reducing water content, thereby intensifying flavor. Conversely, salinity stress can lead to an increase in acidity, which may be detrimental to crops like citrus, where a delicate balance between sweetness and acidity is crucial. Temperature stress similarly influences fruit quality parameters, affecting aspects like color development, texture, and the synthesis of bioactive compounds. Consequently, abiotic stress threatens yield and affects the post-harvest characteristics that determine fruit quality and shelf life, with broader implications for the food supply chain.

Given the complexity of abiotic stress adaptation, there is a growing interest in developing stress-resilient fruit crop varieties through advanced breeding and biotechnological interventions. Traditional breeding efforts have focused on selecting varieties with inherent tolerance traits, such as drought-resistant rootstocks in citrus or salt-tolerant grape genotypes. With recent advances in molecular biology, it is now possible to identify and manipulate specific genes involved in stress responses, paving the way for more targeted approaches. For example, gene editing technologies like

CRISPR/Cas9 offer unprecedented precision in modifying genes associated with osmotic adjustment, ion homeostasis, and antioxidant capacity. Integrating these biotechnological innovations with traditional breeding practices could significantly enhance the resilience of fruit crops, ensuring stable yields and quality under diverse environmental conditions.

1: Adaptive Features for Drought Stress Tolerance

1.1 Morphological Adaptations

Under drought conditions, fruit crops undergo various morphological changes to survive prolonged water scarcity and optimize their water use. Drought stress presents a severe challenge, especially in arid and semi-arid regions where water availability is limited and irregular. The morphological adaptations in fruit crops, including leaf modifications, root system alterations, and stomatal regulation, play a significant role in enabling these plants to thrive in environments where water is a limiting factor. By analyzing each of these adaptations, we can better understand how fruit crops manage water resources, maintain physiological stability, and support growth and productivity during drought stress.

Leaf Modifications

Leaves are essential for photosynthesis, gas exchange, and transpiration, but under drought conditions, these processes become critical areas where plants lose significant amounts of water. To counteract the effects of drought, many fruit crops adapt by altering leaf morphology, which includes a variety of structural modifications aimed at reducing water loss.

1. **Reduced Leaf Area:**
   - The first response to water stress in many fruit crops is to reduce the overall leaf area. A smaller leaf area minimizes the surface exposed to sunlight, which in turn reduces water loss through transpiration. Fruit crops like grapes and olives exhibit this adaptation by either producing smaller leaves or reducing the number of leaves during drought conditions. By curbing leaf growth, the plant conserves water while still allowing enough leaf area for basic photosynthesis.
   - In addition to reducing the number of leaves, some plants may shed older or damaged leaves to focus resources on maintaining the health of younger leaves. Leaf shedding is a strategic approach, as it minimizes the plant's energy requirements and reduces water loss, thus enabling the plant to maintain productivity despite drought stress.
2. **Thickened Leaf Cuticle:**
   - The leaf cuticle, a waxy layer that covers the surface of leaves, acts as a barrier to reduce water evaporation. During drought, many fruit crops like mango and citrus respond by thickening this cuticle layer. A thicker cuticle provides added resistance to water loss, especially under intense sunlight, where dehydration risks are higher.
   - The thickened cuticle also offers additional protection against pathogens and environmental damage, which can become a secondary issue during drought when plants are already weakened. The thicker cuticle helps fruit crops retain their structural integrity, promoting resilience under stress conditions.

- Furthermore, cuticle thickening is often accompanied by an increase in specific wax compounds that further enhance the barrier function. These waxes can reflect sunlight, lowering the leaf's surface temperature, which reduces water loss through transpiration. This adaptation is especially evident in crops grown in regions with high solar radiation.

3. **Trichomes (Hair-like Structures):**

    - Trichomes, tiny hair-like structures that cover the leaf surface, are another effective adaptation seen in drought-tolerant fruit crops. These structures create a microenvironment on the leaf surface by trapping moisture and reducing wind speed near the stomata, which helps lower the rate of transpiration.
    - For example, fig trees develop dense trichomes that act as a protective layer against excessive heat and sunlight, both of which can increase water loss. Trichomes also aid in reflecting sunlight, which prevents leaf tissues from overheating and further mitigates water loss.
    - The presence of trichomes can vary in density depending on the severity of drought. Under extreme water scarcity, certain plants may increase trichome density, thereby creating an insulating layer that minimizes water evaporation.

4. **Waxy Layers**

    - Some fruit crops enhance their drought tolerance by developing an additional waxy coating over the leaf surface. This layer provides an added barrier to water loss by further reducing transpiration. In olive trees, for instance, the waxy layer reflects solar radiation and reduces the surface temperature of leaves, which is critical in minimizing water loss through evaporation.
    - The waxy layer also acts as a defense against environmental pollutants and pathogens, which can exacerbate plant stress during drought. By providing an impermeable layer, the waxy coating helps retain moisture and promotes leaf longevity in arid environments.

Root System Alterations

The root system is fundamental to a plant's ability to access water, and under drought stress, fruit crops exhibit several key root adaptations to maximize water uptake from available resources. These modifications in root structure and growth patterns enhance the plant's drought resilience by enabling it to reach moisture deep within the soil.

1. **Increased Root Depth**

    - Deep rooting is a prominent adaptation in drought-resistant fruit crops, allowing them to access water from deeper soil layers that remain moist even during surface droughts. Crops like pomegranate and olive are known for their extensive root systems that can reach several meters below the soil surface, tapping into groundwater reserves and ensuring continued water availability.
    - This deeper rooting system is particularly beneficial in regions where surface irrigation is limited or inconsistent. By accessing water stored at greater depths, these crops can

maintain cellular hydration, enabling essential physiological processes to continue despite prolonged drought.

2. **Root Density and Thickness**
   - Under drought conditions, many fruit crops increase root density and thickness, particularly in the upper soil layers where moisture may be available after sporadic rainfall. Thickened, fibrous roots improve water absorption efficiency by increasing the contact surface area with soil moisture.
   - Additionally, a denser root system near the soil surface allows the plant to absorb water immediately after rain, maximizing the benefit of limited water resources. For instance, fig trees exhibit an increase in root density and thickness during water-limiting conditions, which enhances their drought resilience.
3. **Hydrotropism**
   - Hydrotropism is the growth of roots toward moisture, an essential adaptive feature that becomes more pronounced during drought. This response is mediated by the root tips, which sense moisture gradients in the soil and adjust root growth accordingly.
   - In fruit crops like mango and citrus, roots actively grow towards moisture pockets, maximizing water uptake. This ability to redirect root growth in response to soil moisture availability allows the plant to optimize water acquisition in challenging environments.
4. **Mycorrhizal Associations**
   - Mycorrhizal fungi form symbiotic relationships with the roots of many drought-resistant fruit crops, enhancing water absorption capabilities. These fungi develop extensive hyphal networks that extend beyond the immediate root zone, reaching water and nutrients located farther from the root surface.
   - In crops such as grapes and citrus, mycorrhizal associations play a crucial role in maintaining water uptake during drought, as the fungal hyphae improve access to soil moisture and nutrients, thus supporting the plant's overall resilience under water stress.

Stomatal Regulation

Stomatal regulation is a critical aspect of drought adaptation in fruit crops, as stomata control both gas exchange and water loss through transpiration. Morphological adjustments in stomatal structure and behavior enable plants to conserve water effectively under drought stress.

1. **Reduced Stomatal Density**
   - During prolonged drought, many fruit crops adapt by reducing stomatal density (the number of stomata per unit leaf area). A lower stomatal density decreases the total area through which water can escape, effectively limiting transpiration rates.
   - This adaptation is common in crops like olives and grapes, where fewer stomata contribute to improved water-use efficiency. By minimizing transpiration, these crops are better equipped to withstand periods of low water availability while still maintaining basic gas exchange for photosynthesis.

2. **Smaller Stomatal Size**
   - Another adaptation observed in drought-tolerant fruit crops is a reduction in stomatal size. Smaller stomata allow for more precise control over the rate of water loss, as the smaller openings limit transpiration more effectively than larger ones.
   - Crops such as mango and citrus exhibit this adaptation, where smaller stomata enable them to conserve water without severely impacting carbon dioxide intake, thereby supporting photosynthesis even under restricted water conditions.
3. **Stomatal Closure**
   - Stomatal closure is a rapid response to drought stress, mediated by plant hormones, particularly abscisic acid (ABA). When water availability is limited, ABA levels increase, signaling the guard cells surrounding each stomatal pore to close, thus preventing water loss.
   - This adaptation is seen in crops like figs and pomegranates, where efficient stomatal closure conserves water while maintaining critical physiological functions. Stomatal closure is balanced with the need for gas exchange, allowing the plant to endure dry conditions without compromising survival.
4. **Temporal Stomatal Opening**
   - Some drought-tolerant fruit crops adopt a strategy known as temporal stomatal opening, in which stomata open primarily during cooler parts of the day, such as early morning or late afternoon, when transpiration rates are lower. This adaptation reduces water loss by avoiding midday heat, which accelerates evaporation.
   - Grapevines are known for this behavior, as they regulate stomatal opening to coincide with lower temperatures, thereby optimizing water use efficiency. By limiting daytime water loss, grapevines and similar crops manage to maintain hydration and resilience during prolonged drought.

1.2 Physiological Mechanisms

Under drought stress, fruit crops activate a range of physiological mechanisms that help them survive, preserve water, and mitigate damage under challenging conditions. These mechanisms, including osmotic adjustment, antioxidant defense systems, and hormonal responses, work at the cellular and molecular levels to sustain cellular structure and function despite limited water availability.

Osmotic Adjustment

Osmotic adjustment is a primary physiological adaptation in plants, allowing cells to retain water by balancing osmotic pressure during drought conditions. This process involves the synthesis and accumulation of osmolytes, which are small, organic molecules that help stabilize cell structures without disrupting metabolic activities. The osmotic adjustment process plays a crucial role in maintaining cellular hydration, preserving cell turgor, and ensuring metabolic continuity during water scarcity.

1. **Accumulation of Proline**
   - **Proline's Multifaceted Role**: Proline, an amino acid, is one of the most frequently observed osmolytes in drought-stressed plants. Proline accumulation helps in various ways: it acts as an osmoprotectant, stabilizing cellular proteins and membranes, and also serves as a compatible solute that helps retain water within cells. This amino acid contributes to the protection of cell structures, particularly enzymes and membranes, from dehydration-induced damage, ensuring that essential cellular functions can continue despite water scarcity.
   - **Energy Source During Recovery**: Beyond its role as an osmoprotectant, proline also functions as an energy and nitrogen source during recovery from drought. Once water availability is restored, plants can break down stored proline to fuel metabolic processes, allowing them to quickly regain normal function. Studies in crops like grapevine and citrus reveal that proline levels rise significantly under drought, which not only supports immediate drought tolerance but also provides resources for recovery post-stress.
   - **Stress Signaling Role**: Recent research also highlights proline's role in stress signaling pathways. Proline acts as a signaling molecule, activating genes associated with drought tolerance. This signaling function enables the plant to coordinate various stress responses, optimizing its physiological processes under challenging conditions. This capacity for both protective and signaling functions underscores the importance of proline accumulation in drought resilience.
2. **Glycine Betaine Accumulation**
   - **Maintaining Turgor and Enzyme Stability**: Glycine betaine is another key osmolyte synthesized by plants during drought conditions. It supports osmotic balance by accumulating within cells to offset the osmotic potential created by water loss, thus preserving cell turgor. Glycine betaine stabilizes enzymes and maintains protein functionality under osmotic stress, which is critical for cellular metabolism in drought-tolerant fruit crops.
   - **Enhanced Photosynthetic Efficiency**: In fruit crops like pomegranate and grapevine, glycine betaine not only helps with osmotic balance but also protects the photosynthetic apparatus, particularly the photosystem II (PSII) complex. By preserving chloroplast structure and photosynthetic function, glycine betaine contributes to continued energy production even under water stress. This allows crops to sustain essential metabolic activities, supporting their resilience against drought.
   - **Membrane Protection**: Glycine betaine plays a significant role in preserving membrane integrity under drought. It stabilizes phospholipids in the cell membrane, preventing leakage and maintaining ion homeostasis within cells. Membrane protection is essential for cell viability during drought, as it minimizes electrolyte leakage, which can otherwise lead to cell death.
3. **Soluble Sugars**
   - **Osmotic Balancing and Signaling**: Soluble sugars such as glucose, fructose, and sucrose

accumulate in response to drought, helping maintain osmotic balance. These sugars act as osmolytes, preserving cell turgor and preventing cellular collapse. Additionally, soluble sugars serve as signaling molecules, activating stress-related pathways and enhancing drought tolerance. In fruit crops like fig and mango, sugar accumulation helps sustain cellular stability under prolonged drought conditions.

- **Role as Energy Reserves**: During drought, soluble sugars also act as energy reserves, supporting metabolic functions when other resources are limited. Upon the return of favorable conditions, these sugars provide the necessary energy for recovery and growth. This dual role as both osmolytes and energy reserves enhances the plant's adaptability, allowing it to endure and recover effectively from drought.
- **Stabilizing Proteins and Membranes**: Soluble sugars are known to interact with proteins and cell membranes, stabilizing them during dehydration. This stabilization protects cellular structures from damage, particularly in extreme drought scenarios where cellular integrity might otherwise be compromised. By preserving the structure and function of proteins and membranes, soluble sugars play a critical role in cellular drought resilience.

Antioxidant Defense System

Drought stress often results in an overproduction of reactive oxygen species (ROS), which can lead to oxidative damage to cellular components. To manage this oxidative stress, drought-tolerant crops activate an extensive antioxidant defense system, composed of enzymatic and non-enzymatic antioxidants. These compounds detoxify ROS, thereby safeguarding cellular structures and ensuring the continuity of metabolic functions.

1. **Enzymatic Antioxidants**
    - **Superoxide Dismutase (SOD)**: SOD is the first line of defense in the antioxidant pathway, converting superoxide radicals (O_2^-) into hydrogen peroxide (H_2O_2), a less harmful molecule. Increased SOD activity is observed in drought-tolerant fruit crops, especially in crops like citrus, where it mitigates oxidative damage. SOD is particularly crucial for maintaining chloroplast health, as it neutralizes ROS generated during photosynthesis, thus preserving photosynthetic efficiency under drought.
    - **Catalase (CAT)**: Catalase further detoxifies hydrogen peroxide by breaking it down into water and oxygen, preventing the buildup of toxic H_2O_2. CAT activity is vital in fruit crops such as mango, where it plays a key role in protecting cells from oxidative stress. This enzyme is found in high concentrations within peroxisomes, where ROS accumulation is most likely to occur during drought stress.
    - **Peroxidase (POD)**: Peroxidases, including ascorbate peroxidase (APX) and glutathione peroxidase (GPX), are essential for removing H_2O_2 from cells. These enzymes use reducing agents such as ascorbate and glutathione to convert H_2O_2 into water, maintaining redox balance and protecting cellular integrity. Enhanced POD activity has been observed in drought-tolerant grapevines, where it contributes significantly to drought resilience by mitigating oxidative damage.

2. **Non-Enzymatic Antioxidants**
   - **Ascorbate (Vitamin C)**: Ascorbate is a potent antioxidant that directly scavenges ROS and supports cellular defense under drought conditions. Ascorbate also serves as a substrate for ascorbate peroxidase, playing a dual role in the antioxidant system. Increased ascorbate levels in crops like olives provide protection against oxidative damage, particularly to chloroplasts, where ROS can severely impair photosynthesis.
   - **Glutathione (GSH)**: Glutathione, another key antioxidant, detoxifies ROS and regenerates ascorbate in the ascorbate-glutathione cycle, ensuring continuous protection against oxidative stress. High glutathione levels in drought-tolerant crops like figs contribute to maintaining cellular stability during drought, providing resilience against oxidative stress-induced damage.
   - **Phenolic Compounds**: Phenolics such as flavonoids and tannins serve as non-enzymatic antioxidants that scavenge ROS and chelate metal ions that catalyze oxidative reactions. During drought, phenolic compounds increase in fruit crops, enhancing the plant's defense against oxidative damage. This increase in phenolic content supports cellular health, contributing to drought resilience.

Hormonal Changes

Hormones act as central regulators in the plant's response to drought, coordinating physiological processes such as water conservation, root growth, and stress signaling. The balance of specific hormones, including abscisic acid (ABA), cytokinins, and ethylene, is modified under drought conditions, allowing the plant to adapt to and mitigate the effects of water scarcity.

1. **Abscisic Acid (ABA)**
   - **Stomatal Regulation and Water Conservation**: ABA is a primary hormone in drought response, increasing in concentration under water stress and inducing stomatal closure to reduce water loss. By signaling guard cells to close stomata, ABA minimizes transpiration, conserving water during drought. In fruit crops like mango, ABA-driven stomatal regulation plays a critical role in sustaining water use efficiency under limited water availability.
   - **Root Growth Enhancement**: Besides controlling stomatal behavior, ABA promotes root growth, particularly the development of lateral and fine roots that enhance water uptake. This dual role of ABA in both water conservation and root expansion is key to drought resilience in citrus and similar drought-tolerant crops.
   - **Signaling Drought Responses**: ABA acts as a signaling molecule that activates various drought-related genes, helping the plant adjust to drought conditions. This signaling function allows fruit crops to coordinate responses at multiple levels, optimizing their adaptation to water scarcity.
2. **Cytokinins**
   - **Resource Allocation for Root Development**: Cytokinins, responsible for promoting cell division and growth, decrease in shoots under drought but maintain higher levels

in roots, promoting root growth over shoot development. This redistribution prioritizes water acquisition and reduces water demand, allowing the plant to conserve resources during drought.

- **Interaction with ABA**: The balance between cytokinins and ABA is essential for drought tolerance, as ABA promotes root growth while reduced cytokinin levels limit shoot growth, supporting resource allocation towards water uptake.

3. **Ethylene**

- **Regulating Leaf Senescence**: Ethylene levels often increase during severe drought, leading to controlled leaf senescence and abscission, which reduces water demand. This response, observed in fig and other drought-tolerant crops, allows the plant to maintain hydration by sacrificing older leaves.
- **Supporting Stomatal Closure**: Ethylene works with ABA to reinforce stomatal closure, adding an extra layer of water conservation. This hormonal synergy is essential for maximizing drought resilience in crops, allowing them to retain water efficiently.

2. Adaptive Features for Salinity Stress Tolerance

2.1 Ion Homeostasis and Exclusion

Salinity stress is a significant abiotic stressor affecting fruit crop productivity, especially in areas with high soil salinity or in regions relying on saline water for irrigation. High levels of sodium (Na^+) and chloride (Cl^-) ions in the soil disrupt the ionic balance within plants, leading to ion toxicity, osmotic stress, and nutrient imbalances. Ion homeostasis and exclusion mechanisms are essential for maintaining cellular ionic equilibrium, enabling fruit crops to function under saline conditions. These mechanisms involve selective ion uptake, which prioritizes beneficial ions over harmful ones, and salt exclusion and compartmentalization strategies that mitigate ion toxicity at cellular and organ levels.

Selective Ion Uptake

Selective ion uptake is a primary mechanism by which salt-tolerant fruit crops maintain cellular functionality in saline environments. This process involves the preferential absorption of essential ions, such as potassium (K^+), over toxic ions like sodium (Na^+), thereby sustaining key physiological processes and avoiding ion toxicity.

1. **Importance of Potassium (K^+) in Ionic Balance**

- Potassium is a vital macronutrient that plays crucial roles in enzyme activation, osmotic regulation, and cellular homeostasis. Under salinity stress, high levels of Na^+ compete with K^+ for uptake sites due to their similar ionic radii and charge. This competition can result in K^+ deficiency, which impairs protein synthesis, disrupts enzymatic reactions, and reduces cellular osmotic potential, leading to stunted growth and reduced productivity.
- In salt-tolerant fruit crops, selective ion channels and transporters enhance K^+ uptake and retention, ensuring that K^+ remains at optimal levels for cellular function. For example, potassium transporters like HAK/KUP and AKT channels actively facilitate K^+ absorption, even under high Na^+ conditions. By maintaining adequate K^+ concentrations,

these crops support photosynthesis, protein synthesis, and cellular energy production, all of which are critical for growth and fruit quality.

2. **High K^+/Na^+ Ratio as an Indicator of Salt Tolerance**

- A high K^+/Na^+ ratio within plant tissues is widely recognized as an indicator of salt tolerance, particularly in fruit crops. Salt-tolerant varieties of grapes, dates, and olives exhibit mechanisms that prioritize K^+ uptake over Na^+, maintaining a favorable ionic balance conducive to metabolic stability.
- This high K^+/Na^+ ratio helps stabilize cell turgor, which is essential for water uptake and nutrient transport. Furthermore, potassium's role in guard cell regulation allows the plant to control stomatal opening efficiently, optimizing gas exchange and minimizing water loss under saline conditions. A balanced K^+/Na^+ ratio also supports the function of essential enzymes and maintains cellular redox balance, both of which are vital for photosynthetic efficiency and growth.

3. **Role of Selective Transporters in Ion Uptake**

- Selective ion transporters play a pivotal role in ensuring that K^+ is preferentially absorbed over Na^+. High-affinity potassium transporters (HKTs) and non-selective cation channels (NSCCs) regulate the movement of K^+ and Na^+ across cell membranes. HKTs, in particular, are known to selectively facilitate K^+ transport while limiting Na^+ influx into cells. These transporters are highly regulated and adjust their activity in response to changes in external salinity, allowing salt-tolerant crops to maintain ionic homeostasis.
- In addition, recent studies have shown that certain salt-tolerant crops express Na^+-specific transporters that pump Na^+ out of root cells, reducing the likelihood of Na^+ entering sensitive tissues. For instance, in dates and grapes, Na^+ transporters located in the root cells actively extrude Na^+ back into the soil, maintaining a high K^+/Na^+ ratio in the root zone and protecting above-ground tissues from ion toxicity.

4. **Calcium (Ca^{2+}) as a Modulator of Ion Selectivity**

- Calcium ions (Ca^{2+}) play an essential role in modulating selective ion uptake under saline conditions. High Na^+ levels disrupt cell membrane stability and function, but adequate Ca^{2+} levels help stabilize membranes and enhance selective K^+ uptake. Calcium also influences the activity of Na^+-selective transporters, promoting Na^+ exclusion from the cytoplasm and reducing its toxic effects on cellular metabolism.
- In fruit crops like pomegranate and citrus, Ca^{2+} supplementation has been shown to enhance salt tolerance by improving cell membrane integrity and boosting the selective uptake of K^+ over Na^+. This interaction underscores the importance of maintaining Ca^{2+} levels in saline conditions to support ion homeostasis and reduce the adverse effects of salinity stress.

Salt Exclusion and Compartmentalization

In addition to selective ion uptake, salt-tolerant crops employ salt exclusion and compartmentalization mechanisms to prevent the toxic accumulation of Na^+ and Cl^- ions within sensitive cellular compartments.

These strategies help isolate harmful ions, allowing the plant to function without interference from excessive Na^+ and Cl^- levels.

1. **Root-Level Salt Exclusion**
   - One of the primary defenses against salinity stress is the exclusion of Na^+ ions at the root level, preventing their entry into the plant's vascular system. Specialized transporters in the root cell plasma membranes actively pump Na^+ out of the cell, effectively blocking it from reaching sensitive tissues in the shoots and leaves.
   - In fruit crops like olive and grape, Na^+-selective transporters (SOS1 transporters) at the root level actively extrude Na^+ ions back into the soil, thereby reducing their accumulation in root cells. This exclusion process is crucial for maintaining ionic balance within the plant, as it protects the plant's photosynthetic tissues from Na^+ toxicity and prevents disruptions to cellular processes.
   - Furthermore, root exclusion mechanisms are often coupled with increased water uptake efficiency, allowing the plant to absorb water from saline soils without a corresponding increase in Na^+ and Cl^- levels. This dual strategy enhances salt tolerance in crops that are frequently exposed to saline environments, enabling them to achieve better growth and productivity under challenging conditions.
2. **Vacuolar Compartmentalization of Na^+ and Cl^-**
   - For Na^+ and Cl^- ions that do enter the root and are transported to the shoot, vacuolar compartmentalization serves as a critical defense mechanism. By sequestering these ions in vacuoles, plants prevent their accumulation in the cytoplasm, where they would interfere with metabolic functions and enzymatic activities.
   - In drought-tolerant fruit crops like dates and citrus, tonoplast (vacuolar membrane) transporters, such as NHX (Na^+/H^+ antiporters), facilitate the active transport of Na^+ into vacuoles. This compartmentalization minimizes the toxic effects of Na^+ on the cytosolic enzymes and metabolic pathways, allowing normal cellular activities to proceed despite the presence of high Na^+ levels in the surrounding environment.
   - Vacuolar sequestration not only reduces Na^+ toxicity but also contributes to osmotic balance within cells. By storing Na^+ in vacuoles, the plant can retain water within its cells, maintaining cell turgor and preventing dehydration under saline conditions. This ability to balance osmotic pressure through vacuolar compartmentalization is especially advantageous in maintaining water status and promoting growth in salt-stressed environments.
3. **Leaf Shedding as a Salt Avoidance Mechanism**
   - In some fruit crops, excessive accumulation of Na^+ and Cl^- in older leaves may lead to programmed leaf senescence, an adaptive response that reduces salt load within the plant. By shedding older, salt-accumulated leaves, the plant minimizes salt toxicity and conserves resources for younger, actively growing tissues.
   - This salt avoidance mechanism is common in crops like mango, where older leaves act as

a buffer, absorbing and sequestering salt ions to protect younger leaves and reproductive organs. Once the leaves accumulate high salt levels, they are shed, effectively reducing the salt burden and protecting the plant's overall health.

4. **Role of Compatible Solutes in Salt Compartmentalization:**

   - Compatible solutes such as proline and glycine betaine assist in maintaining cellular osmotic balance alongside salt compartmentalization strategies. These solutes accumulate in the cytoplasm to counteract the osmotic potential created by Na^+ and Cl^- sequestered within vacuoles.
   - In fruit crops like figs and pomegranates, compatible solutes stabilize proteins and cellular membranes, protecting them from salt-induced damage. These solutes also act as osmoprotectants, preserving enzyme activity and facilitating cellular function under high salinity. By balancing the osmotic pressure in the cytoplasm, compatible solutes work in tandem with vacuolar compartmentalization to maintain cellular stability and enhance salt tolerance.

2.2 Compatible Solute Accumulation

Under salinity stress, high concentrations of sodium (Na^+) and chloride (Cl^-) ions in the soil create osmotic challenges for plants, disrupting water absorption and leading to cellular dehydration. In response, many salt-tolerant fruit crops accumulate compatible solutes—small organic molecules such as proline, glycine betaine, and soluble sugars—that help regulate osmotic pressure without interfering with cellular processes. These solutes act as osmoprotectants, stabilizing proteins and cellular membranes, preserving enzyme functionality, and ensuring the plant's overall resilience under saline conditions.

Role of Proline in Osmotic Adjustment and Protection

Proline is one of the most commonly accumulated compatible solutes in plants under salt stress. As an amino acid, it not only helps maintain osmotic balance within cells but also protects cellular structures, supports enzyme stability, and aids in detoxifying reactive oxygen species (ROS) that accumulate under stress conditions.

1. **Osmotic Balance and Water Retention**

   - Proline accumulation is crucial for osmotic adjustment under salinity stress, as it helps maintain cell turgor by balancing the osmotic potential. This allows the plant to retain water within cells, minimizing dehydration and supporting essential physiological processes even when external salinity levels are high.
   - The high solubility of proline allows it to function effectively as an osmolyte, accumulating in the cytoplasm without disrupting cellular activities. By adjusting the internal osmotic pressure, proline facilitates water uptake despite the adverse osmotic conditions imposed by saline soil, which would otherwise limit water availability to the plant.

2. **Protection of Cellular Structures**

   - Beyond osmotic balance, proline acts as a protective molecule that stabilizes cellular structures, including proteins and membranes. It prevents protein denaturation by forming a protective layer around proteins, helping maintain their three-dimensional structure and functionality even under saline conditions.

- Proline's interaction with cellular membranes is also essential for maintaining membrane integrity. Under salt stress, cell membranes are prone to disruption due to ion imbalance and oxidative stress. Proline helps preserve membrane fluidity and stability, ensuring that cellular compartments remain intact and functional. This protection is vital for maintaining cellular homeostasis, particularly in fruit crops that rely on cellular compartmentalization to manage salt stress.

3. **Antioxidant Properties and ROS Detoxification**

   - Salinity stress often leads to an overproduction of ROS, which can damage lipids, proteins, and nucleic acids. Proline functions as an antioxidant, scavenging ROS and reducing oxidative damage in plant cells. It also works synergistically with other antioxidants, such as ascorbate and glutathione, to bolster the plant's defense against oxidative stress.
   - In addition to directly detoxifying ROS, proline activates genes associated with stress response pathways, enhancing the plant's ability to tolerate high-salinity conditions. This multifaceted role of proline—as an osmolyte, stabilizer, and antioxidant—makes it a critical component in the salinity tolerance of fruit crops like grapes, figs, and olives.

4. **Energy Reserve During Recovery**

   - Upon the alleviation of salt stress, accumulated proline can be broken down to release energy, supporting the plant's recovery and resumption of normal metabolic activities. The catabolism of proline provides energy, carbon, and nitrogen, which are essential for growth and regeneration of tissues damaged by salt stress. This recovery function highlights the adaptability of proline, as it not only supports stress tolerance but also contributes to post-stress rejuvenation in salt-tolerant fruit crops.

Role of Soluble Sugars in Stress Adaptation and Metabolic Stability

Soluble sugars, including glucose, fructose, and sucrose, also play a significant role in salinity stress tolerance by acting as osmolytes, stabilizers of cellular components, and signaling molecules that activate stress responses.

1. **Osmotic Regulation and Water Uptake**

   - Like proline, soluble sugars contribute to osmotic adjustment, allowing the plant to retain water and maintain turgor under saline conditions. By accumulating in the cytoplasm, these sugars balance osmotic pressure between the inside and outside of the cell, facilitating water uptake even in high-salinity environments.
   - The accumulation of soluble sugars helps counteract the osmotic potential created by Na^+ and Cl^- ions, effectively preventing water loss from the cells. This function is particularly beneficial for fruit crops grown in arid and saline regions, as it helps sustain cellular hydration and supports growth under challenging conditions.

2. **Stabilization of Proteins and Membranes**

   - Soluble sugars have a stabilizing effect on proteins and membranes, protecting them from the denaturing effects of high salinity. These sugars interact with proteins, preserving their structure and functionality, which is essential for maintaining metabolic processes under stress.
   - Additionally, sugars stabilize membrane lipids, preventing the leakage of ions and maintaining ion gradients across membranes. This membrane stabilization is crucial for maintaining the functionality of cellular organelles, such as chloroplasts and mitochondria, which are highly sensitive to salinity-induced oxidative stress. In salt-tolerant crops like citrus and pomegranate, the accumulation of soluble sugars helps preserve photosynthetic efficiency and supports cellular respiration, even under salinity stress.

3. **Role as Energy Reserves and Metabolic Support**

   - Soluble sugars also act as an energy source, fueling cellular activities during stress recovery. Upon the reduction of salt stress, these sugars provide the necessary energy for repairing damaged tissues and resuming normal growth. This reserve function is essential for the recovery phase, as it allows the plant to rebuild and restore its physiological balance after prolonged exposure to saline conditions.
   - Moreover, soluble sugars support metabolic stability by serving as substrates for respiration. Under salt stress, respiratory efficiency may decline due to ion imbalances and oxidative stress, but the availability of sugars helps sustain cellular energy production. This metabolic support is critical for salt-tolerant fruit crops, as it enhances their ability to withstand and recover from prolonged saline exposure.

4. **Signaling Role in Stress Response**

   - In addition to their osmotic and stabilizing functions, soluble sugars act as signaling molecules that trigger stress-response pathways. Sugar signaling pathways activate genes involved in antioxidant defense, osmotic regulation, and protein protection, enhancing the plant's overall tolerance to salinity.
   - For example, high sugar levels can signal the activation of antioxidant enzymes, which detoxify ROS generated during salt stress. This signaling function enables fruit crops to respond proactively to salinity stress, adjusting their physiological processes to optimize survival. In salt-tolerant species like olives and dates, sugar signaling plays a vital role in coordinating stress responses and promoting resilience.

Role of Glycine Betaine in Ion Homeostasis and Cellular Stability

Glycine betaine is another prominent compatible solute that accumulates in plants under salinity stress. It plays a key role in osmotic adjustment, cellular protection, and ion homeostasis, particularly in salt-tolerant fruit crops.

1. **Osmotic Adjustment and Water Retention**

   - Glycine betaine functions as an effective osmolyte, helping to balance osmotic pressure and retain water within cells. Its high solubility and stability make it suitable for maintaining cellular hydration without interfering with enzymatic processes. Glycine betaine accumulation is especially beneficial for crops grown in saline soils, as it helps prevent dehydration and supports physiological stability.

- In fruit crops like pomegranate and date palm, glycine betaine accumulation under salinity stress enables these plants to withstand osmotic stress, preserving cell turgor and sustaining metabolic activities even when external water availability is limited.

2. **Stabilization of Enzymes and Protein Structures**

   - Glycine betaine stabilizes enzymes by protecting their structure and function under high-salinity conditions. It forms a protective shell around proteins, preventing them from denaturing due to ion imbalance and osmotic stress. This stabilization is essential for maintaining enzyme activity, especially for metabolic enzymes involved in photosynthesis, respiration, and nutrient assimilation.
   - Furthermore, glycine betaine interacts with cellular membranes, preserving their fluidity and stability. By maintaining membrane integrity, glycine betaine ensures that cellular compartments remain functional, supporting nutrient transport, ion balance, and cellular communication. This role is critical in salt-tolerant crops where cellular stability under saline conditions is necessary for growth and productivity.

3. **Support of Ion Homeostasis**

   - Glycine betaine contributes to ion homeostasis by modulating the movement and compartmentalization of ions. It supports the sequestration of Na^+ and Cl^- ions in vacuoles, preventing their toxic accumulation in the cytoplasm. By promoting ion compartmentalization, glycine betaine minimizes ion toxicity and reduces the detrimental effects of salinity on cellular processes.
   - This role in ion homeostasis is especially important for fruit crops that rely on efficient ion management to survive in saline environments. Glycine betaine's ability to regulate ion distribution within cells allows crops like figs and citrus to maintain metabolic functionality and prevent salt-induced cellular damage.

2.3 Anatomical Adaptations

In addition to physiological and biochemical mechanisms, anatomical adaptations play a crucial role in enhancing salinity tolerance in fruit crops. Salinity-tolerant plants often exhibit structural modifications that reduce water loss, protect cells from ion toxicity, and improve overall resilience to saline conditions. Key anatomical changes include thicker cell walls, a well-developed cuticle, and specialized xylem tissues, all of which contribute to the plant's ability to survive and thrive in high-salinity environments.

Thicker Cell Walls

The cell wall is the outermost structure of plant cells and serves as a protective barrier against environmental stressors. In salinity-tolerant fruit crops, cell walls often become thicker as an adaptation to saline conditions. Thicker cell walls confer several benefits that contribute to the plant's survival in saline environments.

1. **Increased Structural Integrity**
   - A thicker cell wall provides enhanced structural support to plant cells, enabling them to withstand the osmotic stress associated with saline soils. When external salt concentrations are high, water tends to move out of the cells due to osmotic pressure differences. Thicker cell walls help prevent excessive cell shrinkage and maintain cell shape, supporting turgor pressure and preventing wilting.
   - The increased rigidity provided by thicker cell walls allows cells to maintain their structural integrity under the mechanical stresses induced by salinity. This structural adaptation is especially important in fruit crops such as olives and dates, which often face prolonged exposure to saline soils and need to maintain cellular stability for optimal growth and fruiting.
2. **Barrier Against Ion Intrusion**
   - Thicker cell walls act as a physical barrier that limits the movement of Na^+ and Cl^- ions into the cell. By reducing ion permeability, the cell wall prevents excessive accumulation of toxic ions within the cytoplasm, thereby protecting cellular components from ion-induced damage.
   - The composition of the cell wall also plays a role in its effectiveness as a barrier. Salinity-tolerant crops often modify cell wall components, such as lignin and cellulose, to increase resistance to ion penetration. For instance, higher lignin content strengthens the cell wall and makes it more impermeable to harmful ions. This modification helps fruit crops maintain ion homeostasis, which is essential for metabolic function under saline conditions.
3. **Facilitating Ion Sequestration**
   - Thickened cell walls in saline environments may also assist in ion sequestration by trapping Na^+ and Cl^- ions in the apoplast (the cell wall and intercellular spaces), preventing their entry into the cytoplasm. This sequestration limits ion toxicity and allows the plant to manage salt stress more effectively.
   - In crops like grapes and pomegranates, where salt tolerance is critical for productivity, the apoplastic trapping of ions provides an additional line of defense against salt-induced cellular damage. By isolating ions within the apoplastic space, plants can maintain higher K^+/Na^+ ratios in the cytoplasm, which is crucial for cellular processes like enzyme activity and photosynthesis.

Well-Developed Cuticle

The cuticle is a waxy layer that covers the epidermis of leaves and stems, acting as a protective barrier against water loss and environmental stressors. In salt-tolerant fruit crops, the cuticle becomes thicker and more developed, helping to conserve water and reduce the impact of salinity.

1. **Reduction of Transpirational Water Loss**
   - A well-developed cuticle plays a key role in minimizing water loss through transpiration.

In saline environments, where water uptake is limited due to osmotic stress, reducing water loss is essential for maintaining plant hydration. The cuticle acts as a waterproof barrier, preventing excessive evaporation from leaf surfaces and conserving moisture within the plant.

- In crops like olives and citrus, the thicker cuticle is a critical adaptation for water conservation. By minimizing water loss, these plants can retain hydration in saline soils, supporting physiological processes and reducing the risk of dehydration.

2. **Protection Against Ion Penetration**

   - A thicker cuticle also serves as a barrier to salt ions, particularly Na^+ and Cl^-, preventing their penetration into leaf tissues. In saline environments, these ions can accumulate on the leaf surface, potentially entering the plant through stomata or epidermal tissues. A well-developed cuticle limits ion infiltration, reducing the likelihood of ion toxicity and protecting cellular structures.
   - The waxy composition of the cuticle further enhances its effectiveness as a barrier. Salinity-tolerant crops often produce additional cuticular waxes under saline conditions, which not only protect against water loss but also block ion entry. For example, grapevines grown in saline environments exhibit an increased wax deposition on leaves, providing a dual function of water retention and ion exclusion.

3. **Heat and UV Protection**

   - Salinity stress often coexists with other environmental stresses, such as high temperatures and intense sunlight, which can exacerbate water loss and cellular damage. A well-developed cuticle helps shield the plant from excessive heat and ultraviolet (UV) radiation, both of which can compound the effects of salt stress.
   - The cuticle's reflective properties reduce leaf surface temperature, preventing overheating and reducing transpirational demand. This additional protection is particularly advantageous for fruit crops in arid, saline regions where high temperatures are common. By mitigating heat stress, the cuticle supports the plant's ability to cope with multiple stressors simultaneously, enhancing overall resilience.

Modified Xylem Tissues

The xylem is responsible for transporting water and minerals from roots to shoots, and in salt-tolerant crops, xylem tissue undergoes modifications to improve water-use efficiency and reduce the adverse effects of salinity. These adaptations help maintain efficient water transport even when soil salinity disrupts the plant's ability to absorb and utilize water.

1. **Reduction in Xylem Vessel Diameter**

   - Salinity-tolerant fruit crops often exhibit narrower xylem vessels as an adaptation to saline conditions. Smaller vessel diameter reduces the risk of cavitation—air bubble formation within the xylem—which can interrupt water transport. Under saline stress, plants face increased risk of cavitation due to osmotic stress and water scarcity.

- By developing narrower vessels, crops like dates and olives can maintain continuous water flow even in saline soils. This modification enhances the plant's drought resilience and ensures a steady supply of water to leaves, supporting photosynthesis and transpiration balance under challenging conditions.

2. **Increased Xylem Wall Thickness**
   - The thickening of xylem vessel walls is another anatomical adaptation that improves resistance to cavitation and ion intrusion. Thicker vessel walls provide structural stability, reducing the risk of xylem collapse and maintaining water transport capacity under high salt concentrations.
   - In fruit crops like figs and citrus, where water transport is crucial for fruit development, thicker xylem walls help sustain water flow to the leaves and fruits. This adaptation ensures that essential nutrients and water reach the photosynthetic tissues, promoting growth and productivity despite the osmotic challenges posed by saline environments.

3. **Enhanced Xylem Parenchyma Cells**
   - Xylem parenchyma cells, which are responsible for storing water and nutrients within the xylem, are often more developed in salt-tolerant crops. These cells play a vital role in managing ion levels and facilitating osmotic adjustment within the vascular system, particularly under salinity stress.
   - In modified xylem tissues, parenchyma cells help regulate the movement of Na^+ and Cl^- ions, minimizing their accumulation in the xylem sap and preventing transport to sensitive tissues. This selective transport and storage capacity reduces the harmful effects of salinity on water and nutrient flow, supporting the overall health of the plant. Enhanced xylem parenchyma cells contribute to improved ion management in crops like grapes and pomegranates, where ion balance is essential for salt tolerance and fruit quality.

4. **Improved Water-Use Efficiency**
   - Modified xylem tissues contribute to improved water-use efficiency by optimizing water uptake and reducing wastage under saline conditions. This adaptation is vital for fruit crops in saline environments, as it allows them to maximize the benefits of limited water resources.
   - In combination with other anatomical adaptations, modified xylem tissues enable fruit crops to balance water transport with ion management, supporting growth and resilience under high-salinity stress.

3. Adaptive Features for Temperature Stress Tolerance

3.1 High-Temperature Adaptations

High temperatures pose a significant threat to fruit crops, affecting their growth, development, and productivity. Elevated temperatures accelerate metabolic processes, leading to cellular stress, oxidative damage, and, ultimately, reduced crop yields. To survive and thrive under such thermal stress, fruit crops have evolved several adaptive features that enable them to maintain cellular integrity, protect key metabolic processes, and optimize physiological functions. Key adaptations to high-temperature stress include the synthesis of heat shock proteins (HSPs), modifications in lipid composition to maintain membrane fluidity, and alterations in photosynthetic pathways to reduce thermal damage.

Heat Shock Proteins (HSPs)

Heat shock proteins (HSPs) are a group of proteins that play a vital role in protecting cells from temperature-induced stress. They function as molecular chaperones, assisting in protein folding, preventing protein aggregation, and aiding in the repair or degradation of damaged proteins. Under high-temperature stress, the upregulation of HSPs is a critical survival mechanism in fruit crops, helping them maintain cellular stability and protect essential metabolic functions.

1. **Role of HSPs in Protein Folding and Stabilization**
   - High temperatures can cause proteins to denature and lose their functional structure, leading to cellular dysfunction. HSPs stabilize proteins under heat stress by assisting in proper protein folding and preventing aggregation. This function is particularly important for enzymes and structural proteins that are essential for cellular metabolism and integrity.
   - HSPs recognize unfolded or misfolded proteins, binding to them and preventing aggregation, which is irreversible and potentially harmful. By facilitating the refolding of these proteins, HSPs ensure that critical cellular processes continue uninterrupted. This protection is crucial for heat-sensitive crops like grapes and citrus, where metabolic stability directly impacts fruit quality and yield.
2. **Types of Heat Shock Proteins**
   - HSPs are classified into several families based on their molecular weight, including HSP100, HSP90, HSP70, HSP60, and small HSPs (sHSPs). Each family plays a specific role in protecting cells from thermal stress. For example, HSP70 and sHSPs are particularly important in fruit crops as they respond rapidly to temperature increases, stabilizing essential proteins and cellular structures.
   - **HSP70**: This family of proteins is known for its role in protecting cells from heat stress by binding to newly synthesized or denatured proteins, facilitating their proper folding. HSP70 proteins are upregulated during heat stress and play a vital role in thermal tolerance by maintaining the functional conformation of proteins in crops like apples and grapes.
   - **sHSPs**: Small heat shock proteins are smaller in size but highly effective in stabilizing proteins under heat stress. They form complexes with unfolded proteins, preventing irreversible aggregation and facilitating subsequent refolding by other chaperone proteins. The rapid induction of sHSPs in response to heat stress makes them particularly valuable for crops that experience sudden temperature spikes.
3. **Regulation of HSP Expression**
   - The expression of HSPs is tightly regulated by heat shock transcription factors (HSFs). Under normal conditions, HSFs remain inactive, but high temperatures trigger their activation. Once activated, HSFs bind to heat shock elements (HSEs) in the promoter regions of HSP genes, initiating their transcription.

- This rapid response mechanism allows fruit crops to quickly produce HSPs upon exposure to high temperatures. In crops like mango and apricot, the swift upregulation of HSPs through HSF activation enables these plants to counteract heat-induced damage, preserving cellular functions and improving survival rates during heatwaves.

4. **Enhanced Thermotolerance through HSP Accumulation**

   - The accumulation of HSPs confers both immediate and long-term thermotolerance in fruit crops. In grapevines, for instance, high HSP levels during heatwaves help stabilize chloroplasts and other heat-sensitive organelles, protecting photosynthesis and preventing a decline in fruit quality. HSPs also facilitate cellular recovery following heat stress, ensuring that damaged proteins are repaired or degraded as needed.
   - This thermotolerance mechanism is not only crucial for protecting yield and quality but also enables fruit crops to cope with recurrent or prolonged heat stress. By maintaining protein stability and cellular integrity, HSPs allow crops to survive high temperatures without significant metabolic disruption.

Lipid Composition Adjustments

Cellular membranes are particularly vulnerable to high temperatures, which can increase membrane fluidity and lead to leakage, ion imbalances, and loss of cellular integrity. One of the key adaptations to high-temperature stress in fruit crops is the adjustment of lipid composition in cellular membranes, particularly by increasing the proportion of saturated fatty acids. This modification helps stabilize membranes, maintaining cellular fluidity and preventing thermal damage.

1. **Membrane Fluidity and Temperature Sensitivity**

   - Under high temperatures, membranes become more fluid, which can compromise their structural integrity and lead to increased permeability. This loss of integrity disrupts essential cellular functions, including ion balance, transport processes, and enzyme activities.
   - To counteract this, heat-tolerant crops increase the proportion of saturated fatty acids in their membrane lipids. Saturated fatty acids have no double bonds, which allows them to pack closely together, creating a more rigid membrane structure that is less sensitive to temperature fluctuations. This structural adjustment stabilizes the membrane, maintaining its integrity under heat stress.

2. **Role of Saturated Fatty Acids in Membrane Stability**

   - Saturated fatty acids, such as palmitic acid and stearic acid, reduce membrane fluidity by strengthening the hydrophobic interactions between lipid molecules. In heat-tolerant fruit crops like dates and figs, the incorporation of these fatty acids into the membrane bilayer reduces the risk of leakage and ensures that cellular processes continue uninterrupted.
   - In contrast, unsaturated fatty acids contain double bonds that introduce kinks in their structure, preventing tight packing and increasing fluidity. Under normal conditions, a balance of saturated and unsaturated fatty acids maintains membrane flexibility. However, under heat stress, a higher saturated-to-unsaturated fatty acid ratio helps stabilize the membrane and reduce the potential for thermal damage.

3. **Chloroplast Membrane Stability and Photosynthesis Protection**
   - The chloroplast membrane is particularly susceptible to heat-induced damage, as it houses the photosynthetic machinery. By modifying the lipid composition of chloroplast membranes, heat-tolerant fruit crops protect photosynthesis from thermal disruption.
   - Enhanced membrane stability prevents the leakage of essential ions and metabolites, preserving the photosynthetic capacity even under high temperatures. In crops like olives and citrus, this adaptation ensures that photosynthesis remains efficient, supporting growth and productivity during heat stress.
4. **Enhanced Expression of Desaturase Enzymes**
   - In addition to increasing the proportion of saturated fatty acids, some fruit crops regulate the activity of desaturase enzymes, which modify unsaturated fatty acids to optimize membrane fluidity under fluctuating temperatures. This flexibility in lipid composition enables crops to adapt to both high and low temperatures, promoting resilience in diverse environments.
   - By balancing desaturation and saturation processes, crops like pomegranate can fine-tune membrane fluidity, enhancing their ability to cope with sudden heat spikes or prolonged warm periods. This adaptive lipid modification strategy contributes to the overall thermal tolerance in these fruit crops.

Altered Photosynthetic Pathways

Photosynthesis is highly sensitive to temperature changes, as elevated temperatures can disrupt electron transport, reduce enzyme efficiency, and lead to photo-oxidative damage. High-temperature adaptations in photosynthetic pathways allow heat-tolerant fruit crops to optimize energy production, maintain chlorophyll stability, and minimize the adverse effects of thermal stress.

1. **Adjustments in Photorespiration**
   - Photorespiration is a process where oxygen competes with carbon dioxide at the enzyme ribulose-1,5-bisphosphate carboxylase/oxygenase (RuBisCO), leading to a reduction in photosynthetic efficiency. Under high temperatures, photorespiration rates increase, which can limit carbon fixation and reduce growth.
   - To counter this, heat-tolerant crops like grapes and mangoes may adjust their photorespiratory pathways to minimize energy loss and enhance carbon assimilation. This adaptation allows the plant to maintain efficient photosynthesis, even under high temperatures, by reducing the impact of photorespiration on energy production and carbon fixation.
2. **Increased Chlorophyll Stability**
   - High temperatures can degrade chlorophyll, impairing the plant's ability to capture light and perform photosynthesis. Heat-tolerant fruit crops protect chlorophyll molecules from thermal damage by upregulating chlorophyll-binding proteins and other protective factors that stabilize chlorophyll structure.

- Enhanced chlorophyll stability allows plants to maintain a high photosynthetic rate during heat stress, supporting growth and productivity. For example, in crops like citrus, which require consistent photosynthetic activity for fruit development, maintaining chlorophyll integrity is essential for withstanding prolonged exposure to high temperatures.

3. **Modifications in Chlorophyll Fluorescence and Light Utilization**

- Chlorophyll fluorescence, a measure of the light energy emitted by chlorophyll molecules, is often used as an indicator of photosynthetic efficiency under stress. Heat-tolerant crops adjust their chlorophyll fluorescence, optimizing light absorption and minimizing excess energy that could lead to photo-oxidative damage.
- By modifying fluorescence and light utilization pathways, crops like figs and apricots reduce the risk of photoinhibition—a process where excessive light energy damages the photosynthetic apparatus. This adjustment enables these plants to use available light more efficiently, preventing oxidative stress and promoting resilience under high-temperature conditions.

4. **Enhanced Activity of Antioxidant Enzymes in Chloroplasts**

- High temperatures increase the production of reactive oxygen species (ROS) within chloroplasts, leading to oxidative damage. Heat-tolerant fruit crops enhance the activity of antioxidant enzymes, such as superoxide dismutase (SOD) and catalase (CAT), within chloroplasts to neutralize ROS and protect the photosynthetic machinery.
- This antioxidant response is critical for maintaining photosynthetic efficiency, as it prevents the degradation of key components like thylakoid membranes and photosystem II. By neutralizing ROS, antioxidant enzymes help sustain photosynthesis, supporting energy production and carbon fixation even under extreme heat.

3.2 Low-Temperature Adaptations

Low-temperature stress is a significant challenge for fruit crops, particularly during sensitive growth stages such as flowering and fruit setting. Exposure to cold can impair cellular processes, disrupt enzyme activity, and lead to ice formation within cells, all of which pose risks to crop health, yield, and quality. To withstand cold conditions, many fruit crops have developed specialized adaptations that protect cellular structures, maintain membrane stability, and optimize metabolic processes for survival at low temperatures. Key adaptations for cold tolerance include the accumulation of cryoprotective compounds, adjustments in cell membrane fluidity, and modifications in photosynthetic and respiratory pathways.

Cryoprotective Compounds

Cryoprotective compounds play a crucial role in safeguarding plant cells against cold-induced damage. These compounds, which include antifreeze proteins, sugars, and amino acids like proline, prevent ice formation, stabilize cellular structures, and protect enzymes from denaturation under low temperatures.

1. **Antifreeze Proteins**

- Antifreeze proteins (AFPs) are specialized proteins that inhibit ice crystal formation and growth within plant tissues, preventing cellular injury. By binding to ice nuclei and inhibiting their expansion, AFPs reduce the likelihood of ice damage within cells, which is particularly valuable for crops exposed to sudden cold snaps or frost.

- In fruit crops such as apples and cherries, AFPs accumulate in response to cold stress, acting as a molecular barrier against ice formation. These proteins help maintain cell integrity, enabling crops to endure freezing temperatures without suffering severe cellular damage. AFPs are especially beneficial in maintaining the function of sensitive structures like the vascular system, which is crucial for nutrient and water transport even under cold conditions.

2. **Role of Proline as a Cryoprotectant**

- Proline, an amino acid commonly associated with drought and salt stress, also plays a vital role in cold tolerance. As a cryoprotectant, proline stabilizes proteins and cellular membranes, preventing them from denaturing under low temperatures. Its highly soluble nature enables it to accumulate within cells without interfering with metabolic activities, protecting without disrupting normal physiological processes.
- In crops like citrus and grapes, proline levels rise significantly during cold stress, helping to preserve cellular structures and reduce the impact of freezing temperatures. Proline also acts as an osmolyte, balancing the osmotic potential in cells and preventing dehydration during cold stress, where water availability may become limited due to ice formation outside the cell.

3. **Accumulation of Soluble Sugars**

- Soluble sugars, including sucrose, glucose, and fructose, are crucial cryoprotectants that prevent ice formation and stabilize cellular structures. By accumulating in the cytoplasm, these sugars lower the freezing point of cellular fluids, reducing the risk of ice formation within cells. This antifreeze effect is particularly beneficial for crops in regions that experience early frosts or unpredictable temperature drops.
- In fruit crops such as pomegranates and figs, soluble sugars act as osmoprotectants, preserving cell turgor and preventing dehydration. Additionally, these sugars stabilize proteins and membranes, ensuring that enzymes remain active and cellular structures remain intact despite low temperatures. This accumulation of sugars helps crops survive transient cold events, minimizing damage and maintaining productivity.

4. **Function of Glycine Betaine as a Protective Solute**

- Glycine betaine is another compatible solute that functions as a cryoprotectant, accumulating in response to cold stress and helping to protect cellular components. Like proline, glycine betaine stabilizes proteins and membranes, supporting structural integrity under low temperatures.
- Glycine betaine also helps maintain ion balance within cells by preventing excessive ion leakage, which can occur under cold stress due to changes in membrane permeability. In salt-tolerant crops such as olives, glycine betaine provides dual protection against both salinity and cold stress, enhancing the crop's resilience in challenging environments.

Cell Membrane Fluidity

Cell membranes are highly susceptible to cold-induced damage, as low temperatures cause lipids to become more rigid, potentially leading to membrane leakage, ion imbalances, and cellular damage. To counteract these effects, cold-tolerant fruit crops modify their membrane lipid composition, increasing the proportion of unsaturated fatty acids to maintain membrane fluidity and integrity at low temperatures.

1. **Role of Unsaturated Fatty Acids in Membrane Fluidity**
    - Unsaturated fatty acids contain one or more double bonds, which introduce kinks into the fatty acid chains, preventing them from packing tightly together. This structural feature enhances membrane fluidity, enabling the membrane to remain flexible even at low temperatures.
    - By increasing the proportion of unsaturated fatty acids, cold-tolerant crops stabilize their cellular membranes, preventing them from becoming rigid or brittle under cold stress. This adaptation is essential for maintaining ion gradients, nutrient transport, and enzyme activity within the cell. For example, crops like apples and apricots show an increase in unsaturated fatty acids during cold stress, which helps maintain cellular function and supports survival in cold environments.
2. **Increased Activity of Desaturase Enzymes**
    - Desaturase enzymes, which introduce double bonds into fatty acids, are upregulated in response to cold stress, allowing the plant to increase the proportion of unsaturated fatty acids in its membranes. This enzymatic adjustment is critical for adapting membrane fluidity to fluctuating temperatures, particularly for crops exposed to sudden drops in temperature.
    - Enhanced desaturase activity has been observed in crops like grapes, where it helps maintain membrane flexibility and prevents cold-induced cellular damage. By continuously adjusting the lipid composition, crops can optimize membrane fluidity in response to changing temperatures, improving their resilience to cold stress.
3. **Maintenance of Chloroplast Membrane Integrity**
    - The chloroplast membrane is especially vulnerable to cold-induced rigidity, which can impair photosynthesis and reduce energy production. Cold-tolerant crops adjust the lipid composition of chloroplast membranes, ensuring that photosynthetic machinery remains functional even under low temperatures.
    - By increasing unsaturated lipid content in chloroplast membranes, crops like citrus protect their photosystems from thermal damage. This adjustment prevents the leakage of ions and stabilizes the thylakoid membranes, allowing efficient electron transport and energy production during cold periods.
4. **Antifreeze Lipid Compounds**
    - Some cold-tolerant crops produce antifreeze lipid compounds that further enhance membrane stability under freezing conditions. These lipids, which contain specific

molecular structures that resist freezing, integrate into cell membranes and contribute to their flexibility and resistance to cold-induced damage.

- This adaptation is beneficial for fruit crops exposed to extreme cold, as it reduces the risk of membrane rupture and ion leakage. By combining unsaturated fatty acids and antifreeze lipids, crops enhance their cold tolerance, ensuring cellular functionality in adverse conditions.

Photosynthetic and Respiratory Adjustments

Low temperatures slow down metabolic reactions, which can reduce energy production, impair growth, and delay development in fruit crops. To adapt, cold-tolerant plants modify their photosynthetic and respiratory processes, balancing energy conservation with warmth generation to optimize metabolic efficiency under cold stress.

1. **Reduced Photosynthetic Rates**
   - At low temperatures, photosynthetic enzymes become less efficient, which can limit carbon fixation and reduce energy production. Cold-tolerant fruit crops adjust their photosynthetic pathways to minimize energy loss, often reducing photosynthetic rates as a strategy to conserve energy.
   - By slowing down photosynthesis, crops like pears and cherries avoid excessive accumulation of ROS, which can lead to oxidative stress and cellular damage. This reduction in photosynthetic activity is accompanied by increased activity of antioxidant enzymes, which detoxify ROS and protect cellular components from oxidative damage.
2. **Enhanced Photorespiration**
   - Photorespiration, a process that consumes oxygen and releases carbon dioxide, increases in some cold-tolerant crops as a way to generate additional heat and prevent chilling injury. The heat generated by photorespiration helps maintain cellular temperature, providing warmth to sensitive tissues during cold stress.
   - In crops like apples and grapes, increased photorespiration under low temperatures helps protect chloroplasts from photo-oxidative damage and maintains a stable temperature within leaf tissues. This adaptation allows these crops to endure chilling stress without compromising their photosynthetic machinery.
3. **Increased Respiration for Heat Generation**
   - In addition to photosynthetic adjustments, cold-tolerant fruit crops often increase their respiratory activity under low temperatures. Respiration generates heat as a byproduct, which helps warm up cells and prevent chilling damage in cold environments.
   - By upregulating respiratory rates, crops like citrus and apricot maintain a higher internal temperature, which supports enzymatic activity and metabolic function. This thermal adaptation enables crops to withstand cold nights and frosty conditions, especially during critical growth stages.

4. **Accumulation of Sugars to Support Respiration**
   - Cold-tolerant crops accumulate sugars as an energy reserve to fuel respiration under low temperatures. These sugars provide a readily available source of energy, ensuring that the plant can maintain warmth through respiration even when photosynthesis is limited.
   - Sugars like sucrose and glucose support metabolic processes by fueling the respiratory pathway, providing the energy needed to counteract cold stress. In grapes and figs, sugar accumulation helps stabilize cellular structures and supports cold tolerance by maintaining a steady supply of energy.

4. Interaction among Different Stresses

Abiotic stresses rarely occur in isolation; their combined effect often exacerbates the impact on crop physiology and productivity. Some key interactions include:

4.1 Drought and Heat Stress

Drought and heat stress are two of the most challenging environmental factors that fruit crops face, often occurring simultaneously during hot, dry periods. The combined impact of drought and heat stress can be more severe than either stress alone, as the stresses intensify each other's effects, leading to profound physiological disturbances. This combination often results in accelerated water loss, impaired photosynthesis, increased oxidative stress, and an overall decline in crop productivity. The adaptive mechanisms fruit crops employ under these dual stressors, such as enhanced antioxidant activity, ABA-induced stomatal closure, and other physiological adjustments, is critical for improving drought and heat resilience in fruit species like apple, citrus, grape, and pomegranate.

Physiological Challenges of Combined Drought and Heat Stress

1. **Severe Dehydration and Water Deficit**
   - When both drought and heat stress occur, fruit crops experience rapid water loss due to increased transpiration, which significantly depletes internal water reserves. High temperatures accelerate evaporation from leaf surfaces, while drought conditions limit the availability of soil moisture, compounding the plant's struggle to maintain hydration.
   - The combined stress leads to severe dehydration, as plants are unable to replace lost water. This water deficit reduces cell turgor, affecting cell expansion and overall growth. In crops like apple and citrus, where cell hydration is essential for fruit development, the impact of this dehydration is visible as reduced fruit size and quality.
2. **Accelerated Senescence and Leaf Drop**
   - The stress combination accelerates cellular aging or senescence in fruit crops, leading to early leaf drop and premature yellowing. This response is often a survival mechanism, as shedding leaves reduces water demand and conserves resources, but it comes at a significant cost to photosynthetic capacity and productivity.
   - In citrus and apple trees, early leaf drop under combined drought and heat stress reduces the canopy's ability to capture light and perform photosynthesis, which directly impacts

fruit set and yield. The premature loss of leaves also limits the plant's ability to store carbohydrates, affecting fruit quality and leading to smaller, less flavorful produce.

3. **Impaired Photosynthesis**

   - Both drought and heat stress disrupt photosynthesis, the process by which plants convert sunlight into chemical energy. High temperatures denature photosynthetic enzymes and damage photosynthetic structures, while drought limits water availability, which is essential for the photosynthetic electron transport chain.
   - Combined drought and heat stress cause a decline in chlorophyll content, inhibit the activity of the enzyme ribulose-1,5-bisphosphate carboxylase/oxygenase (RuBisCO), and reduce carbon fixation, all of which diminish the plant's ability to produce energy. As a result, growth is stunted, and fruit development is impaired, which is especially problematic for crops like grapes and pomegranates, where high sugar content and flavor complexity are essential.

Enhanced Antioxidant Response

The simultaneous presence of drought and heat stress triggers an overproduction of reactive oxygen species (ROS), which can cause severe oxidative damage to cellular components such as lipids, proteins, and DNA. To counteract this oxidative stress, fruit crops enhance their antioxidant defense system, a vital adaptive response for managing ROS levels and protecting cellular integrity.

1. **Role of Enzymatic Antioxidants**

   - Enzymatic antioxidants, including superoxide dismutase (SOD), catalase (CAT), and ascorbate peroxidase (APX), are upregulated under combined drought and heat stress. These enzymes work together to detoxify ROS, preventing oxidative damage to cell membranes and other vital structures.
   - SOD converts superoxide radicals (O_2^-) into hydrogen peroxide (H_2O_2), which is then broken down into water and oxygen by CAT and APX. In fruit crops like apples and citrus, increased levels of these antioxidants protect the chloroplasts from oxidative damage, allowing photosynthesis to continue at reduced levels even under stress.

2. **Accumulation of Non-Enzymatic Antioxidants**

   - Non-enzymatic antioxidants, such as ascorbate (vitamin C), glutathione, and flavonoids, also accumulate in response to the oxidative stress caused by combined drought and heat stress. These compounds directly scavenge ROS, preventing them from causing cellular damage.
   - In crops like grapes and figs, which are known for their high levels of flavonoids, these compounds act as powerful antioxidants that protect cellular structures and contribute to resilience under environmental stress. By preserving cell membranes and stabilizing proteins, non-enzymatic antioxidants reduce the impact of oxidative stress, supporting growth and survival even in extreme conditions.

3. **Synergistic Role of Antioxidants in Protecting Photosynthesis**

   - Antioxidants play a critical role in preserving photosynthetic efficiency under combined stress. By detoxifying ROS, they protect chloroplast membranes and photosynthetic machinery, allowing the plant to maintain energy production and carbohydrate synthesis, which are essential for fruit development.
   - In heat-sensitive crops like citrus, the antioxidant system helps protect photosystem II from photo-oxidative damage, ensuring that light energy can be converted into chemical energy. This protection supports fruit set and yield, even though overall photosynthetic rates may be reduced.

ABA-Driven Stomatal Closure

Abscisic acid (ABA) is a plant hormone that plays a pivotal role in the response to drought and heat stress. Under combined stress, ABA levels increase significantly, leading to rapid stomatal closure, reduced water loss, and enhanced root growth, which collectively help the plant conserve water and improve stress resilience.

1. **Mechanism of ABA-Induced Stomatal Closure**

   - In response to water deficit and heat stress, ABA is synthesized in roots and transported to leaves, where it signals the guard cells surrounding stomata to close. Stomatal closure minimizes water loss by reducing transpiration, which is crucial for maintaining hydration under combined stress.
   - Stomatal closure is a vital survival mechanism in crops like pomegranate and grape, where water conservation is critical during hot, dry periods. However, this response also limits CO_2 intake, reducing photosynthesis and potentially impacting growth and yield. Despite this trade-off, ABA-driven stomatal closure is essential for survival under severe drought and heat.

2. **Promotion of Root Growth and Water Uptake**

   - In addition to regulating stomatal closure, ABA promotes root growth and enhances the plant's ability to extract water from deeper soil layers. By extending root length and increasing root surface area, ABA-driven root expansion allows the plant to access moisture reserves that may be unavailable to shallower roots.
   - This adaptation is especially valuable for fruit crops in arid regions, where soil moisture is limited to deeper layers. In apple and citrus, increased root growth under ABA signaling improves water uptake, reducing the negative impact of drought on crop health and yield.

3. **Interaction with Other Hormones**

   - ABA interacts with other plant hormones, such as ethylene and cytokinins, to coordinate the plant's response to combined drought and heat stress. For instance, elevated ABA levels inhibit cytokinin production in shoots, slowing vegetative growth and reducing water demand. Ethylene, which promotes leaf senescence, may also increase under combined stress, leading to controlled leaf drop and further reducing water loss.

- This hormonal interplay allows fruit crops to prioritize water conservation and root expansion over shoot growth, a strategy that enhances survival during prolonged drought and heat stress. In crops like figs and olives, which often endure hot, dry summers, this hormonal balance is crucial for maintaining productivity and fruit quality.

Impacts on Growth and Productivity

The combined effects of drought and heat stress have substantial implications for fruit crop growth, development, and yield. The stress combination not only reduces overall biomass but also negatively affects fruit set, quality, and marketability, posing significant challenges for fruit producers.

1. **Reduction in Fruit Set and Yield**
   - Combined drought and heat stress often lead to reduced fruit set due to flower abortion, impaired fertilization, and premature fruit drop. High temperatures and water scarcity create unfavorable conditions for pollination and seed development, directly impacting the number of fruits that mature to harvest.
   - In crops like apples and citrus, early fruit drop and lower fruit set under combined stress conditions result in a significant yield reduction. This decline in yield has economic implications, especially in regions where drought and heatwaves are frequent.
2. **Impairment of Fruit Quality**
   - Heat and drought stress can lead to smaller fruit size, lower sugar content, and reduced nutritional value. These factors not only affect the taste and quality of the fruit but also diminish its market value and consumer appeal.
   - In grapes, for instance, combined drought and heat stress can lead to excessive berry shrinkage, reducing juice content and concentration. In crops like citrus, sugar accumulation and acidity balance may be compromised, affecting the flavor profile and overall fruit quality.
3. **Challenges in Post-Harvest Management**
   - Fruits harvested from stressed plants often have a shorter shelf life and are more prone to post-harvest disorders due to compromised structural integrity and altered nutrient composition. This poses additional challenges for storage and transportation, especially for fruits intended for export.
   - Post-harvest quality management becomes more complex under these conditions, as fruits may deteriorate faster or show signs of skin wrinkling and flavor degradation. Addressing these challenges requires careful post-harvest handling practices and may necessitate additional cooling or protective measures to maintain quality.

4.2 Salinity and High-Temperature Stress

Salinity and high-temperature stress are two environmental stressors that, when occurring together, pose a formidable challenge to fruit crops. High temperatures increase water evaporation rates, leading to higher soil salt concentrations as moisture decreases. This intensified salinity under high temperatures disrupts ion uptake by roots, causes severe osmotic and oxidative stress, and impairs plant metabolic functions. Together, these stresses result in stunted growth, reduced fruit quality, and decreased yield in crops like grapes and tomatoes. Understanding the physiological effects and adaptations required to manage these stresses is crucial for enhancing resilience in fruit crops.

Increased Evaporation Rates and Soil Salt Concentration

High temperatures accelerate evaporation, causing soil moisture to decrease rapidly. When evaporation rates exceed the rate of replenishment from irrigation or rainfall, salt levels in the soil rise. This increase in soil salinity, compounded by high temperatures, has profound effects on fruit crops by exacerbating water scarcity and intensifying osmotic stress.

1. **Intensified Osmotic Stress**
   - High salinity conditions create a gradient that draws water out of plant roots, making it difficult for plants to absorb moisture even when water is present in the soil. As temperatures rise, water within plant cells evaporates more quickly, leaving cells dehydrated and unable to maintain turgor pressure.
   - This dehydration limits cellular expansion and growth, causing stunted development in fruit crops. In tomatoes and grapes, for example, osmotic stress from compounded salinity and heat results in visibly reduced plant size, decreased leaf area, and inhibited root elongation.
2. **Concentration of Toxic Ions in the Root Zone**
   - When soil moisture decreases under high temperatures, Na^+ and Cl^- ions become concentrated in the root zone, increasing the risk of ion toxicity in plants. This high salt concentration hinders the plant's ability to selectively absorb essential nutrients like potassium (K^+) and calcium (Ca^{2+}), leading to imbalances that disrupt cellular processes.
   - In crops like grapes, the accumulation of toxic ions in the root zone can interfere with nutrient uptake, affecting photosynthesis, respiration, and overall growth. This ion imbalance also triggers the accumulation of compatible solutes to counteract the osmotic stress, diverting resources away from growth and reproductive processes.

Impact on Root Ion Uptake

The simultaneous occurrence of salinity and high-temperature stress severely impairs the root system's ability to absorb water and nutrients. This disruption in ion uptake hampers nutrient transport within the plant, resulting in deficiencies that affect metabolic activities, enzyme function, and overall plant health.

1. **Reduced Water Uptake and Root Functionality**
   - Under saline conditions, water uptake by roots is already challenging due to osmotic pressure created by high salt levels. When combined with high temperatures, the water demand within the plant rises, but the ability to absorb water decreases. This imbalance places substantial strain on root tissues, leading to dehydration and, in severe cases, root necrosis.
2. In tomatoes, where root functionality is critical for supporting rapid growth and fruit production, the compounded stress limits water absorption, affecting both vegetative and

reproductive stages. Reduced water uptake under salinity and heat can lead to diminished leaf turgor, wilting, and an overall decline in plant vitality.

3. **Selective Ion Absorption Challenges**
   - Roots under salinity stress normally rely on selective ion channels and transporters to absorb beneficial ions like K^+ while excluding toxic ions like Na^+. However, high temperatures compromise this selective uptake by increasing membrane permeability, making it harder for roots to filter out Na^+ and Cl^- ions.
   - In crops like grapes, where K^+ is essential for maintaining cellular osmotic balance and enzyme activation, reduced K^+ uptake under salinity and heat stress affects metabolic efficiency. The inability to regulate ion uptake impairs physiological processes, causing deficiencies in essential nutrients that impact growth and fruit quality.
4. **Disruption of Mycorrhizal Associations**
   - Mycorrhizal fungi, which often form beneficial symbiotic relationships with fruit crop roots, are sensitive to both salinity and temperature. These fungi assist in nutrient and water absorption, especially under saline conditions. However, high temperatures disrupt fungal growth and function, weakening this symbiotic relationship and reducing the plant's ability to cope with stress.
   - The disruption of mycorrhizal associations reduces the plant's access to phosphorus and other micronutrients, compounding the effects of nutrient deficiency under combined salinity and heat. This weakened symbiosis limits the stress tolerance of crops like tomatoes, where mycorrhizae are integral to nutrient uptake and resilience.

Intensified Oxidative Damage

Both salinity and high temperatures independently induce oxidative stress, which can lead to an overproduction of reactive oxygen species (ROS) like superoxide radicals, hydrogen peroxide, and hydroxyl radicals. The combined impact of these stressors amplifies ROS generation, causing severe oxidative damage to cellular structures, proteins, lipids, and nucleic acids.

1. **Lipid Peroxidation and Membrane Damage**
   - High temperatures increase the fluidity of cellular membranes, while salinity destabilizes membrane structure by disrupting lipid composition and increasing Na^+ levels. Together, these stresses promote lipid peroxidation, where ROS attack the unsaturated fatty acids in membrane lipids, leading to cell membrane deterioration.
   - In crops like grapes and olives, lipid peroxidation under combined salinity and heat stress results in membrane leakage, ion imbalance, and loss of cellular integrity. This damage impacts the cell's ability to retain essential ions and nutrients, causing further physiological disruptions that impair growth and productivity.
2. **Damage to Photosynthetic Apparatus**
   - The photosynthetic machinery, especially photosystem II (PSII), is highly susceptible to oxidative damage. High temperatures increase ROS production within chloroplasts,

while salinity disrupts the chloroplast membrane potential. The compounded stress damages PSII and reduces chlorophyll content, leading to impaired photosynthesis.

- In tomatoes, which are particularly sensitive to changes in photosynthetic efficiency, the combined stress reduces carbon assimilation and energy production, resulting in stunted growth and diminished fruit development. The loss of chlorophyll and reduction in photosynthetic rate directly impact fruit yield and quality.

3. **Activation of Antioxidant Defense Mechanisms**
   - To mitigate oxidative damage, fruit crops upregulate their antioxidant defense systems, including both enzymatic and non-enzymatic antioxidants. Enzymes like superoxide dismutase (SOD), catalase (CAT), and glutathione peroxidase (GPX) play a crucial role in detoxifying ROS, while compounds like ascorbate, glutathione, and carotenoids act as ROS scavengers.
   - In crops such as olives and grapes, enhanced antioxidant activity under combined salinity and heat stress helps protect cellular structures from oxidative damage. However, maintaining this defense system requires significant energy and metabolic resources, which may divert resources from growth and reproduction, further affecting yield.

Impact on Growth and Fruit Quality

The compounded effects of salinity and high-temperature stress reduce overall plant growth, impair flowering and fruit set, and lead to poor fruit quality, impacting both productivity and economic value.

1. **Stunted Growth and Reduced Biomass:**
   - Under combined salinity and heat stress, plants prioritize survival over growth, leading to reduced biomass accumulation. This stunted growth is a result of limited water uptake, nutrient deficiencies, and impaired photosynthesis, all of which restrict cellular expansion and vegetative growth.
   - In crops like tomatoes, the reduced growth under these conditions manifests as shorter plants, smaller leaves, and thinner stems, which limit the plant's ability to produce and support fruits. This decline in vegetative growth affects the overall yield potential and limits the plant's ability to recover from stress.
2. **Poor Fruit Set and Reduced Yield**
   - Combined salinity and heat stress disrupt flowering and fruit set, as both processes require optimal water availability, nutrient supply, and temperature conditions. High temperatures during flowering cause pollen sterility, while salinity affects the development of reproductive organs, leading to a decrease in the number of fruits set per plant.
   - For crops like grapes, poor fruit set under combined stress conditions translates to fewer clusters and a significant reduction in harvest. This impact on yield is particularly challenging for commercial production, where reduced fruit numbers directly affect profitability.

3. **Diminished Fruit Quality**

   - Salinity and heat stress negatively impact fruit quality by reducing sugar content, altering acidity, and diminishing flavor. High temperatures accelerate respiration, depleting sugar reserves, while salinity limits the uptake of potassium, which is crucial for sugar and flavor synthesis.

   - In grapes, the dual stress reduces berry size and sugar concentration, affecting the quality of wine and table grapes. Similarly, in tomatoes, increased salinity and temperature stress can lead to smaller, less flavorful fruits with altered color and texture. These changes reduce the fruit's appeal to consumers, affecting marketability.

4. **Increased Susceptibility to Post-Harvest Disorders**

   - Fruits from crops exposed to combined salinity and high temperatures often have a reduced shelf life and are more prone to post-harvest disorders, such as skin wrinkling, rapid spoilage, and susceptibility to fungal infections. This vulnerability is due to compromised cell wall structure, reduced antioxidant levels, and altered water content in the fruit.

   - These post-harvest challenges are particularly concerning for fruits intended for export or long storage periods. Maintaining fruit quality during and after harvest becomes more complex, requiring additional storage, handling, and preservation techniques to ensure that fruits reach the market in acceptable condition.

4.3 Drought and Salinity Stress

Drought and salinity are two major environmental stressors that frequently occur together, especially in arid and semi-arid regions where water scarcity restricts leaching. The lack of sufficient water results in the accumulation of salts in the soil, which compounds the impact of drought by creating osmotic stress and ion toxicity for plants. Under this combined stress, fruit crops face challenges in water uptake, ionic balance, and cellular stability. To survive these harsh conditions, fruit crops like olives and figs employ a range of adaptive strategies, including enhanced osmotic adjustment, ion compartmentalization, antioxidant capacity, and efficient ion exclusion mechanisms. These adaptations are essential for managing water and ion balance, maintaining cellular integrity, and sustaining growth and productivity in high-salinity, low-water environments.

Impact of Combined Drought and Salinity Stress

1. **Increased Osmotic Stress**

   - Under drought conditions, limited soil moisture already creates osmotic stress, making it difficult for plants to absorb water. When combined with high salinity, the osmotic potential of the soil solution decreases even further, intensifying the water scarcity experienced by plants.

   - High concentrations of salt in the soil amplify osmotic stress, creating a situation where the plant's internal water content drops, leading to cellular dehydration. This results in reduced turgor pressure, which inhibits cell expansion and overall growth. In crops like figs, reduced turgor causes wilting, leaf curling, and stunted growth, directly impacting photosynthetic capacity and productivity.

2. **Ion Toxicity and Nutrient Imbalance**
   - Salinity introduces high levels of sodium (Na^+) and chloride (Cl^-) ions into the root zone, which can be toxic to plants. These ions compete with essential nutrients like potassium (K^+) and calcium (Ca^{2+}) for uptake, leading to nutrient deficiencies that further weaken the plant.
   - Ion toxicity disrupts metabolic processes, particularly in sensitive tissues like leaves and young roots. In drought and salinity-tolerant crops like olives, the high levels of Na^+ and Cl^- under combined stress conditions require effective mechanisms to exclude these ions from critical cellular compartments to prevent metabolic disruption.
3. **Decreased Photosynthetic Efficiency**
   - Both drought and salinity negatively affect photosynthesis. Drought reduces water availability for photosynthetic reactions, while salinity causes ion toxicity in chloroplasts, where photosynthesis takes place. The combined stress can lead to chlorophyll degradation, reduced photosynthetic enzyme activity, and lower carbon assimilation rates.
   - In olives and figs, this stress combination results in a decline in photosynthetic efficiency, as chlorophyll content and RuBisCO activity decrease, limiting the plant's ability to capture and convert light energy. This reduction in photosynthesis has a direct impact on growth, biomass accumulation, and fruit development, diminishing yield and fruit quality.

Osmotic Adjustment as an Adaptive Mechanism

Osmotic adjustment is a crucial adaptation that allows plants to maintain cell turgor and prevent dehydration under drought and salinity stress. By accumulating compatible solutes, or osmolytes, such as proline, glycine betaine, and soluble sugars, plants balance the osmotic potential within cells, allowing them to retain water even in low-moisture, high-salt environments.

1. **Role of Proline in Osmotic Balance**
   - Proline, an amino acid, is one of the most common osmolytes accumulated by plants under combined drought and salinity stress. Proline acts as an osmoprotectant, stabilizing proteins, membranes, and other cellular structures while balancing osmotic potential within cells.
   - In drought-tolerant crops like figs, high levels of proline help maintain cell hydration, ensuring that essential metabolic processes can continue despite limited water availability. Proline also functions as an antioxidant, scavenging reactive oxygen species (ROS) generated by stress, protecting cells from oxidative damage.
2. **Glycine Betaine and Soluble Sugars**
   - Glycine betaine and soluble sugars (such as sucrose, glucose, and fructose) are other key osmolytes that accumulate in response to drought and salinity. These solutes protect cellular structures and enzymes from dehydration, maintaining cellular integrity and function under stress.
   - In olives, the accumulation of glycine betaine and sugars provides osmotic balance and

supports cell turgor. This adaptation is particularly beneficial during prolonged dry periods with high soil salinity, as it helps the plant sustain growth and physiological activities even when external water is scarce.

3. **Increased Production of Compatible Solutes**
   - Under combined drought and salinity stress, the synthesis of compatible solutes is significantly upregulated, allowing plants to adapt to the osmotic challenges posed by low water availability and high salt concentration. The production of these solutes is energy-intensive, but it is essential for survival.
   - In both olives and figs, increased compatible solute production enables these crops to endure extended periods of stress by maintaining cellular hydration and protecting cellular components from dehydration and ionic stress.

Ion Compartmentalization and Exclusion

To cope with the toxic effects of excess Na^+ and Cl^- ions under salinity, fruit crops use ion compartmentalization and exclusion mechanisms. These strategies help sequester harmful ions in specific compartments, preventing them from interfering with cellular processes, or exclude them altogether from sensitive tissues.

1. **Vacuolar Compartmentalization of Na^+ and Cl^-**
   - Ion compartmentalization involves sequestering excess Na^+ and Cl^- ions in vacuoles, where they cannot interfere with cytoplasmic functions. By storing these ions in vacuoles, plants maintain low concentrations of toxic ions in the cytoplasm, protecting enzymes and metabolic pathways from disruption.
   - In drought and salt-tolerant crops like olives, the vacuolar compartmentalization of Na^+ is facilitated by Na^+/H^+ antiporters located on the tonoplast (vacuolar membrane). These transporters actively pump Na^+ into vacuoles, ensuring that cytoplasmic ionic balance is preserved, even under high salinity.
2. **Selective Ion Transport and Exclusion at the Root Level**
   - Selective ion transport mechanisms enable plants to absorb essential ions like K^+ while excluding or minimizing the uptake of Na^+ and Cl^-. In combined drought and salinity stress, selective ion channels and transporters in the root cells are crucial for maintaining ionic homeostasis.
   - In figs, for example, root cells express Na^+-specific transporters that pump Na^+ out of the root cells and back into the soil, reducing the risk of Na^+ toxicity in above-ground tissues. This selective ion exclusion mechanism allows crops to sustain nutrient uptake, preventing nutrient deficiencies that would otherwise exacerbate the effects of combined stress.
3. **Role of Calcium in Ion Homeostasis**
   - Calcium (Ca^{2+}) is essential for stabilizing cell membranes and counteracting the toxic effects of Na^+. Under combined drought and salinity stress, calcium uptake is enhanced to help maintain membrane stability and support selective ion transport.

- In olives, higher Ca^{2+} levels improve the plant's ability to exclude Na^+ from sensitive tissues, reducing the impact of salinity. Calcium ions also contribute to the structural integrity of cell walls, providing additional resilience against osmotic and ionic stress.

Antioxidant Capacity Enhancement

The combined stresses of drought and salinity lead to the overproduction of ROS, which can cause oxidative damage to lipids, proteins, and nucleic acids. To mitigate oxidative damage, fruit crops strengthen their antioxidant systems, which include both enzymatic and non-enzymatic components.

1. **Upregulation of Enzymatic Antioxidants**
   - Enzymatic antioxidants, such as superoxide dismutase (SOD), catalase (CAT), and peroxidase (POD), play a vital role in detoxifying ROS under stress conditions. These enzymes work together to neutralize harmful free radicals, preventing oxidative damage to cellular structures.
   - In olives and figs, increased activity of these antioxidants under drought and salinity stress helps maintain cellular integrity, ensuring that photosynthesis, respiration, and other metabolic processes continue. By protecting chloroplasts from oxidative damage, enzymatic antioxidants support energy production, which is critical for stress resilience.
2. **Accumulation of Non-Enzymatic Antioxidants**
   - Non-enzymatic antioxidants, including ascorbate (vitamin C), glutathione, and phenolic compounds, act as ROS scavengers. These compounds directly neutralize ROS, preventing them from damaging cellular components and aiding in stress tolerance.
   - In figs, high levels of phenolic compounds provide an additional layer of protection against oxidative stress. These antioxidants not only protect cellular membranes and enzymes but also contribute to improved fruit quality, as they are associated with enhanced flavor and nutritional value.
3. **Antioxidant Protection of Photosynthetic Machinery**
   - Drought and salinity stress can cause significant damage to photosynthetic machinery, particularly photosystem II (PSII). Antioxidants help protect PSII from oxidative damage by neutralizing ROS generated during photosynthesis.
   - In crops like olives, a robust antioxidant system preserves chlorophyll content and prevents photo-oxidative damage, allowing the plant to maintain photosynthetic efficiency even under combined stress. This protection is essential for sustaining growth and productivity in water-scarce, saline environments.

5: Impact on Crop Growth and Productivity

The physiological responses and adaptive features developed by fruit crops under abiotic stress are crucial for survival but often come at a cost to growth and productivity. The impact on crop performance varies based on the type, intensity, and duration of the stress, as well as the crop's inherent tolerance.

5.1 Impaired Photosynthesis and Growth

Abiotic stresses such as drought, salinity, extreme temperatures, and nutrient deficiencies impose substantial challenges to the growth and productivity of fruit crops. Photosynthesis, the primary process through which plants convert light energy into chemical energy, is particularly vulnerable to these stressors. Photosynthetic impairment under abiotic stress not only affects the immediate energy supply but also impacts growth rates, biomass accumulation, and ultimately the yield and quality of fruit crops. This section explores how abiotic stresses disrupt photosynthesis through mechanisms such as stomatal closure, chlorophyll degradation, oxidative damage to chloroplasts, and other related physiological responses, and examines their cascading effects on plant growth and fruit yield.

Mechanisms of Photosynthetic Impairment Under Abiotic Stress

Abiotic stress affects photosynthesis at multiple levels, from the opening and closing of stomata to the structural integrity of chloroplasts, and even down to the biochemical processes within the Calvin cycle. Each of these disruptions limits the plant's ability to capture light energy, assimilate carbon dioxide (CO_2), and synthesize carbohydrates, which are fundamental to growth and development. This section examines the various mechanisms by which abiotic stress impairs photosynthesis.

1. Stomatal Closure and CO_2 Assimilation

Stomata, tiny pores on the leaf surface, play a crucial role in regulating gas exchange between the plant and its environment. These pores allow CO_2 to enter the leaf for photosynthesis and release water vapor as part of the transpiration process. Stomatal behavior is highly responsive to environmental conditions, particularly under drought and high-temperature stress, as plants close stomata to conserve water. While this response helps mitigate water loss, it also limits the availability of CO_2 for photosynthesis, leading to a reduction in the rate of carbon fixation.

Mechanism of Stomatal Closure in Response to Abiotic Stress

» Drought stress leads to the production of abscisic acid (ABA), a plant hormone that signals guard cells to close the stomata. Guard cells respond to ABA by losing water, which causes them to collapse, thereby closing the stomatal pores. This reduction in stomatal aperture minimizes water loss but restricts CO_2 entry, reducing photosynthetic rate.

» High temperatures can also promote stomatal closure as part of a protective response to prevent excessive water loss. However, with the stomata closed, the internal concentration of CO_2 declines, impeding photosynthesis.

» In fruit crops like grapes and olives, where stomatal closure is a common response to both drought and high heat, reduced CO_2 assimilation directly impacts carbohydrate production. These plants face reduced energy availability, slowing down growth and decreasing overall productivity.

Reduction in CO_2 Assimilation and Photosynthetic Efficiency

» With limited CO_2 availability due to stomatal closure, the plant's capacity for carbon assimilation diminishes. This reduced CO_2 assimilation rate negatively affects the Calvin cycle, where CO_2 is fixed into sugars, ultimately impacting the energy reserves needed for growth.

» In crops such as apples and citrus, where fruit size and quality are closely tied to photosynthetic performance, limited CO_2 uptake reduces carbohydrate availability. This reduction in carbohydrate synthesis impacts vegetative and reproductive growth, leading to smaller fruit size and reduced overall yield.

Implications for Biomass Accumulation

» Reduced photosynthesis due to stomatal closure translates to lower biomass accumulation. In crops like pomegranates and mangoes, diminished biomass is visible as reduced leaf area, weaker stems, and shallower root systems, all of which impair the plant's ability to support fruit development.

» A smaller biomass limits the plant's photosynthetic capacity further, creating a feedback loop that constrains growth, productivity, and resilience under stress. Consequently, yield losses in fruit crops under drought and heat stress are often substantial, with a marked decline in both fruit quantity and quality.

2. Chlorophyll Degradation and Light Absorption

Chlorophyll, the green pigment in leaves, is essential for capturing light energy and initiating photosynthesis. However, abiotic stresses such as drought, salinity, and extreme temperatures can trigger chlorophyll degradation, leading to a reduction in the plant's light-harvesting capability and photosynthetic efficiency.

Causes of Chlorophyll Degradation Under Stress

» Abiotic stresses increase the production of reactive oxygen species (ROS) in plant cells, especially in the chloroplasts, where photosynthesis occurs. These ROS molecules cause oxidative damage to cellular components, including chlorophyll molecules. ROS-induced chlorophyll degradation reduces the plant's ability to absorb light efficiently.

- In saline soils, excess sodium (Na^+) and chloride (Cl^-) ions interfere with chlorophyll biosynthesis and trigger chlorophyll breakdown. High salt levels can disrupt the availability of essential nutrients, particularly magnesium, a central component of the chlorophyll molecule. Magnesium deficiency leads to chlorosis (yellowing of leaves) and further reduces the plant's photosynthetic capacity.

Impact on Light Absorption and Energy Conversion

» Chlorophyll degradation limits the plant's ability to capture light energy, impacting the entire photosynthetic process. In heat-stressed crops like grapes, a decrease in chlorophyll content reduces the light-absorbing capacity of the chloroplasts, leading to lower energy production.

» Reduced light absorption affects the efficiency of the electron transport chain, where light energy is converted into ATP and NADPH. These molecules are essential for the Calvin cycle, where carbon is fixed into sugars. Lower energy availability from the electron transport chain hinders the Calvin cycle's efficiency, slowing down photosynthesis and limiting the plant's ability to produce carbohydrates.

Visible Symptoms and Impact on Fruit Quality

- Chlorophyll degradation is often visibly noticeable as leaf yellowing or chlorosis, which is a common sign of abiotic stress. In fruit crops such as tomatoes and peppers, chlorosis reduces leaf area and light absorption, impacting the plant's photosynthetic efficiency and productivity.
- Degraded chlorophyll content under stress conditions also affects the nutritional and aesthetic qualities of the fruit. For example, in high-value crops like apples and grapes, chlorophyll degradation leads to a reduction in sugar accumulation, affecting flavor, sweetness, and color. This impact on fruit quality diminishes the crop's marketability, directly affecting farmers' profitability.

3. Oxidative Damage to Chloroplasts and Photosynthetic Machinery

Abiotic stress, especially in the form of drought, salinity, and high temperatures, leads to excessive ROS production within plant cells. ROS molecules cause oxidative damage to cellular structures, including chloroplasts, where photosynthesis takes place. Damage to the photosynthetic machinery reduces the plant's ability to produce energy, impacts carbon fixation, and stunts overall plant growth.

Generation of Reactive Oxygen Species (ROS) Under Abiotic Stress

- Under normal conditions, ROS levels are regulated within plant cells. However, abiotic stresses disrupt this balance, leading to an overproduction of ROS, which then damage cell structures and proteins, impairing cellular function.
- High temperatures increase metabolic rates and ROS production, while drought and salinity exacerbate oxidative stress by disrupting the electron transport chain in chloroplasts. The excess ROS generated under these conditions causes oxidative damage, leading to the breakdown of vital cellular components, including chlorophyll and photosynthetic enzymes.

Damage to Photosystem II and Photosynthetic Enzymes

- Photosystem II (PSII), a critical component of the photosynthetic electron transport chain, is highly sensitive to oxidative stress. Excess ROS production leads to lipid peroxidation within the thylakoid membrane where PSII resides, which disrupts electron flow and reduces ATP and NADPH synthesis.
- Key enzymes in the Calvin cycle, such as RuBisCO, are also susceptible to oxidative damage. RuBisCO is essential for carbon fixation, and oxidative modifications to this enzyme reduce its efficiency, lowering the rate of photosynthesis and energy production.

Antioxidant Defense Mechanisms to Mitigate Oxidative Damage

- To counteract ROS damage, plants activate an antioxidant defense system that includes enzymatic antioxidants (e.g., superoxide dismutase, catalase, peroxidase) and non-enzymatic antioxidants (e.g., ascorbate, glutathione, carotenoids). These antioxidants neutralize ROS, reducing oxidative stress and protecting the photosynthetic machinery.
- In drought-tolerant crops like olives and figs, the upregulation of antioxidant activity under stress conditions helps preserve chloroplast integrity, allowing photosynthesis to continue

at reduced levels. However, maintaining this defense system requires significant metabolic resources, which can limit growth and reproduction under prolonged stress.

Effects on Growth and Biomass Accumulation

Reduced photosynthesis due to stomatal closure, chlorophyll degradation, and oxidative damage has far-reaching impacts on plant growth and biomass accumulation. Without adequate photosynthesis, plants lack the energy and carbon compounds needed for cell division, cell expansion, and overall development. This results in stunted growth, reduced leaf area, and impaired root development, all of which hinder the plant's ability to support fruit growth and achieve full yield potential.

Reduction in Leaf Area and Canopy Size

» Abiotic stress-induced reduction in photosynthesis leads to smaller leaves and reduced canopy size, limiting the plant's light-capturing ability and compounding the decline in photosynthetic efficiency.

» In crops like pomegranate and mango, a reduced canopy size decreases the plant's productivity, as fewer leaves translate to less overall energy production. This reduction in canopy size also limits the plant's ability to shade its fruit from excessive sunlight, potentially exposing fruits to sunburn damage in high-temperature environments.

Impact on Root Growth and Nutrient Uptake

» Photosynthesis supplies the energy required for root development, and reduced photosynthesis under stress limits root growth, leading to a less extensive root system. This compromised root system is less effective at absorbing water and nutrients, further exacerbating the effects of drought and salinity stress.

» In fruit crops like grapes and citrus, which rely on deep root systems for water access, reduced root growth under abiotic stress conditions limits the plant's resilience to environmental changes, affecting growth and yield.

Effects on Fruit Set, Development, and Quality

» Photosynthesis is vital for fruit set and development, as it provides the energy and carbon skeletons needed for cellular growth and the synthesis of sugars, acids, and other compounds that contribute to fruit quality.

» Reduced photosynthetic capacity under abiotic stress leads to smaller fruits with lower sugar content and diminished flavor. In high-value fruit crops like apples and grapes, this reduction in quality can significantly lower market value, affecting profitability.

5.2 Disrupted Reproductive Processes

The reproductive phase of fruit crops—encompassing flowering, pollination, fertilization, fruit set, and early fruit development—is highly sensitive to environmental stresses. Conditions like drought, high temperatures, and salinity disrupt these reproductive processes, leading to decreased fruit production, lower fruit quality, and diminished yield potential. During critical stages, such as flowering and fruit set, the effects of abiotic stressors can have irreversible consequences on fruit crops. Drought and heat stress often lead to floral abortion and premature fruit drop in crops like mango and citrus, while salinity stress adversely affects pollen viability and fertilization in pomegranate.

The Importance of Reproductive Processes in Fruit Crops

Reproduction is the foundational phase that determines the yield and quality of fruit crops. Successful reproduction involves a series of stages, beginning with the development of floral buds, followed by pollination, fertilization, and finally fruit set. During this process, plants allocate significant energy and resources to ensure successful seed and fruit development. However, environmental stress can impair these processes at various stages, severely limiting a plant's reproductive success.

1. **Flowering and Pollination**
    - Flowering is the first step in the reproductive process, and its success is critical for fruit production. Environmental stresses during this phase can lead to poor floral development, reduced flower numbers, and decreased pollen viability, all of which reduce the likelihood of successful pollination and fertilization.
    - Pollination—the transfer of pollen from male to female reproductive organs—is another sensitive stage. Effective pollination is essential for fertilization, fruit set, and seed development. Stress conditions disrupt pollen production and reduce pollen viability, resulting in lower fertilization rates and compromised fruit set.
2. **Fruit Set and Early Development**
    - Once fertilization occurs, the plant transitions to the fruit set stage, where young fruits begin to develop. This is another critical period that requires optimal conditions for cellular growth, nutrient allocation, and hormonal balance. Abiotic stress can lead to early fruit drop, nutrient deficiency, and slower fruit growth, directly affecting yield and fruit quality.
    - For crops like mango, citrus, and pomegranate, stress-induced disruptions in fruit set and early development lead to reduced fruit number and size, with far-reaching consequences for overall productivity.

Impacts of Drought Stress on Reproductive Success

Drought stress is a major threat to reproductive success in fruit crops, as it limits water availability, disrupts nutrient transport, and induces cellular dehydration. During drought conditions, plants prioritize survival over reproduction, which often results in reduced flowering, floral abortion, and fruit drop. These effects have significant implications for yield, particularly in water-intensive crops like citrus, mango, and pomegranate.

1. **Reduced Flowering and Floral Bud Development**
    - Under drought stress, plants experience limited water availability, which impacts floral bud initiation and development. The energy needed for flower production is redirected to survival mechanisms, reducing the number of flowers produced.
    - In mango, for example, drought stress during bud initiation limits energy and water resources for flower formation, resulting in fewer buds and a reduction in the total

number of flowers. Consequently, the potential for successful pollination and fruit set is diminished, reducing yield from the outset.

2. **Floral Abortion and Premature Flower Drop**
   - Drought-induced water deficit often triggers floral abortion and early flower drop as a survival mechanism. In crops such as citrus, mango, and pomegranate, drought stress during flowering causes plants to shed flowers prematurely, as they prioritize water conservation over reproduction.
   - This response significantly reduces the number of flowers available for pollination, leading to a lower number of potential fruits. In regions prone to frequent drought, such flower loss translates directly into lower fruit yields, impacting both quantity and quality of produce.
3. **Decreased Pollen Viability and Germination**
   - Pollen viability and germination are highly sensitive to water availability. Drought stress affects pollen development by reducing water content within pollen grains, leading to desiccation and lower viability. In crops like pomegranate, drought conditions reduce pollen viability, impacting successful fertilization rates and ultimately lowering fruit set.
   - Drought stress also impacts pollen tube growth, which is essential for fertilization. Insufficient water availability limits pollen hydration and reduces the length of pollen tubes, decreasing the chances of successful fertilization and resulting in fewer fertilized flowers and reduced fruit set.
4. **Impact on Fruit Set and Early Fruit Drop**
   - Drought stress affects the fruit set stage by limiting water and nutrient transport to young fruits, causing poor cell expansion and smaller fruit size. In addition, drought-stressed plants often experience early fruit drop as a protective mechanism to conserve resources.
   - In citrus, early fruit drop under drought stress results in lower yield potential, as fewer fruits remain on the tree to reach maturity. This early fruit abscission, or shedding, significantly impacts overall productivity and is a common response in fruit crops grown in arid and semi-arid regions.

Effects of Heat Stress on Flowering, Pollination, and Fruit Set

High temperatures during flowering and fruit set stages present unique challenges for fruit crops, as heat stress damages reproductive tissues, impairs pollen function, and disrupts hormonal regulation. Heat stress is particularly problematic for crops like mango and citrus, where optimal fruit set requires stable, moderate temperatures. When temperatures rise beyond optimal levels, the reproductive process is adversely affected, leading to reduced fruit set and early fruit drop.

1. **Floral Morphology and Abnormalities**
   - Heat stress can cause structural abnormalities in flowers, such as incomplete formation of stamens and pistils. These abnormalities make it difficult for flowers to achieve successful fertilization, as reproductive organs may not be fully developed.

- In mango, high temperatures during flowering can lead to malformed flowers with non-functional reproductive parts, limiting the chances of successful pollination. This effect is particularly noticeable in mango orchards exposed to heatwaves during the flowering season, where abnormal floral morphology contributes to reduced yield potential.

2. **Pollen Viability, Desiccation, and Pollination Efficiency**
   - High temperatures can desiccate pollen grains, reducing their viability and ability to fertilize ovules. Heat stress also accelerates pollen aging, shortening the lifespan of pollen grains and lowering the likelihood of successful pollination.
   - In citrus, for example, pollen desiccation under high temperatures decreases the chances of successful fertilization, as viable pollen grains are limited in number and may fail to reach the stigma. As a result, pollination efficiency is compromised, leading to a lower fruit set.
3. **Disruption of Pollination and Fertilization**
   - Successful fertilization relies on the growth of pollen tubes to transport sperm cells to the ovule. High temperatures disrupt this process by inhibiting pollen tube growth, which reduces fertilization success.
   - In pomegranate, heat stress during flowering and pollination affects the growth rate of pollen tubes, leading to incomplete fertilization and lower fruit set. This reduced fertilization rate is a direct consequence of disrupted pollen tube elongation, which is critical for successful fertilization.
4. **Increased Ethylene Production and Early Fruit Drop**
   - Heat stress increases ethylene production, a hormone associated with fruit ripening and abscission. Elevated ethylene levels under high temperatures induce early fruit drop, as the plant sheds immature fruits in response to unfavorable conditions.
   - In mango, elevated ethylene levels cause premature fruit drop, reducing the number of fruits that reach maturity. This heat-induced fruit abscission is a significant issue in regions with frequent high-temperature fluctuations, as it leads to reduced yield and lower quality fruit.

Influence of Salinity Stress on Reproductive Success

Salinity stress, caused by high soil salt concentrations, poses unique challenges to reproductive processes in fruit crops. Excessive salt levels disrupt water absorption, create ion imbalances, and induce osmotic stress, all of which interfere with flower development, pollen viability, and fruit set. Salinity is particularly detrimental to salt-sensitive crops like pomegranate, where reproductive processes are highly vulnerable to ion toxicity.

1. **Impact on Flower Development and Floral Organ Differentiation**
   - Salinity stress interferes with floral bud initiation and development, leading to fewer flowers per plant and compromised floral morphology. In pomegranate, for instance, high salt levels in the soil restrict water uptake and create osmotic stress, leading to poorly developed flowers.

- The reduction in flower quality and quantity under salinity stress limits the reproductive potential of the plant, resulting in fewer pollination opportunities and lower fruit set. This decline in floral productivity is particularly challenging in saline-prone regions where water scarcity and soil salinization are prevalent.

2. **Reduced Pollen Viability, Germination, and Fertilization**
   - Salinity stress affects pollen viability by disrupting ionic balance and nutrient availability. High levels of sodium (Na^+) and chloride (Cl^-) ions compete with essential nutrients, leading to deficiencies in potassium and calcium, which are crucial for pollen development and germination.
   - Pomegranate pollen grains exposed to saline conditions exhibit lower viability, reduced germination rates, and impaired pollen tube growth, which hinder successful fertilization. As a result, fewer flowers are successfully fertilized, and fruit set declines significantly.

3. **Impaired Fertilization and Early Fruit Abscission**
   - Even if fertilization occurs under salinity stress, developing fruits may experience osmotic stress and ion toxicity, leading to early fruit drop. Salinity induces hormonal imbalances that trigger abscission processes, causing the plant to shed fruits prematurely.
   - In citrus, salinity stress frequently leads to the early shedding of young fruits as the plant reallocates resources to cope with the harsh conditions. This early fruit drop reduces the number of mature fruits, impacting both yield and marketability.

4. **Combined Impact of Drought and Salinity on Reproductive Success**
   - In regions where drought and salinity often co-occur, the combined stress effects amplify reproductive challenges. Water scarcity limits flower hydration, while high salt concentrations create osmotic stress, ion toxicity, and nutrient imbalances, severely reducing the likelihood of successful pollination and fertilization.
   - For example, in pomegranate, combined drought and salinity stress causes severe reductions in flower and pollen viability, leading to widespread floral abortion and limited fruit set. This cumulative impact of drought and salinity significantly decreases the crop's reproductive success, posing a substantial threat to yield.

5.3 Decline in Fruit Quality

The quality of fruit is defined by multiple parameters that go beyond size and yield. While yield is essential for economic returns, fruit quality determines marketability, consumer satisfaction, and nutritional value. Key quality attributes include sugar content, acidity, texture, flavor, and vitamin and antioxidant levels, all of which are influenced by environmental conditions. Abiotic stresses, particularly drought, salinity, and extreme temperatures, significantly alter these quality traits, sometimes leading to both positive and negative outcomes. For instance, while drought stress can increase sugar concentration, it may also reduce fruit size and alter flavor balance, impacting overall acceptability.

Influence of Drought Stress on Fruit Quality

Drought stress, or water deficit, affects nearly every aspect of fruit quality. As plants conserve water, nutrient transport, and photosynthesis rates are altered, leading to changes in sugar content, acidity, size, and texture of fruits. Although drought stress can improve certain quality parameters in specific crops, prolonged or severe drought often negatively impacts overall quality, flavor, and nutritional content.

1. **Impact on Sugar Content and Flavor Profile**
   - Drought stress often leads to an increase in sugar concentration within fruits, as water availability decreases. This concentration effect occurs because lower water content reduces fruit size, concentrating sugars within the remaining water content.
   - In grapes, for instance, controlled water deficit is sometimes employed to enhance sugar accumulation, improving the sweetness and flavor profile desired for winemaking. However, if drought stress becomes too intense, it can lead to excessive sugar levels, which can unbalance the flavor, leading to wines with overly high alcohol content. Additionally, an imbalance between sugars and organic acids can affect the overall taste profile, making the fruit overly sweet without sufficient acidity for balance.
2. **Reduction in Fruit Size and Texture Changes**
   - Drought stress generally results in smaller fruit size, as limited water availability restricts cellular expansion and overall fruit growth. Smaller fruits may have higher concentrations of sugars, acids, and other compounds, but they are less commercially desirable due to reduced visual appeal and weight.
   - The texture of drought-stressed fruits can also change, becoming firmer or even tougher in certain cases. In some crops, like apples, drought stress can lead to an increase in fruit firmness, which may be beneficial for storage and transportation. However, in fruits like melons and berries, increased firmness can reduce the juicy quality and affect consumer acceptance.
3. **Alterations in Acidity and Organic Acid Composition**
   - Drought stress influences organic acid metabolism, often resulting in higher acidity levels due to increased concentrations of malic, citric, or tartaric acids. In tomatoes, water deficit during fruit development can lead to an increase in acidity, affecting flavor balance. The increased acidity may be desirable in some culinary contexts, but it can also lead to a less palatable taste if excessive.
   - In grapes, organic acid metabolism is highly sensitive to drought, affecting wine acidity and taste. Drought-stressed grapes often have higher tartaric acid levels but lower malic acid, which impacts the wine's sensory profile and stability.
4. **Effect on Phenolic Compounds and Antioxidant Levels:**
   - Drought stress can lead to an increase in phenolic compounds and antioxidants, as plants produce these compounds to protect themselves from oxidative stress. These changes are particularly notable in grapes, where polyphenols, flavonoids, and anthocyanins increase under drought stress, enhancing the color and antioxidant capacity of the fruit.
   - The elevated levels of phenolic compounds not only improve the health benefits of the fruit but also enhance the complexity and color stability in grape-derived products

like wine. However, excessive accumulation of certain phenolics can lead to a bitter or astringent taste, which may not be desirable for fresh consumption.

Effects of Salinity Stress on Fruit Quality

Salinity stress, caused by excessive salt levels in the soil, poses significant challenges to fruit quality, particularly in salt-sensitive crops. High salinity affects water uptake and creates ionic imbalances, leading to nutrient deficiencies and ion toxicity. These conditions alter sugar levels, acidity, vitamin content, and texture, often resulting in reduced fruit quality and marketability.

1. **Decrease in Fruit Size and Juiciness**
   - Salinity stress limits water absorption by creating an osmotic gradient that restricts the flow of water into plant roots. This results in smaller fruits, as cellular expansion is compromised, and in some cases, the fruits may also become less juicy.
   - In tomatoes, high salinity levels lead to reduced fruit size and dry texture. While smaller fruit size may increase the concentration of certain flavor compounds, it generally reduces consumer appeal and market value. Additionally, reduced juiciness impacts the texture and overall eating experience, particularly in fresh market varieties.
2. **Changes in Sugar and Acidity Balance**
   - Salinity stress often results in altered sugar and acid profiles, with the potential to either enhance or detract from flavor. For example, salinity can increase sugar content in certain crops by inducing osmotic adjustment, as plants accumulate soluble sugars to balance osmotic stress. However, this increase in sugar concentration can be accompanied by changes in acid levels, affecting overall flavor.
   - In citrus fruits, salinity stress tends to increase acidity, as the accumulation of chloride ions leads to higher levels of organic acids in the juice. This results in a tangier or more acidic flavor profile, which can be undesirable in varieties intended for sweet or mild flavor profiles.
3. **Reduction in Vitamin C and Nutritional Value**
   - Salinity stress adversely affects the synthesis of ascorbic acid (vitamin C), particularly in crops like tomatoes and citrus. High salinity levels disrupt nutrient uptake, leading to deficiencies in essential minerals required for vitamin synthesis. As a result, the vitamin C content in fruits decreases, reducing the nutritional value of the produce.
   - In salt-sensitive crops, lower vitamin C levels impact the fruit's appeal to health-conscious consumers, as vitamin C is a key antioxidant with recognized health benefits. Reduced nutritional value under salinity stress poses a significant challenge for fruit producers, especially in regions where soil salinization is common.
4. **Increase in Phenolic Compounds and Antioxidant Activity**
   - Similar to drought, salinity stress induces oxidative stress, prompting plants to increase the synthesis of phenolic compounds and other antioxidants as a defense mechanism. These compounds enhance the fruit's antioxidant profile, which can be a positive aspect in terms of health benefits.

- In pomegranate, for example, salinity stress is associated with increased levels of polyphenols, flavonoids, and tannins, which contribute to the fruit's color, bitterness, and antioxidant activity. While this boost in antioxidants is beneficial for health, excessive bitterness from phenolic compounds can affect the taste, making the fruit less appealing for direct consumption.

Impact of High Temperature on Fruit Quality

High temperatures, especially during the maturation phase, exert considerable influence on fruit quality traits, including sugar content, acidity, color, and texture. Heat stress accelerates ripening, alters metabolic pathways, and increases respiration rates, leading to changes in flavor, aroma, and shelf life.

1. **Accelerated Ripening and Overripening**
   - High temperatures speed up the ripening process, causing fruits to mature quickly and sometimes leading to overripening. Rapid ripening increases sugar accumulation in the short term but may result in a decline in quality if the fruit becomes overly soft or loses its desired texture.
   - In fruits like tomatoes, high temperatures can lead to a mushy texture as the fruit ripens too quickly, reducing firmness and shortening shelf life. Accelerated ripening under heat stress also affects harvest timing, as fruits may need to be picked earlier to prevent quality decline.
2. **Reduction in Acidity and Flavor Balance**
   - High temperatures tend to reduce acidity in many fruits, as organic acids degrade more quickly under heat. This decline in acidity affects flavor balance, leading to a less tangy or mildly flavored fruit, which may not be desirable in certain crops like citrus and tomatoes.
   - For citrus, lower acidity under heat stress results in a sweeter but often less balanced flavor, affecting consumer expectations, especially in varieties prized for their tangy taste.
3. **Changes in Color and Pigment Composition**
   - Heat stress can alter pigment composition, particularly anthocyanins and carotenoids, which contribute to fruit color. High temperatures may cause uneven or poor coloration, affecting the visual appeal of fruits like apples, peppers, and grapes.
   - In red grapes, for example, high temperatures may reduce anthocyanin accumulation, leading to lighter-colored berries. This change in color affects the aesthetic and market appeal, especially in grape varieties used for winemaking, where color intensity is a valued trait.
4. **Impact on Aroma and Volatile Compounds**
   - High temperatures influence the production of volatile compounds responsible for aroma and flavor. These compounds are critical for sensory quality, as they contribute to the unique aroma profiles of fruits.
   - In strawberries, for example, heat stress reduces the production of volatile esters, leading to a diminished aroma and flavor profile. This reduction in volatile compounds affects consumer perception, making the fruit less appealing in both fresh and processed forms.

Cumulative Impact of Abiotic Stress on Fruit Shelf Life and Post-Harvest Quality

The combined effects of drought, salinity, and heat stress not only affect the quality of fruit at harvest but also influence post-harvest characteristics like shelf life, susceptibility to spoilage, and storability. Stressed fruits often have compromised cell integrity, lower antioxidant levels, and altered pH, all of which impact their ability to withstand storage and transport.

1. **Reduced Shelf Life and Increased Susceptibility to Spoilage**
   - Abiotic stress compromises cell wall structure and membrane integrity, making fruits more prone to spoilage and microbial attack. Fruits grown under drought or heat stress may have weaker cell walls, leading to increased susceptibility to bruising and decay.
   - This issue is particularly problematic for fruits like grapes and tomatoes, where cell wall integrity is essential for firmness and shelf life. Stressed fruits require more careful handling and may have a shorter storage period, affecting supply chain logistics.
2. **Altered Nutritional Composition and Antioxidant Stability**
   - The antioxidant content of fruits, including vitamins and phenolic compounds, can decline over time, especially if the fruit has been exposed to abiotic stress. This decline affects nutritional quality and reduces the health benefits associated with fruit consumption.
   - In citrus, vitamin C levels degrade more rapidly in fruits affected by salinity and heat stress, leading to lower nutritional value during storage. This degradation presents a challenge for producers aiming to maintain quality in fresh markets, where nutrient content is a key selling point.
3. **Challenges in Marketability and Consumer Acceptance**
   - Fruits affected by abiotic stress may exhibit external and internal quality defects, including color irregularities, texture changes, and flavor imbalances. These factors influence consumer preference, as visually and texturally inconsistent fruits may be perceived as lower quality.
   - In commercial settings, stressed fruits may fail to meet quality standards, impacting their marketability. This is especially relevant for high-value crops like apples, grapes, and citrus, where appearance, taste, and texture are crucial for consumer acceptance.

6. Potential for Genetic and Biotechnological Interventions

To enhance the resilience of fruit crops to multiple abiotic stresses, advances in genetics and biotechnology play a critical role. Traditional breeding and modern biotechnological approaches are employed to develop varieties with enhanced stress tolerance, focusing on traits like deep rooting, osmolyte accumulation, and antioxidant capacity.

6.1 Breeding for Stress-Tolerant Varieties

The need for stress-tolerant varieties in fruit crops has become increasingly critical due to climate change and environmental variability, which introduce a range of abiotic stresses such as drought, salinity, extreme temperatures, and nutrient deficiencies. Developing varieties that can withstand these stressors is essential for sustaining yield, improving fruit quality, and ensuring food security. Conventional breeding has long focused on selecting genotypes with favorable stress-adaptive traits. However, recent advancements, including marker-assisted selection (MAS), genomic selection, and gene-editing technologies, have revolutionized the breeding process, making it more precise, efficient, and targeted.

The Importance of Stress-Tolerant Varieties in Fruit Crops

Abiotic stresses pose significant threats to fruit crops, impacting productivity, quality, and sustainability. Fruit crops are particularly vulnerable due to their perennial nature and high sensitivity to environmental fluctuations. Developing stress-tolerant varieties provides a sustainable solution to managing these challenges, allowing crops to maintain growth, yield, and quality under unfavorable conditions.

1. **Maintaining Yield and Quality Under Stress**
   - Stress-tolerant varieties can better maintain physiological processes, such as photosynthesis and nutrient uptake, under adverse conditions, reducing yield losses and maintaining fruit quality. For instance, drought-tolerant mango varieties can withstand water scarcity without sacrificing fruit size or sugar content.
   - Breeding for stress tolerance also addresses market demands for high-quality fruit with consistent texture, flavor, and nutritional value, even when grown under challenging environmental conditions. This consistency is essential for both fresh and processed fruit markets, where consumer expectations for quality are high.
2. **Ensuring Resilience in Perennial Crops**
   - Fruit crops are generally perennial and remain in the field for several years, making them more susceptible to cumulative environmental stresses over time. Developing varieties that can adapt to long-term climatic shifts, such as rising temperatures and soil salinization, is essential for the sustainability of orchards and vineyards.
   - Stress-tolerant varieties allow farmers to manage resource inputs more efficiently, reducing the need for excessive irrigation or soil amendments, which are often required to mitigate stress in sensitive varieties. This resilience reduces costs and conserves natural resources, making fruit production more sustainable.

Conventional Breeding for Stress Tolerance

Conventional breeding methods have laid the foundation for developing stress-tolerant fruit crops, relying on the selection of genotypes that exhibit adaptive traits under specific stress conditions. This process involves screening diverse germplasm for traits such as drought resistance, salt tolerance, and heat resilience, followed by cross-breeding to combine favorable traits into a single cultivar.

1. **Trait Selection and Germplasm Screening**
   - The first step in conventional breeding is identifying traits that contribute to stress tolerance. For instance, traits like deep rooting systems, efficient water use, and osmotic adjustment are associated with drought tolerance, while ion exclusion and compartmentalization are crucial for salinity tolerance.

- Breeders screen existing germplasm collections, including wild relatives and landraces, to identify sources of stress tolerance. Wild relatives of crops often exhibit superior stress-adaptive traits, as they have evolved in natural environments with limited human intervention. For example, wild grape species from arid regions often exhibit greater drought tolerance than cultivated varieties.

2. **Hybridization and Selection**
   - After identifying promising genotypes, breeders use hybridization to cross plants with desirable traits. Through successive generations, plants are selected based on their ability to thrive under stress conditions, gradually accumulating stress-tolerant traits in the progeny.
   - This process is time-consuming and requires extensive field trials under various stress conditions to confirm the stability and effectiveness of the selected traits. In perennial fruit crops, which have long life cycles, breeding can take several years or even decades to develop a stable, stress-tolerant variety. However, the result is often a robust cultivar with a strong genetic foundation for stress adaptation.

3. **Limitations of Conventional Breeding**
   - Conventional breeding is limited by the complexity of stress tolerance traits, which are often governed by multiple genes and influenced by environmental interactions. This complexity makes it challenging to achieve consistent results, as breeders cannot directly control the expression of specific genes.
   - The long generation time of fruit crops further slows down the breeding process, delaying the release of improved varieties. Additionally, conventional breeding relies on natural genetic variation, which may not always provide the specific traits needed to combat emerging stressors associated with climate change.

Advancements in Marker-Assisted Selection (MAS)

Marker-assisted selection (MAS) has transformed breeding for stress tolerance by allowing breeders to target specific genetic markers associated with desirable traits. MAS accelerates the breeding process by enabling early selection, reducing the reliance on phenotypic evaluations, and increasing the precision of trait incorporation.

1. **Identification of Genetic Markers Linked to Stress Tolerance**
   - MAS relies on identifying molecular markers (such as single nucleotide polymorphisms, or SNPs) that are closely associated with genes conferring stress tolerance. These markers serve as proxies for the desired traits, allowing breeders to track the inheritance of stress-tolerance genes in breeding populations.
   - For example, in salt-tolerant tomato breeding, markers linked to Na^+ and Cl^- exclusion mechanisms enable breeders to select for genotypes that effectively manage ionic stress. This targeted approach increases the efficiency of selecting salt-tolerant individuals, reducing the need for extensive field trials.

2. **Application of MAS in Early-Stage Selection**
   - With MAS, breeders can conduct early-stage selection by analyzing seedlings or young plants for the presence of markers linked to stress tolerance traits. This early selection accelerates the breeding process by identifying potential candidates before stress conditions are applied, saving time and resources.
   - In drought-tolerant grape breeding, for instance, markers associated with root architecture traits are used to select individuals with deep-rooting tendencies at an early stage, improving water access and resilience to drought. MAS allows breeders to select for these traits even before field trials, expediting the development of drought-tolerant grape varieties.
3. **Integration of MAS with Conventional Breeding**
   - MAS is often used in conjunction with conventional breeding to increase the efficiency of trait selection. After initial hybridization, MAS enables the identification of individuals carrying the desired stress-tolerance genes, focusing resources on promising candidates.
   - In citrus breeding, for example, MAS is integrated with traditional breeding to develop varieties with combined tolerance to drought and salinity. By selecting individuals with markers linked to water-use efficiency and salt exclusion, breeders can accelerate the development of resilient citrus cultivars suited for semi-arid and saline environments.

Genomic Selection and Its Role in Breeding for Stress Tolerance

Genomic selection (GS) is a more advanced technique that leverages genome-wide data to predict the breeding value of individual plants. Unlike MAS, which targets specific markers, GS uses genome-wide markers to create a predictive model for complex traits, enabling breeders to make more informed selection decisions.

1. **Genome-Wide Prediction Models**
   - Genomic selection involves creating a training population where both genotypic and phenotypic data are collected. Using this data, statistical models are developed to predict the performance of untested individuals based on their genetic profile.
   - In drought-tolerant apple breeding, GS models are used to predict water-use efficiency, leaf water potential, and root architecture based on genome-wide marker data. These predictions enable breeders to select individuals with high predicted tolerance levels, even without extensive phenotypic testing under drought conditions.
2. **Efficiency in Breeding for Polygenic Traits**
   - Many stress-tolerance traits are polygenic, meaning they are controlled by multiple genes with small effects. Genomic selection is particularly useful for these traits, as it considers the cumulative effect of numerous markers across the genome, providing a more accurate prediction of stress tolerance.
   - In salinity tolerance breeding for pomegranate, GS enables breeders to account for multiple physiological traits, such as ion exclusion, osmotic adjustment, and antioxidant

activity, which collectively contribute to salinity resilience. This approach allows breeders to select for overall stress tolerance rather than individual traits, resulting in more resilient varieties.

3. **Accelerating Breeding Cycles**
   - By reducing the need for extensive field trials and accelerating selection, GS shortens the breeding cycle, which is especially valuable for perennial fruit crops with long juvenile phases. This shortened cycle allows breeders to introduce new stress-tolerant varieties more rapidly in response to emerging environmental challenges.
   - For crops like mango, where conventional breeding takes several years due to long juvenile phases, GS enables breeders to make faster progress in developing drought and heat-tolerant varieties. This accelerated breeding process is critical in adapting fruit production to climate change.

Gene Editing and Transgenic Approaches for Stress Tolerance

Gene-editing technologies like CRISPR/Cas9 have introduced new possibilities for directly modifying genes associated with stress tolerance. These tools offer unprecedented precision, allowing breeders to target specific genes and enhance stress-resilience traits without relying on natural variation.

1. **CRISPR/Cas9 and Targeted Gene Editing**
   - CRISPR/Cas9 enables breeders to edit genes with precision, allowing for the knockout, insertion, or modification of specific stress-tolerance genes. This technique has shown promise in improving stress tolerance by targeting genes involved in water-use efficiency, ion transport, and antioxidant production.
   - In citrus, CRISPR/Cas9 has been used to modify genes related to drought tolerance, enhancing water-use efficiency without affecting growth or fruit quality. Such targeted modifications allow breeders to introduce stress tolerance without undesirable side effects, making gene editing a valuable tool in fruit crop breeding.

2. **Development of Transgenic Varieties**
   - Transgenic approaches involve the introduction of genes from other species to confer stress tolerance. While these methods are more complex and face regulatory challenges, they offer the potential for substantial improvements in stress resilience.
   - For example, transgenic tomato varieties have been developed with genes for salinity tolerance, enabling these plants to maintain growth and productivity in saline soils. The introduction of genes from halophytes (salt-tolerant plants) has proven effective in enhancing salinity tolerance, although consumer acceptance and regulatory hurdles remain a challenge.

3. **Ethical and Regulatory Considerations**
   - Gene editing and transgenic methods raise ethical and regulatory considerations, particularly concerning consumer acceptance and environmental impact. However, these technologies offer a promising avenue for developing stress-tolerant varieties,

especially as climate change introduces new challenges that traditional breeding may struggle to address.

- In regions facing severe water scarcity and soil degradation, the adoption of gene-edited and transgenic varieties could provide a sustainable solution, though careful consideration of biosafety and public perception is essential.

6.2 Genetic Engineering and CRISPR Technology

The field of genetic engineering has advanced significantly, providing researchers with powerful tools to introduce or modify specific genes responsible for stress tolerance in plants. Traditional genetic engineering involves the insertion of genes into a plant's genome to enhance stress response capabilities, such as increased water-use efficiency, ion regulation, or oxidative stress mitigation. More recently, CRISPR (Clustered Regularly Interspaced Short Palindromic Repeats) technology has revolutionized genetic engineering by allowing precise, targeted editing of genes associated with stress response. CRISPR technology is enabling the development of fruit crop varieties that can withstand multiple stressors simultaneously, potentially transforming fruit production in areas prone to drought, salinity, and temperature extremes.

The Role of Genetic Engineering in Stress Tolerance

Genetic engineering introduces exogenous genes or modifies endogenous genes associated with stress tolerance, giving plants the tools they need to survive in challenging environments. By enhancing pathways related to stress response, such as abscisic acid (ABA) synthesis and antioxidant production, genetic engineering helps fruit crops withstand adverse conditions that would otherwise reduce yield and quality.

1. **Insertion of Stress-Tolerance Genes**
   - Genetic engineering can introduce specific genes from other organisms that enhance a plant's ability to cope with environmental stress. For example, genes from halophytes (salt-tolerant plants) can be transferred to salt-sensitive crops, improving their ability to exclude sodium ions and tolerate high-salinity conditions.
   - In citrus, transgenic varieties have been developed by introducing genes related to osmotic adjustment and ion transport, enabling the plant to manage water and salt balance more effectively. These transgenes help the plant maintain cellular stability in saline soils, preserving growth and fruit quality.
2. **Enhancing Abscisic Acid (ABA) Synthesis Pathways**
   - ABA is a plant hormone that regulates stress responses, particularly in relation to drought. Genetic engineering can enhance ABA synthesis pathways, enabling plants to respond more effectively to water deficit by closing stomata, reducing transpiration, and conserving water.
   - In grapevine, for instance, transgenic approaches have been used to upregulate genes involved in ABA biosynthesis, resulting in improved drought tolerance. By closing stomata during water scarcity, these genetically engineered grapevines reduce water loss and sustain productivity even in dry conditions.

3. **Improving Antioxidant Production for Oxidative Stress Management**

   - Abiotic stressors like heat, salinity, and drought increase the production of reactive oxygen species (ROS) in plants, causing oxidative damage to cells. Genetic engineering can introduce genes that boost antioxidant enzyme activity, such as superoxide dismutase (SOD), catalase (CAT), and ascorbate peroxidase (APX), helping plants detoxify ROS and protect cellular structures.
   - Transgenic tomato plants with enhanced SOD and CAT enzyme activity have shown increased tolerance to oxidative stress under drought and high temperatures. This improvement in ROS detoxification reduces cellular damage, supporting better growth and yield under stress.

CRISPR Technology: A Precision Tool for Gene Editing

CRISPR technology represents a major leap in genetic engineering, providing unprecedented precision in gene editing. Unlike traditional transgenic approaches that introduce foreign genes, CRISPR allows researchers to edit the plant's own genes by targeting specific DNA sequences for modification. This precision enables the creation of stress-tolerant crops without introducing foreign DNA, which often reduces regulatory barriers and enhances consumer acceptance.

1. **Mechanism of CRISPR/Cas9 and Targeted Gene Editing**

   - CRISPR technology uses a guide RNA to direct the Cas9 nuclease to a specific DNA sequence within the plant's genome. Once bound to the target, Cas9 makes a cut in the DNA, which can then be repaired to introduce a mutation, delete a sequence, or insert a new gene.
   - In fruit crops like tomatoes and apples, CRISPR/Cas9 has been used to knock out genes associated with excessive water loss or susceptibility to salinity. By targeting these genes, researchers can suppress undesirable traits while enhancing the plant's natural resilience to environmental stress.

2. **Editing Genes Associated with Drought Tolerance**

   - CRISPR technology enables targeted modification of genes that regulate water-use efficiency, root architecture, and ABA synthesis. For instance, editing genes that control stomatal density and aperture can reduce water loss by limiting transpiration, a critical adaptation for drought tolerance.
   - In citrus, CRISPR has been used to edit genes that enhance ABA sensitivity, allowing the plant to respond more effectively to drought by closing stomata earlier and conserving water. This modification supports the plant's survival and productivity under prolonged drought conditions.

3. **Enhancement of Antioxidant Pathways Using CRISPR**

   - Targeted editing of antioxidant pathway genes helps plants manage oxidative stress more efficiently. CRISPR can upregulate genes involved in the production of antioxidant enzymes, allowing plants to better neutralize ROS generated under high salinity, heat, or drought.

- For example, in tomato, CRISPR editing has been applied to increase the expression of genes involved in glutathione and ascorbate synthesis, which play a vital role in ROS scavenging. By enhancing the plant's antioxidant defenses, CRISPR technology contributes to greater resilience against combined abiotic stresses.

Applications of Genetic Engineering and CRISPR in Developing Stress-Tolerant Fruit Crops

Genetic engineering and CRISPR have been applied to a variety of fruit crops, each with unique stress-tolerance goals tailored to specific environmental challenges. The flexibility of these technologies enables the precise enhancement of desired traits, providing fruit crops with the ability to withstand multiple stressors simultaneously.

1. **Drought-Tolerant Grapes and Citrus**
   - Drought stress is a primary concern for grape and citrus production in arid and semi-arid regions. Genetic engineering has introduced drought-response genes in these crops, enhancing their ability to retain water and minimize damage during dry periods.
   - In grapes, both traditional transgenic approaches and CRISPR have been used to improve root system architecture, allowing the plant to access deeper soil moisture. Enhanced root depth improves the plant's ability to sustain growth under water-limited conditions, supporting yield and quality even in prolonged drought.
2. **Salt-Tolerant Tomatoes and Pomegranates**
   - Salinity is a significant issue in coastal areas and regions with poor-quality irrigation water. Genetic engineering has introduced salt tolerance genes in tomatoes, such as those responsible for Na^+ and Cl^- exclusion, allowing these plants to thrive in saline soils.
   - In pomegranate, CRISPR has been employed to increase the expression of genes that regulate ion compartmentalization, a mechanism that helps prevent toxic levels of sodium from accumulating in photosynthetically active tissues. By sequestering sodium in vacuoles, these pomegranate varieties can tolerate higher soil salinity, reducing yield losses in salt-affected regions.
3. **Heat-Resilient Strawberries and Apples**
   - Rising temperatures during the growing season can impair fruit quality, ripening, and overall productivity. In strawberries and apples, genetic engineering has been used to introduce heat-shock proteins and other molecular chaperones that protect cellular structures from heat-induced damage.
   - CRISPR technology enables researchers to fine-tune heat tolerance by editing genes involved in temperature-sensitive metabolic pathways. For example, in apples, CRISPR has been applied to reduce the expression of ethylene-responsive genes, helping to delay overripening under high temperatures and extending the fruit's marketable life.
4. **Multi-Stress Resilience Using CRISPR Stacking**
   - One of the significant advantages of CRISPR technology is its ability to stack multiple gene edits, allowing for the development of crops that can tolerate several abiotic stresses

simultaneously. By targeting multiple stress-response genes, CRISPR enables breeders to create fruit varieties that can handle complex environmental challenges.

- In citrus, for example, CRISPR stacking has been used to edit genes related to both drought and salinity tolerance. This multi-stress resilience allows citrus trees to thrive in regions that experience both high soil salinity and intermittent drought, providing a sustainable solution for fruit production in challenging climates.

Ethical, Regulatory, and Consumer Considerations

While genetic engineering and CRISPR hold great potential for developing stress-tolerant fruit crops, there are several ethical, regulatory, and consumer considerations that influence the adoption of these technologies.

1. **Regulatory Hurdles and Approval Processes**
    - Traditional genetically engineered crops face stringent regulatory processes in many countries, primarily due to concerns about environmental impact, food safety, and the introduction of foreign DNA. However, CRISPR-edited crops, which do not necessarily involve foreign genes, have been subject to more lenient regulations in certain regions, such as the United States.
    - In fruit crops, regulatory acceptance varies widely depending on the country and the specific gene-editing approach used. This regulatory variability affects the global availability and commercialization of genetically modified and gene-edited fruit varieties.
2. **Consumer Acceptance and Public Perception**
    - Consumer acceptance of genetically modified and gene-edited foods remains mixed. While CRISPR technology is often perceived more favorably due to its precision and lack of foreign DNA, there is still skepticism among some consumers regarding the safety and ethics of genetic modifications.
    - In high-value crops like apples and strawberries, consumer acceptance is particularly important, as these fruits are often sold directly to consumers in fresh markets. Transparent labeling and consumer education regarding the benefits of CRISPR and genetic engineering in enhancing stress tolerance may improve public perception and acceptance.
3. **Environmental and Ethical Considerations**
    - The introduction of genetically modified and gene-edited crops raises ethical questions about biodiversity and the potential impact on natural ecosystems. There is concern that widespread use of these crops could reduce genetic diversity and contribute to the homogenization of agricultural landscapes.
    - Ethical considerations also include the rights of farmers to save and exchange seeds, as genetically engineered and gene-edited varieties are often patented. Ensuring fair access to these technologies, particularly for small-scale farmers, is crucial for the equitable distribution of the benefits associated with stress-tolerant varieties.

Chapter - 5

Mitigation Measures and Conservation Practices - Relevance to Abiotic Stresses

Mitigation Measures and Conservation Practices

Abiotic stresses, such as extreme temperatures, drought, and salinity, are exacerbated by environmental factors such as the greenhouse effect and methane emissions. These factors contribute to climate change, impacting fruit crop productivity and quality. Addressing these challenges requires a combination of mitigation measures and conservation practices tailored to the specific stress factors affecting crop growth. Key approaches in stress management include understanding the greenhouse effect and methane emissions in the context of crop stress, using antitranspirants to reduce water loss, and applying plant growth regulators (PGRs) to enhance stress resilience.

A. The Greenhouse Effect and Methane Emissions Concerning Abiotic Stress

The greenhouse effect is a crucial factor in regulating Earth's climate, allowing the planet to maintain a temperature conducive to life. This effect occurs as greenhouse gases (GHGs) like carbon dioxide (CO_2) and methane (CH_4) trap heat within the atmosphere, preventing it from escaping into space. However, human activities—such as fossil fuel combustion, industrial processes, and deforestation—have significantly increased GHG levels, intensifying the greenhouse effect and leading to global warming. This rise in atmospheric temperatures has profound implications for agriculture, particularly in the form of increased abiotic stresses such as heat, drought, and irregular precipitation patterns. These stresses impact fruit crops, which are often sensitive to environmental changes, making it essential to understand the greenhouse effect's role in agricultural stress management and to explore strategies for mitigating its impact.

1. Understanding the Greenhouse Effect and Its Impact on Abiotic Stress

The greenhouse effect is a natural process that warms the Earth's surface. When the Sun's energy reaches the Earth's atmosphere, some of it is reflected to space, while the rest is absorbed by the surface, warming it. The Earth then radiates this heat in the form of infrared energy, but greenhouse gases trap a portion of this energy, keeping the planet warm. This process is vital for life, as it maintains temperatures that allow ecosystems to thrive. However, increased GHG levels have amplified this natural effect, resulting in an overall warming trend known as global warming.

DOI: 10.1201/9781003717409-5

1. **The Role of Greenhouse Gases (GHGs) in Climate Regulation**
   - Greenhouse gases, including CO_2, CH_4, nitrous oxide (N_2O), and water vapor, are responsible for trapping heat in the Earth's atmosphere. Each of these gases has a different capacity for trapping heat, with CH_4 being over 25 times more effective than CO_2 at retaining heat over 100years.
   - While GHGs occur naturally and are necessary for climate stability, their concentrations have risen sharply due to human activities. CO_2 levels, for example, have increased by approximately 40% since the Industrial Revolution, largely due to fossil fuel burning and deforestation. Methane levels have also doubled, primarily from agriculture, livestock production, and waste decomposition.
   - As GHG concentrations rise, more heat is trapped, causing an increase in global temperatures. This warming effect disrupts climatic patterns, contributing to phenomena such as prolonged droughts, more intense storms, and irregular rainfall. These altered patterns of temperature and precipitation present significant challenges to agricultural production, particularly for fruit crops that are sensitive to temperature changes and water availability.
2. **Human Activities Contributing to the Enhanced Greenhouse Effect**
   - **Fossil Fuel Combustion**: The burning of fossil fuels—such as coal, oil, and natural gas—releases large amounts of CO_2, contributing to over 75% of global GHG emissions. These fuels are used extensively for energy production, transportation, and industry, making them a significant contributor to global warming.
   - **Deforestation and Land Use Changes**: Forests act as carbon sinks, absorbing CO_2 from the atmosphere. However, deforestation and land conversion for agriculture release this stored carbon, reducing the Earth's capacity to offset emissions. The loss of forest cover also disrupts local rainfall patterns and soil stability, which exacerbates abiotic stresses for crops.
   - **Agricultural Practices**: Agriculture is a major source of methane and nitrous oxide, both potent GHGs. Livestock production, particularly ruminants like cattle, produces methane during digestion. Rice paddies and improper manure management also contribute to methane emissions. Additionally, the use of nitrogen-based fertilizers releases nitrous oxide, which has a global warming potential 298 times that of CO_2.
3. **Impact of Global Warming on Abiotic Stress Factors in Agriculture**
   - **Increased Temperature and Heat Stress**: As global temperatures rise, fruit crops experience more frequent and intense heat waves. High temperatures can disrupt photosynthesis, respiration, and transpiration, leading to reduced growth, poor fruit set, and lower yields. Heat stress is particularly damaging during flowering and fruiting stages, where even slight temperature increases can affect crop productivity and quality.
   - **Altered Precipitation Patterns and Drought**: Climate change is altering precipitation patterns, resulting in longer dry spells and irregular rainfall distribution. Drought stress

impairs water uptake, reduces nutrient availability, and inhibits root growth, making it difficult for fruit crops to maintain cellular hydration. Crops such as grapes, mangoes, and apples are highly susceptible to water scarcity, which affects their growth, fruit size, and sugar content.

- **Increased Frequency of Extreme Weather Events**: Climate change has led to a rise in extreme weather events, such as storms, floods, and cyclones. These events damage crops, erode soil, and disrupt farming operations, leading to substantial economic losses. Fruit crops, especially those with perennial systems like orchards, suffer from prolonged recovery times after storm damage.

4. **Implications of Greenhouse Gases for Salinity Stress and Soil Degradation**

- The greenhouse effect contributes to sea level rise, a phenomenon driven by the melting of polar ice caps and the thermal expansion of seawater as it warms. Rising sea levels lead to saltwater intrusion in coastal areas, increasing soil salinity in agricultural lands. Salinity stress impacts crop growth by creating osmotic imbalances and ion toxicity, both of which impair root function and nutrient absorption.
- In salt-sensitive fruit crops such as citrus, high soil salinity disrupts root uptake, reduces leaf turgor, and inhibits photosynthesis, leading to stunted growth and lower fruit yield. This effect is particularly pronounced in coastal regions, where groundwater becomes saline, reducing the availability of fresh water for irrigation.
- Soil degradation, including increased salinization, erosion, and nutrient loss, is exacerbated by climate change. As soil quality declines, its capacity to support healthy crop growth diminishes, making it challenging to sustain productivity in fruit crops. Conservation practices and soil amendments are therefore essential to mitigate the negative impacts of soil degradation on fruit production.

5. **Methane's Role in Climate Change and Agricultural Stress**

- Methane (CH_4), though present in lower concentrations than CO_2, has a global warming potential that is significantly higher. Agriculture, particularly livestock production, is one of the primary sources of methane emissions. Methane is released during the digestive process in ruminants and from the anaerobic decomposition of organic matter in waterlogged soils, such as rice paddies.
- The increase in methane contributes directly to warming, which heightens the impact of abiotic stress on crops. Higher temperatures resulting from methane emissions create a feedback loop, intensifying heat and drought stress on fruit crops. Additionally, methane-driven warming affects ecosystem dynamics, potentially altering pest and disease pressures in agricultural systems.
- The cumulative effect of methane emissions, combined with CO_2 and other GHGs, accelerates climate-induced stress factors that challenge fruit crops, pushing them beyond their optimal growing conditions. This added pressure necessitates adaptive management practices to protect yields and ensure sustainable production.

6. **Climate Change-Induced Abiotic Stress and Crop Vulnerability**

    - Rising temperatures, driven by the greenhouse effect and methane emissions, lead to a greater incidence of abiotic stresses such as heat, drought, and salinity. These stressors affect fruit crops by disrupting their physiological functions and reducing their ability to absorb water and nutrients.
    - For example, high temperatures increase the rate of transpiration, causing water loss and dehydration in crops like citrus, mango, and apple. In drought-prone regions, reduced rainfall exacerbates water scarcity, further stressing these crops and impacting fruit size, quality, and yield.
    - Increased abiotic stress from climate change also influences the timing of growth stages, such as flowering, pollination, and fruit set. As temperatures fluctuate, crops may experience shifts in their development cycles, leading to poor synchronization with pollinators, reduced fruit set, and lower yields.

Higher Temperatures and Their Impact on Heat and Drought Stress in Fruit Crops

The greenhouse effect has led to an increase in global temperatures, which directly impacts agricultural systems by intensifying heat and drought stress. These elevated temperatures increase evaporation rates, reducing soil moisture levels and making crops more vulnerable to water scarcity. As a result, fruit crops face heightened challenges in maintaining physiological functions essential for growth, such as photosynthesis, respiration, and nutrient transport. The increased incidence of heat and drought stress due to higher temperatures under the greenhouse effect disrupts crop productivity and quality, posing significant threats to sustainable fruit production.

1. **Increased Evaporation and Soil Moisture Loss**

    - Higher temperatures accelerate the rate of evaporation from both soil and plant surfaces. In areas experiencing prolonged heat, soil moisture depletion occurs rapidly, creating conditions that limit the availability of water for fruit crops. This intensified evaporation cycle is particularly problematic in regions where water resources are already scarce or seasonal.
    - In fruit orchards, such as those for mango, citrus, and apple, water scarcity due to increased evaporation leads to reduced root-zone moisture. This limited water availability stresses the plant's root system, impairing the uptake of nutrients and water essential for growth. Reduced soil moisture limits the plant's ability to maintain cell turgor and carry out basic metabolic activities, contributing to slower growth, smaller fruits, and lower yields.

2. **Increased Transpiration Rates and Water Deficit in Plants**

    - Elevated temperatures drive higher rates of transpiration, the process by which plants lose water vapor through stomata. While transpiration is a necessary mechanism for cooling and nutrient transport, excessive water loss due to high temperatures leads to water deficit within the plant.
    - As plants lose water faster than they can absorb it from the soil, they enter a state of

water stress. This is particularly detrimental to fruit crops that require steady hydration to support fruit development, such as grapes, apples, and citrus. Under severe heat, plants may close their stomata to reduce water loss, but this also limits CO_2 uptake, disrupting photosynthesis and reducing energy production.

3. **Disruption of Photosynthesis and Reduced Carbon Assimilation**

   - Photosynthesis is highly sensitive to temperature, and elevated temperatures under greenhouse conditions disrupt the efficiency of this critical process. Optimal photosynthesis typically occurs within a specific temperature range, and higher temperatures can inhibit the functioning of photosynthetic enzymes and reduce the stability of chloroplast membranes.
   - Heat stress can lead to a reduction in chlorophyll content, affecting the plant's ability to capture sunlight and produce energy. Additionally, high temperatures interfere with the Calvin cycle, the phase of photosynthesis responsible for carbon fixation. As a result, the plant's ability to convert atmospheric CO_2 into sugars is compromised, reducing the energy available for growth and fruit development.
   - In fruit crops like tomatoes and strawberries, heat stress-induced photosynthetic inhibition leads to lower sugar production, affecting the sweetness, flavor, and overall quality of the fruits. Reduced photosynthetic efficiency directly impacts yield, as plants struggle to maintain productivity under high temperatures.

4. **Increased Respiration Rates and Energy Loss**

   - High temperatures accelerate respiration, the process by which plants break down sugars to release energy for growth and maintenance. While respiration is essential for cellular functions, elevated temperatures cause plants to respire at higher rates, consuming more of the sugars produced during photosynthesis.
   - In fruit crops, the energy cost of increased respiration under heat stress can surpass the energy generated through photosynthesis, leading to a net energy deficit. This energy imbalance weakens the plant, reducing its ability to produce new leaves, flowers, and fruits. Increased respiration under heat stress is particularly detrimental during fruit ripening, as the plant allocates more energy to respiration than to building up sugars and other compounds that contribute to fruit quality.

5. **Impact on Flowering, Pollination, and Fruit Set**

   - Elevated temperatures from the greenhouse effect can disrupt reproductive stages such as flowering, pollination, and fruit set. High temperatures during flowering can lead to poor pollen viability, reduced pollen tube growth, and incomplete fertilization, resulting in fewer fruits per plant.
   - For instance, in crops like mango and citrus, heat stress during the flowering season can cause flower drop, reducing the number of flowers available for pollination. The heat also affects the timing and viability of pollen, impacting pollination success and ultimately reducing fruit set.

- Reduced fruit set leads to lower yield potential, with significant economic implications for growers. Additionally, the fruits that do develop may be smaller or misshapen due to inadequate nutrient and water supply during critical growth stages.

6. **Vulnerability to Secondary Abiotic Stresses**
   - Elevated temperatures often co-occur with drought, creating compounded stress that exacerbates crop vulnerability. Fruit crops experiencing both heat and water stress struggle to maintain cellular hydration and nutrient transport, leading to reduced resilience against other abiotic stresses.
   - In regions where soil salinity is a concern, heat stress further intensifies salt accumulation in the root zone due to increased evaporation, making it more difficult for plants to access water and nutrients. This combination of heat, drought, and salinity stress can be particularly damaging in coastal orchards, where fruit crops are exposed to saline groundwater or soil.

7. **Overall Decline in Crop Productivity and Fruit Quality**
   - The cumulative effects of elevated temperatures on soil moisture, photosynthesis, respiration, and reproductive success lead to a decline in both yield and fruit quality. Heat-stressed fruit crops tend to produce smaller, less flavorful fruits with lower sugar content and diminished nutritional value.
   - In high-value crops like apples, grapes, and citrus, these changes in fruit quality impact marketability and profitability, as consumers may perceive heat-stressed fruits as less desirable. Lower yields and reduced fruit quality under heat and drought stress present economic challenges for growers, particularly in regions where climate change has led to more frequent and intense temperature fluctuations.

Impact of Temperature Fluctuations During Critical Growth Stages in Fruit Crops

Temperature plays a crucial role in the growth and development of fruit crops, especially during sensitive stages such as flowering, pollination, and fruit set. These stages are highly susceptible to environmental changes, and even minor temperature fluctuations can disrupt physiological and reproductive processes, leading to significant impacts on yield and quality. Higher temperatures, as a result of the greenhouse effect, often coincide with these critical growth periods, creating additional stress for fruit crops and reducing their productivity and marketable quality. In particular, crops like apples and grapes experience lower fruit sets and compromised fruit quality when exposed to high temperatures during flowering and fruit set stages.

1. **Sensitivity of Flowering and Fruit Set Stages to Temperature**
   - The flowering and fruit set stages are highly sensitive to environmental conditions, as they involve complex processes such as pollen development, pollination, and fertilization. Optimal temperatures are required for pollen viability, stigma receptivity, and successful fertilization, all of which contribute to fruit set and subsequent development.
   - Temperature fluctuations, particularly increases, can disrupt these processes by affecting

pollen structure and function, leading to lower pollen viability and poor pollination success. When temperatures exceed the crop's tolerance threshold, the chances of successful fertilization diminish, resulting in fewer fruits per plant and lower overall yield.

- In perennial fruit crops, such as apples and grapes, which produce flowers and fruits once per growing season, any disruption during flowering and fruit set can lead to a season-long impact on productivity. This makes temperature management crucial during these growth stages to ensure a healthy crop and optimal fruit yield.

2. **Effect of High Temperatures on Pollen Viability and Pollination Success**

- High temperatures can adversely affect pollen development, viability, and function, which are essential for successful pollination and fruit set. In apples, for example, high temperatures during flowering reduce pollen viability, as heat damages pollen cells, impairing their ability to germinate and form pollen tubes.
- Grapes also experience reduced pollination success under high temperatures, as the heat disrupts pollen-stigma interactions, decreasing the likelihood of fertilization. The reduced viability and functionality of pollen due to high temperatures lead to incomplete fertilization and fewer seeds, resulting in smaller fruit set and lower yield.
- Pollination success is further impacted when high temperatures shorten the window of stigma receptivity, the period during which the stigma can accept and support pollen germination. Under heat stress, this window becomes limited, reducing the number of successful fertilization events and the potential fruit set.

3. **Impact on Fruit Set and Yield Reduction**

- High temperatures during flowering and fruit set stages lead to lower fruit set, as fewer flowers are successfully pollinated and fertilized. This is particularly evident in apple orchards, where a decrease in fruit set due to heat stress results in fewer apples per tree, directly reducing yield.
- In grapes, high temperatures during the flowering stage result in fewer grape clusters and smaller berries, as heat interferes with cell division and expansion within the developing fruit. This reduction in fruit set and berry size not only affects yield but also impacts grape quality, especially in wine production, where berry size and composition are critical for flavor and aroma.
- Lower fruit set and yield due to high temperatures are further compounded in regions where temperature spikes are frequent during flowering and fruit set. For crops that have a limited flowering period, such as apples, a single heat event can severely impact the entire season's production, underscoring the importance of temperature management strategies.

4. **Implications for Fruit Quality and Marketability**

- High temperatures during fruit set stages can influence the size, texture, and composition of fruits that do manage to develop. In grapes, for example, smaller berry size under heat stress leads to higher skin-to-pulp ratios, affecting the concentration of phenolic compounds, sugars, and acids in the fruit. These changes impact flavor, texture, and quality, which are critical attributes in wine grape varieties.

- Apples exposed to high temperatures during flowering and early fruit development may experience irregular growth patterns, resulting in misshapen or unevenly colored fruits. Such quality issues affect the marketability of apples, as consumers generally prefer fruits with uniform size, shape, and appearance.
- Reduced fruit size and altered composition also affect storage potential and shelf life, particularly for fresh-market fruits. Smaller, heat-stressed fruits may have reduced water content and nutritional quality, making them less desirable for consumers and reducing their overall value in the market.

5. **Long-Term Impacts on Crop Productivity and Economic Returns**
   - Repeated temperature fluctuations and high heat events during flowering and fruit set can have cumulative effects on perennial fruit crops. Over time, these stresses can weaken the plants, reducing their resilience to other stress factors such as drought and pests, ultimately lowering productivity.
   - For growers, reduced yield and compromised fruit quality translate to economic losses, especially for high-value crops like apples and grapes. Poor pollination, lower fruit set, and reduced fruit quality can affect profitability and increase the risk of crop failure in years with extreme temperature fluctuations.
   - Furthermore, the need for additional measures to counteract heat stress, such as shading or misting systems, increases production costs, impacting the overall sustainability of fruit crop farming in regions prone to climate variability.
6. **Adaptive Strategies to Manage Temperature Fluctuations During Flowering and Fruit Set**
   - Given the sensitivity of flowering and fruit set stages to temperature, adaptive management strategies are essential to mitigate the impact of heat stress. Shading techniques, such as the use of shade nets, can lower canopy temperatures and protect flowers and developing fruits from direct sunlight.
   - Installing misting or evaporative cooling systems in orchards helps reduce canopy temperatures, maintaining a favorable environment for flowering and fruit set even during high heat periods. These systems are particularly beneficial for high-value fruit crops like apples and grapes, where maintaining fruit quality is crucial.
   - Breeding and selecting heat-tolerant varieties are also essential long-term solutions. Varieties with heat-resilient pollen and enhanced fruit set stability under variable temperatures can help fruit crops maintain productivity despite temperature fluctuations. In grapes, for instance, certain cultivars exhibit better heat tolerance during flowering, making them more suitable for warmer climates.

2. Methane Emissions and Their Contribution to Climate-Induced Stress

Methane (CH_4) is a highly potent greenhouse gas, with a global warming potential that is over 25 times greater than carbon dioxide (CO_2) over a 100-year period. Despite its lower concentration in the atmosphere compared to CO_2, methane plays a disproportionately large role in contributing to global warming and climate change. Methane emissions originate from various natural and anthropogenic sources, including agriculture, livestock production, rice paddies, and wetlands. As methane levels rise, they contribute to the greenhouse effect, leading to increased atmospheric temperatures. This warming exacerbates heat stress, water scarcity, and irregular rainfall patterns, which have profound implications for fruit crop production. These climate-induced stresses challenge the growth, yield, and quality of fruit crops, necessitating strategies to manage their impact on agriculture.

Sources of Methane Emissions

1. **Agriculture and Livestock Production**
   - Agriculture is one of the largest sources of methane emissions, with livestock production being a major contributor. Ruminant animals, such as cattle, sheep, and goats, produce methane as part of their digestive process (enteric fermentation). This methane is released into the atmosphere when the animals exhale or burp.
   - Manure management also contributes to methane emissions. When manure is stored or treated in anaerobic conditions, such as in lagoons or pits, methane is produced as a byproduct of microbial decomposition.
   - Methane emissions from livestock and manure are a significant concern, especially as global demand for animal products increases. The rise in methane emissions from livestock has added to the greenhouse effect, exacerbating climate-induced stresses in agricultural regions.
2. **Rice Paddies and Wetlands**
   - Methane is also produced in waterlogged soils, where anaerobic conditions promote the activity of methanogenic bacteria. Rice paddies, which are flooded for long periods, create ideal conditions for methane production. The bacteria decompose organic material in the absence of oxygen, releasing methane as a byproduct.
   - Wetlands, both natural and artificial, are another source of methane. While natural wetlands play a vital role in ecosystems, they contribute to global methane emissions, especially as temperatures increase and wetland areas expand or shift due to climate change.
3. **Other Sources of Methane Emissions**
   - Other sources of methane include landfills, where organic waste decomposes anaerobically, and fossil fuel extraction, where methane is released during oil and gas production. These sources, combined with agricultural emissions, contribute significantly to atmospheric methane levels and intensify the greenhouse effect.
   - Methane leakage from natural gas systems is another critical source, as methane is the

primary component of natural gas. Leaks can occur during extraction, transportation, and processing, further increasing methane's contribution to climate change.

Methane's Role in the Greenhouse Effect and Contribution to Global Warming

Methane is an effective heat-trapping gas that significantly amplifies the greenhouse effect, contributing to global warming. Although methane has a shorter atmospheric lifetime than CO_2 (approximately 12 years compared to CO_2's hundreds of years), its impact on climate is intense during that period.

1. **Amplifying the Greenhouse Effect**
   - Methane traps heat in the Earth's atmosphere by absorbing infrared radiation, which would otherwise escape into space. As a potent greenhouse gas, methane has a substantial impact on the Earth's energy balance, creating an enhanced greenhouse effect that drives global warming.
   - As methane emissions increase, more heat is trapped in the atmosphere, leading to higher global temperatures. These rising temperatures create a feedback loop, where warming triggers further methane release from sources like permafrost and wetlands, intensifying climate change.
2. **Implications for Agriculture and Fruit Crop Production**
   - The enhanced greenhouse effect due to methane emissions increases atmospheric temperatures, which has direct implications for agriculture. Higher temperatures lead to heat stress in crops, particularly during critical growth stages. In fruit crops, heat stress can damage flowers, reduce fruit set, and impair fruit quality, impacting both yield and marketability.
   - For perennial fruit crops, such as apples, grapes, and citrus, sustained exposure to high temperatures over multiple seasons can reduce long-term productivity and affect the lifespan of orchards. Heat stress also influences fruit ripening, often accelerating the process and resulting in fruits that are smaller, less flavorful, and lower in nutritional content.

Increased Heat Stress and Water Scarcity in Fruit Crops

1. **Heat Stress and Physiological Challenges**
   - As temperatures rise due to methane-driven warming, fruit crops experience greater levels of heat stress, which affects photosynthesis, respiration, and water balance. Photosynthesis is highly temperature-sensitive, and excessive heat can inhibit enzyme function, reduce chlorophyll content, and disrupt the Calvin cycle, which is crucial for carbon fixation.
   - High temperatures also increase the rate of respiration, leading to a faster breakdown of sugars within the plant. This imbalance between photosynthesis and respiration reduces the energy available for growth and fruit development, weakening the plant's ability to produce high-quality fruits.
   - Heat stress during flowering and fruit set stages can lead to poor pollination, reduced fruit set, and increased fruit drop. In crops like citrus and grapes, high temperatures

impair pollen viability and reduce the receptivity of the stigma, leading to incomplete fertilization and fewer fruits per plant.

2. **Water Scarcity and Increased Drought Conditions**
   - Methane emissions contribute indirectly to changes in atmospheric patterns, which affect precipitation and water availability. Irregular rainfall and prolonged dry periods create water scarcity conditions that exacerbate drought stress in fruit crops.
   - As water becomes scarce, fruit crops struggle to maintain cellular hydration and nutrient transport, essential for physiological processes and growth. Drought conditions force plants to conserve water by closing stomata, which limits CO_2 uptake and reduces photosynthesis, leading to stunted growth and poor fruit quality.
   - In addition to affecting yield, water scarcity influences fruit composition, often leading to higher concentrations of sugars and acids as the plant compensates for dehydration. While this may enhance certain quality traits, it typically reduces fruit size and impacts the overall marketability of the produce.

Impact of Methane Emissions on Rainfall Patterns and Drought Frequency

Methane emissions contribute not only to increased temperatures but also to shifts in precipitation patterns, which can have significant effects on water availability for fruit crops.

1. **Changes in Rainfall Patterns**
   - Methane-driven global warming affects atmospheric circulation, which in turn influences rainfall distribution. Many regions have experienced shifts in their typical rainfall patterns, with some areas seeing reduced precipitation and others facing an increase in the intensity of storms.
   - Irregular rainfall creates challenges for fruit crop irrigation management, as water availability becomes unpredictable. Inconsistent rainfall disrupts the moisture levels required for stable growth, particularly in sensitive fruit crops that rely on specific water conditions for quality fruit development.
2. **Prolonged Dry Periods and Increased Drought Frequency**
   - Methane emissions contribute to prolonged dry spells and increased drought frequency in many agricultural regions. For fruit crops, these extended periods of drought increase water stress and force the plants to rely on limited soil moisture reserves, which are quickly depleted under high temperatures.
   - Drought conditions impact the long-term sustainability of fruit orchards, particularly those with perennial crops like apples, mangoes, and grapes, which require consistent water availability throughout their life cycle. Prolonged drought stress not only reduces current yields but also weakens the plant's resilience, making it more susceptible to future stress events.
   - In regions where drought conditions are intensified by methane-driven climate change, fruit growers face higher production costs due to increased irrigation needs. Water scarcity

also poses challenges for small-scale farmers who may not have access to advanced irrigation systems, reducing their ability to manage drought and heat stress effectively.

3. **Cumulative Impact on Crop Productivity and Quality**
   - The combined effects of heat stress, water scarcity, and irregular rainfall patterns due to methane emissions lead to a cumulative impact on fruit crop productivity and quality. Repeated exposure to these stressors weakens fruit plants, affecting their ability to recover between growing seasons.
   - In crops like citrus, apples, and berries, stress-induced reductions in fruit size, flavor, and texture make the produce less appealing to consumers, impacting market value. For crops used in processing, such as wine grapes, altered sugar and acid levels under stress conditions affect the final product's flavor profile and quality.
   - Additionally, methane-driven climate change increases the likelihood of extreme weather events, such as heatwaves and droughts, further compounding the challenges of fruit production in affected regions. Growers must adapt to these changing conditions by implementing climate-resilient practices, such as efficient irrigation systems, mulching, and selecting drought-tolerant crop varieties.

3. Impact of Climate Change on Salinity and Soil Degradation

Climate change, driven by the greenhouse effect and methane emissions, has significant implications for soil quality and salinity levels, especially in coastal areas. One of the most concerning impacts of climate change is the gradual rise in sea levels, a consequence of melting polar ice and the thermal expansion of seawater due to global warming. As sea levels rise, coastal regions face increased risks of saltwater intrusion, leading to the salinization of soil and groundwater resources. This heightened salinity poses a serious threat to agriculture, particularly for salt-sensitive fruit crops such as citrus, grapes, and strawberries, which suffer from osmotic stress and ion toxicity under saline conditions. The accumulation of salt in soil reduces water availability, disrupts root function, impairs nutrient uptake, and inhibits growth, adding another layer of stress for crops already challenged by heat and drought.

Sea Level Rise and Saltwater Intrusion in Coastal Soils

1. **Connection Between Climate Change, Sea Level Rise, and Soil Salinity**
   - Sea level rise is one of the most direct consequences of climate change, driven by factors such as melting glaciers and ice sheets, as well as the thermal expansion of ocean water as it warms. As sea levels rise, saltwater intrusion into coastal areas becomes more prevalent, leading to increased salinity in both soil and groundwater.
   - Saltwater intrusion occurs when seawater infiltrates freshwater aquifers and coastal soils, replacing freshwater resources with saline water. This process is especially pronounced during high tides and storm surges, where seawater can penetrate further inland, affecting farmland located close to the coast.
   - Coastal regions with high agricultural activity, such as those cultivating fruit crops, are particularly vulnerable to this intrusion. In these areas, saltwater intrusion results in

salinized soil and reduced access to freshwater for irrigation, compounding the challenges faced by farmers and impacting the productivity and quality of their crops.

2. **Implications for Salt-Sensitive Fruit Crops**
   - Salt-sensitive crops, such as citrus, grapes, and strawberries, are highly susceptible to the detrimental effects of increased soil salinity. These crops struggle to survive in saline environments because their root systems are less efficient at excluding sodium (Na^+) and chloride (Cl^-) ions, which are prevalent in seawater.
   - In saline conditions, fruit crops experience osmotic stress, a condition where high salt concentrations in the soil solution limit the plant's ability to absorb water through its roots. This osmotic imbalance creates a water deficit within the plant, causing dehydration and impairing vital physiological processes.
   - Salt-sensitive crops also experience ion toxicity due to the accumulation of Na^+ and Cl^- ions within their tissues. These ions interfere with cellular function, disrupt enzyme activity, and cause nutrient imbalances, leading to stunted growth, leaf chlorosis, and reduced photosynthetic efficiency. Over time, these effects contribute to poor fruit development, reduced yield, and lower crop quality.

Increased Soil Salinity and Reduced Water Availability

1. **Osmotic Stress and Water Uptake Challenges:**
   - Increased soil salinity intensifies osmotic stress, making it difficult for plants to take up water from the soil. As the concentration of salt in the soil solution rises, the osmotic pressure around the plant's roots increases, creating a water gradient that favors water movement away from the roots instead of into them.
   - For fruit crops already struggling with heat and drought, increased osmotic pressure exacerbates water scarcity, leading to a state of dehydration. To cope, plants may close their stomata to reduce water loss, but this also limits CO_2 uptake, reducing photosynthesis and energy production, which are crucial for growth and fruit formation.
   - In crops like grapes and strawberries, reduced water uptake due to high soil salinity affects berry size, sugar content, and overall fruit quality. Smaller fruit size, lower sugar accumulation, and poor flavor reduce the marketability of these crops, impacting both fresh and processed fruit industries.

2. **Disruption of Root Function and Ion Imbalance**
   - High salinity disrupts root function by causing ionic imbalances and toxicity. Sodium ions (Na^+), in particular, can displace essential nutrients like potassium (K^+) and calcium (Ca^{2+}) from root cells, leading to deficiencies in these vital nutrients.
   - Potassium is crucial for maintaining cellular function, enzyme activation, and osmotic regulation. When sodium displaces potassium, the plant's metabolic processes are disrupted, and cellular function deteriorates. This ionic imbalance affects nutrient transport, leaf turgor, and cellular integrity, which are essential for healthy plant growth and fruit development.

- In citrus crops, salt-induced potassium deficiency weakens the plant's resistance to stress, reducing its ability to produce high-quality fruit. Additionally, salt toxicity in the root zone damages root hairs and reduces root elongation, limiting the plant's capacity to explore the soil for water and nutrients.

3. **Inhibition of Nutrient Uptake and Growth**

- Soil salinity negatively impacts nutrient uptake by interfering with the transport of essential nutrients, such as nitrogen, phosphorus, and magnesium. As sodium and chloride ions accumulate in the root zone, they compete with other nutrients for uptake sites, leading to deficiencies that impair growth and yield.
- Nutrient deficiencies induced by salinity stress are particularly problematic for fruit crops that require balanced nutrient levels to achieve optimal fruit quality. For instance, magnesium deficiency affects chlorophyll synthesis, leading to leaf chlorosis and reduced photosynthesis, which are detrimental to the plant's energy production.
- In strawberries, which are highly sensitive to salinity, nutrient uptake issues result in smaller berries, poor color, and suboptimal flavor. Magnesium and calcium deficiencies, common under high salinity, contribute to fruit softness and reduced shelf life, affecting the market appeal and profitability of these crops.

Soil Degradation and Its Long-Term Impacts on Agricultural Productivity

1. **Soil Structure Degradation Due to Salt Accumulation**

- Increased salinity contributes to soil structure degradation, which further exacerbates challenges for fruit crop production. As sodium accumulates in the soil, it disrupts soil structure by displacing calcium ions that hold soil particles together. This process, known as sodification, leads to soil dispersion and compaction.
- Compacted soil has poor drainage, reduced porosity, and limited root penetration, making it difficult for plants to access water and nutrients. In fruit orchards, where deep root systems are essential for stability and nutrient uptake, degraded soil structure restricts root growth, reducing the plant's ability to sustain itself during stress periods.
- Poor soil structure also increases susceptibility to erosion, as compacted soil is less effective at retaining moisture and organic matter. Over time, soil degradation from salinity impacts the long-term productivity of agricultural lands, necessitating costly interventions to restore soil health.

2. **Loss of Soil Fertility and Organic Matter**

- Salinity-induced soil degradation often leads to a decline in soil fertility and a reduction in organic matter. Salt buildup reduces microbial activity in the soil, as beneficial soil organisms are sensitive to high salinity levels. This loss of microbial diversity impacts nutrient cycling, leading to lower levels of available nitrogen, phosphorus, and other essential nutrients.
- Reduced organic matter and soil fertility hinder the soil's capacity to support healthy

plant growth, as nutrients become less available to crops. In fruit crops that depend on nutrient-rich soils for high yield and quality, such as strawberries and citrus, soil fertility decline poses a major challenge for productivity.

- The decline in soil fertility due to salinity also affects soil carbon sequestration, as organic matter plays a key role in storing carbon within the soil. Reduced carbon sequestration has broader implications for climate change, as soils lose their ability to act as carbon sinks, contributing to increased atmospheric CO_2 levels.

3. **Impact on Crop Sustainability and Agricultural Viability**

- Soil salinity and degradation reduce the long-term viability of agricultural lands, especially in regions with limited access to freshwater resources. As soils become increasingly saline and degraded, the cost of maintaining productivity rises, placing economic strain on farmers and reducing the profitability of fruit crops.
- For salt-sensitive crops, the cumulative impact of salinity, nutrient deficiencies, and soil degradation leads to lower yields and diminished fruit quality. These factors make it challenging for farmers to meet market standards, impacting their income and threatening the sustainability of fruit crop production in affected regions.
- The financial burden of soil management practices, such as gypsum application to displace sodium, leaching to remove excess salts, and soil amendments to improve structure, increases operational costs. In areas facing climate-induced salinity, these measures become necessary but may not always be feasible for small-scale farmers, making long-term sustainability a concern.

Adaptive Strategies to Manage Soil Salinity and Degradation

1. **Improved Irrigation Practices to Prevent Salt Accumulation**

- Efficient irrigation methods, such as drip irrigation, help minimize salt accumulation in the root zone by delivering water directly to the plant's roots and reducing evaporation. Controlled irrigation also reduces the leaching of essential nutrients, preserving soil fertility.
- Salt flushing techniques, where fields are periodically flooded with freshwater to wash out excess salts, can help reduce soil salinity. However, this approach requires substantial freshwater resources, which may not be feasible in all regions, particularly those already experiencing water scarcity.

2. **Soil Amendments and Organic Matter Addition**

- Adding organic matter, such as compost or manure, can improve soil structure, enhance water retention, and increase microbial activity. Organic amendments help buffer the soil against salinity by improving nutrient availability and supporting root health.
- Gypsum (calcium sulfate) is commonly used to displace sodium ions in sodic soils, improving soil structure and drainage. Applying gypsum helps prevent soil compaction and allows for better root growth, particularly in orchards where soil health is crucial for perennial crop sustainability.

3. **Breeding Salt-Tolerant Crop Varieties**

   - Developing and cultivating salt-tolerant fruit crop varieties is a long-term strategy to address soil salinity challenges. Breeding programs aim to enhance traits such as sodium exclusion, osmotic adjustment, and root resilience, allowing crops to thrive in saline conditions.
   - Salt-tolerant rootstocks for citrus, grapes, and strawberries can improve crop resilience by reducing ion uptake and maintaining growth under saline conditions. These rootstocks provide farmers with options to sustain productivity and protect yield in salinity-affected areas.

B. Use of Antitranspirants in Stress Management

Antitranspirants are compounds applied to plant foliage to reduce water loss through transpiration, a process in which water vapor exits the plant through small pores called stomata. During drought or high heat conditions, water conservation becomes critical, particularly for fruit crops that require stable hydration levels to support growth, flowering, fruit set, and quality maintenance. By limiting transpiration, antitranspirants help plants retain moisture, making them a valuable tool in managing abiotic stress. Antitranspirants function either by forming a protective film on the leaf surface or by influencing stomatal behavior, thereby reducing water vapor loss and helping crops cope with water scarcity and high temperatures.

1. Types of Antitranspirants and Their Mode of Action

Antitranspirants come in several types, each designed to reduce water loss in different ways. The primary categories include film-forming, stomatal closing, and reflective antitranspirants. Each type has a unique mode of action that targets specific aspects of the plant's water conservation mechanisms.

Film-Forming Antitranspirants

1. **Mode of Action**

   - Film-forming antitranspirants create a thin, water-impermeable layer on the leaf surface, physically blocking water loss by preventing water vapor from escaping through the stomata. This layer acts as a barrier between the plant and the atmosphere, effectively reducing the rate of transpiration.
   - These compounds often include natural or synthetic substances like waxes, latex, or polymers, which can adhere to the leaf surface without significantly impacting photosynthesis or gas exchange. The film is designed to be permeable to CO_2 and O_2, allowing for minimal disruption to the plant's metabolic processes while blocking water vapor from exiting the leaf.

2. **Examples and Applications**

   - Film-forming antitranspirants include products made from carnauba wax, paraffin wax, and certain latex compounds. Synthetic polymers, such as polyvinyl alcohol and polyethylene glycol, are also used for this purpose due to their durability and flexibility.

- In fruit crops like apples, grapes, and citrus, film-forming antitranspirants are applied during dry periods to reduce water loss and help the plants maintain hydration. These antitranspirants are particularly useful in orchards and vineyards located in arid or semi-arid regions, where water availability is often limited.
- By conserving moisture, film-forming antitranspirants support essential physiological functions, reducing the risk of leaf wilting and dehydration. This enables fruit crops to survive extended drought periods without severe stress impacts on growth or productivity.

3. **Benefits and Considerations**
   - **Benefits**: Film-forming antitranspirants are effective in reducing water loss, especially during periods of high evapotranspiration. They are easy to apply and provide immediate protection, helping fruit crops conserve moisture during critical growth stages.
   - **Considerations**: While generally effective, excessive application of film-forming antitranspirants may reduce gas exchange and interfere with photosynthesis if the film is too thick or impermeable. Additionally, the layer may degrade under intense sunlight or rainfall, requiring reapplication in certain conditions.

Stomatal Closing Antitranspirants

1. **Mode of Action**
   - Stomatal closing antitranspirants work by chemically inducing stomatal closure, reducing transpiration by limiting the opening of stomata. These compounds affect the stomatal guard cells, which regulate the opening and closing of stomata based on environmental cues.
   - Abscisic acid (ABA) is a natural plant hormone widely used in stomatal closing antitranspirants. ABA triggers stomatal closure in response to water stress, reducing water loss and allowing the plant to conserve moisture. Synthetic ABA analogs and other compounds that mimic the effects of ABA are also used in antitranspirant formulations.
2. **Examples and Applications**
   - Stomatal closing antitranspirants commonly include natural or synthetic ABA. Other substances, such as potassium salts and certain oils, have also been used to induce stomatal closure.
   - These antitranspirants are especially beneficial for fruit crops during midday heat, when temperatures peak, and transpiration rates are highest. By reducing stomatal opening during hot and dry conditions, these compounds help conserve water, allowing crops to maintain hydration even in intense heat.
   - Stomatal closing antitranspirants are often used in crops like citrus, mangoes, and avocados, where water conservation is critical during flowering and fruit set stages. Reducing water loss at these stages minimizes the risk of stress-related fruit drop and promotes a higher fruit set rate.

3. **Benefits and Considerations**
   - **Benefits**: Stomatal closing antitranspirants are effective in reducing water loss during periods of high temperature and low humidity, providing a natural mechanism for water conservation. They help fruit crops withstand midday heat without risking dehydration.
   - **Considerations**: Excessive or prolonged stomatal closure can restrict CO_2 uptake, impacting photosynthesis and reducing energy production. Additionally, overuse of ABA-based antitranspirants can lead to a delay in stomatal reopening, which may affect gas exchange and overall plant health if the compound remains active for too long.

Reflective Antitranspirants

1. **Mode of Action**
   - Reflective antitranspirants contain materials that reflect sunlight, reducing leaf temperature and lowering water loss due to heat. By decreasing the amount of solar radiation absorbed by the leaf, reflective antitranspirants prevent overheating and reduce transpiration.
   - These compounds typically include kaolin clay, calcium carbonate, or mineral-based formulations that create a reflective coating on the leaf surface. This coating not only reflects sunlight but also provides a cooling effect, helping the plant maintain optimal leaf temperature and reduce heat stress.
2. **Examples and Applications**
   - Kaolin clay and calcium carbonate are common ingredients in reflective antitranspirants. These materials are finely ground and mixed with water to create a suspension that can be sprayed onto the leaves.
   - Reflective antitranspirants are widely used in crops like grapes, apples, and berries, where excessive heat can lead to fruit sunburn, reduced sugar accumulation, and poor color development. By reducing leaf and fruit temperatures, these antitranspirants protect against heat stress and promote uniform ripening.
   - In addition to reducing transpiration, reflective antitranspirants help prevent sunburn on fruits, which is a common issue in regions with intense sunlight. By reflecting solar radiation, they protect fruit quality, ensuring consistent color and preventing blemishes that affect marketability.
3. **Benefits and Considerations**
   - **Benefits**: Reflective antitranspirants provide dual protection by lowering leaf temperature and reducing water loss. They are especially beneficial in high-value fruit crops where sunburn and heat stress can negatively impact fruit appearance and quality. These antitranspirants are also environmentally friendly, as they rely on natural materials like clay.
   - **Considerations**: Reflective antitranspirants may alter the appearance of the plant by leaving a white residue on the leaves and fruits. This can be unappealing for fresh market produce and may require washing before sale. Additionally, reflective coatings may need reapplication if they are washed off by rain.

2. Practical Application and Benefits of Antitranspirants

Antitranspirants are widely recognized for their role in helping plants conserve water during periods of limited water availability. These compounds are particularly valuable for fruit crops, which often have high water requirements to support growth, flowering, and fruit set. By slowing down water loss through transpiration, antitranspirants help plants withstand drought and heat stress, thereby supporting critical physiological functions and improving overall crop productivity. Their application is especially beneficial in regions prone to low rainfall, high temperatures, or drought conditions, where water scarcity is a recurrent challenge.

Targeted Application During Drought-Prone Periods

1. **Timing and Situational Use**

   - Antitranspirants are strategically applied during drought-prone periods or times when water availability is low, helping plants retain moisture and minimize stress during critical growth stages. By forming a barrier or influencing stomatal behavior, antitranspirants reduce water vapor loss and allow plants to sustain growth with less water.
   - In arid and semi-arid regions with limited rainfall, crops like grapes, olives, and almonds benefit greatly from antitranspirant applications. These crops are typically grown in climates where water availability is inconsistent, and the soil may not retain enough moisture to meet the plant's needs throughout the growing season. By applying antitranspirants, growers can mitigate the effects of drought and help these crops maintain productivity during dry spells.

2. **Case Examples in Grapes and Olives**

   - **Grapes**: In vineyards, water availability is crucial for maintaining grapevine health, promoting fruit set, and ensuring high-quality grape production. During dry periods, antitranspirants are applied to grapevines to reduce transpiration and preserve soil moisture. This practice is especially useful in wine-producing regions where water scarcity can impact grape yield, size, and sugar content. By conserving water, antitranspirants allow grapevines to allocate resources to berry development, resulting in more consistent yields and improved wine quality.
   - **Olives**: Olive trees, though drought-tolerant, also benefit from antitranspirant applications during extended dry periods. When applied to olives, antitranspirants help the trees retain water, maintain leaf turgor, and reduce leaf drop, which can occur when water stress is severe. By preventing excessive water loss, antitranspirants support stable olive production and ensure that the trees remain healthy and productive even under challenging environmental conditions.

3. **Enhanced Water Conservation in Low-Rainfall Regions**

   - In regions with limited access to freshwater resources, such as Mediterranean and arid climates, water conservation is essential for sustainable fruit crop production. Antitranspirants help optimize water usage, allowing crops to thrive despite low rainfall or irrigation constraints.

- For instance, in citrus orchards located in low-rainfall areas, applying antitranspirants during the dry season conserves water, supports fruit set, and reduces the risk of fruit drop. These benefits translate into more reliable harvests, as water-stressed trees are better able to retain their fruit and sustain growth throughout the season.

Reduction of Midday Water Loss

1. **Managing Transpiration Peaks During High Heat**
   - Transpiration rates typically peak during midday, especially in hot climates, when high temperatures and low humidity drive water loss. Midday water loss can rapidly deplete the plant's available water, causing physiological stress, wilting, and reduced productivity.
   - Antitranspirants are particularly effective in managing these midday transpiration peaks by reducing the rate of water vapor loss from the leaf surface. By limiting midday transpiration, antitranspirants extend the period over which water remains available to the plant, enabling it to maintain essential physiological functions even during the hottest part of the day.
   - In fruit crops like mangoes and avocados, which are sensitive to heat and drought stress, antitranspirants help conserve moisture, preventing leaf scorch and wilting. This ensures that the trees can continue photosynthesis and nutrient transport throughout the day, supporting healthy growth and reducing heat-induced stress.
2. **Supporting Critical Physiological Functions**
   - By slowing down water loss, antitranspirants help support critical physiological functions such as photosynthesis, nutrient transport, and cell division. Water conservation allows the plant to sustain these processes for longer periods, even under limited water conditions.
   - For example, in apple orchards, midday application of antitranspirants reduces water stress, allowing trees to continue producing carbohydrates through photosynthesis. This sustained energy production supports fruit development and improves apple quality, even in high-temperature conditions.
   - In addition to photosynthesis, antitranspirants support turgor maintenance in leaves and stems. Leaf turgor is essential for cell expansion, which directly influences growth rates and yield potential. By maintaining turgor pressure, antitranspirants reduce the risk of leaf wilting and ensure steady growth during stress periods.
3. **Longer Period of Water Availability**
   - The application of antitranspirants extends the duration over which water remains available to the plant. This extended period of hydration is especially beneficial during drought or dry periods, as it allows the plant to endure longer stretches without supplemental water.
   - In crops like strawberries and peaches, which are susceptible to drought stress, the use of antitranspirants enables prolonged water retention, supporting fruit size, texture, and flavor. For strawberries, consistent hydration is crucial for producing juicy, well-formed fruits, while in peaches, water conservation during fruit development promotes optimal flavor and sweetness.

Improved Water Retention, Yield, and Fruit Quality

1. Maintaining Leaf Turgor and Reducing Wilting

- In fruit crops like citrus, antitranspirants have been shown to reduce wilting by helping plants maintain leaf turgor. Leaf turgor is essential for supporting growth, as it keeps the leaves upright and exposed to sunlight, maximizing photosynthetic efficiency.
- By preserving leaf turgor, antitranspirants reduce the risk of premature leaf drop, which can weaken the plant and reduce its ability to photosynthesize. This turgor maintenance is particularly valuable during high-temperature periods, as it helps plants retain their foliage and sustain growth despite the challenges posed by heat and drought stress.
- In citrus orchards, applying antitranspirants during dry periods has been found to improve canopy health and reduce leaf wilting, allowing trees to retain their foliage and optimize fruit production. This not only benefits yield but also contributes to overall orchard productivity by ensuring that trees have sufficient leaf area to support future growth.

2. Improved Water-Use Efficiency

- Antitranspirants enhance water-use efficiency by enabling plants to make better use of available water. By reducing unnecessary water loss, these compounds allow crops to thrive with less water, which is especially beneficial in water-scarce regions or during prolonged droughts.
- In pomegranate orchards, for instance, applying antitranspirants increases water-use efficiency by conserving moisture during dry periods. This conserved moisture supports steady fruit development, resulting in higher fruit set and improved quality. By making efficient use of limited water resources, antitranspirants help farmers maintain crop productivity while reducing their reliance on irrigation.
- Enhanced water-use efficiency also benefits crop resilience, as plants are better able to withstand short-term water deficits and recover more quickly from temporary drought conditions. This resilience is important for crops that are exposed to variable weather patterns, ensuring that they can continue producing quality fruit even in challenging climates.

3. Prevention of Early Fruit Drop and Improved Fruit Set

- Water stress, especially during flowering and fruit set stages, can cause early fruit drop, reducing yield potential. Antitranspirants help prevent this by conserving water, supporting fruit set, and reducing the likelihood of stress-induced fruit drop.
- In fruit crops like citrus and grapes, where early fruit drop can significantly impact yield, antitranspirants reduce the risk by maintaining hydration during sensitive growth stages. By providing moisture stability, these compounds promote higher fruit set, resulting in more consistent and productive harvests.
- Improved fruit set and reduced fruit drop contribute directly to increased yield, benefiting both crop profitability and marketability. Consistent yield is especially valuable in commercial fruit production, where fluctuations in crop output can impact supply chains and pricing.

4. **Enhanced Fruit Quality and Market Value**
   - Water retention supported by antitranspirants contributes to enhanced fruit quality by promoting proper hydration, flavor, texture, and size. Fruits that develop with consistent moisture levels are often larger, juicier, and have higher sugar content, making them more appealing to consumers.
   - In crops like mangoes and apples, antitranspirant applications during fruit development result in improved texture, flavor, and color, all of which are crucial for marketability. The increased sugar accumulation in fruits grown under water-stable conditions contributes to a more desirable taste profile, enhancing their market value.
   - For high-value crops where fruit quality directly impacts revenue, antitranspirants serve as a practical tool to optimize crop quality, ensure customer satisfaction, and increase profitability. By reducing the negative effects of drought and heat stress on fruit quality, antitranspirants help farmers produce consistent, high-quality harvests that meet market demands.

3. Limitations and Considerations in the Use of Antitranspirants

While antitranspirants are valuable tools for managing drought and heat stress in fruit crops, their application comes with certain limitations and considerations. These compounds provide only temporary relief from water stress and do not replace the need for consistent irrigation. Moreover, improper application of antitranspirants, particularly film-forming types, can interfere with essential plant processes, including photosynthesis and gas exchange. Reflective antitranspirants, although effective in reducing heat stress, may also affect the appearance of fruit, which can influence marketability. Understanding these limitations is crucial for growers to maximize the benefits of antitranspirants while minimizing potential negative effects on plant health and crop quality.

Temporary Relief and Need for Adequate Irrigation

1. **Temporary Nature of Antitranspirant Protection**
   - Antitranspirants provide temporary protection against drought stress by conserving water, but their effects are not permanent. These compounds help reduce transpiration for a limited period, after which they may wear off due to environmental factors such as rain, high winds, or degradation under intense sunlight.
   - As antitranspirants gradually break down or are washed off, their effectiveness diminishes, requiring reapplication to maintain their water-conserving benefits. For growers, this reapplication adds to labor and material costs, particularly in areas where water scarcity is chronic and long-term water management is essential.
   - Given the temporary nature of antitranspirants, they are most effective when used as part of a broader water management strategy that includes adequate irrigation, mulching, and soil moisture conservation practices.
2. **Irrigation as a Necessary Complement**
   - Although antitranspirants help conserve water, they do not eliminate the need for

irrigation. Regular irrigation is essential for providing consistent moisture to support plant growth, especially for fruit crops with high water requirements.

- In drought-prone areas, reliance solely on antitranspirants without supplemental irrigation can lead to water stress if the plants deplete their internal moisture reserves. This is particularly true during critical growth stages, such as flowering and fruit set, when water availability directly influences fruit development and yield.
- For optimal results, growers should use antitranspirants in conjunction with efficient irrigation practices, such as drip irrigation, to ensure that plants receive sufficient water to meet their physiological needs. This combination allows crops to benefit from both reduced transpiration and consistent soil moisture, enhancing drought resilience without sacrificing growth.

Potential Impact on Photosynthesis and Plant Metabolism

1. **Interference with Photosynthesis Due to Reduced CO_2 Uptake**
   - Some antitranspirants, particularly film-forming and stomatal-closing types, can limit the amount of carbon dioxide (CO_2) entering the leaf. Since CO_2 is a critical component for photosynthesis, reduced CO_2 uptake can impact the plant's ability to produce energy and grow.
   - When stomata are closed or partially obstructed by a film, gas exchange is restricted, leading to lower photosynthetic efficiency. As a result, the plant's energy production may decrease, affecting growth rates, biomass accumulation, and, ultimately, fruit development.
   - For fruit crops like apples and citrus, where growth and productivity depend heavily on photosynthesis, any reduction in CO_2 uptake can slow down growth and limit yield potential. Therefore, applying antitranspirants in excess or during active growth periods may hinder crop productivity by interfering with essential metabolic processes.
2. **Balancing Dosage to Avoid Negative Impacts on Gas Exchange**
   - The effectiveness of antitranspirants depends on proper dosage and timing. Overuse of film-forming antitranspirants may lead to an excessive barrier on the leaf surface, which can further restrict gas exchange beyond what is optimal for water conservation.
   - A thick or improperly applied film can limit both CO_2 intake and oxygen release, causing metabolic stress and possibly resulting in oxygen buildup within the leaf. This oxygen buildup can lead to photorespiration, a process that reduces photosynthetic efficiency and consumes energy, ultimately impacting plant growth and fruit development.
   - To avoid these issues, growers need to carefully follow manufacturer recommendations regarding dilution, application timing, and frequency. It is generally advisable to apply antitranspirants during periods of high heat or drought stress, rather than during active growth stages when high photosynthetic activity is essential for crop development.

Aesthetic and Marketability Considerations

1. **Effect of Reflective Antitranspirants on Fruit Appearance**
   - Reflective antitranspirants, often made from kaolin clay or calcium carbonate, leave a visible residue on the plant's leaves and fruit. While this coating is effective in reducing heat absorption and protecting against sunburn, it can alter the appearance of the fruit.
   - For fruits intended for fresh markets, such as apples, citrus, and grapes, a white or dusty coating can make them less visually appealing to consumers, potentially impacting marketability. Shoppers may perceive the residue as dirt or an indication of lower quality, which can affect their purchasing decisions.
   - Washing the fruit before sale may be necessary to remove the residue, adding to labor and post-harvest handling costs. This consideration is especially important for high-value fruits that rely on visual appeal to attract consumers in retail settings.
2. **Implications for High-Value Fruit Crops**
   - High-value crops, such as table grapes, berries, and certain tropical fruits, are particularly sensitive to market perceptions of quality. Any alteration in appearance, even if it does not affect fruit quality, can impact the product's appeal and, subsequently, its price in the market.
   - Reflective antitranspirants are most suitable for crops that are less dependent on appearance for their market value or for fruits that will undergo processing, where appearance is less critical. However, for fresh-market fruits, growers must weigh the benefits of heat and water stress reduction against the potential aesthetic impacts.
   - In situations where aesthetic concerns are paramount, growers may opt for alternative antitranspirant types, such as film-forming or stomatal-closing antitranspirants, which are less likely to alter the visual quality of the fruit. Choosing the right type based on crop value, intended market, and environmental conditions is essential for maximizing benefits without compromising marketability.

Environmental Considerations and Cost Implications

1. **Environmental Impact of Repeated Applications**
   - Depending on environmental conditions, such as rainfall and wind, antitranspirants may need reapplication to maintain their effectiveness. Frequent reapplication not only increases production costs but also raises environmental concerns regarding the cumulative effects of antitranspirant residues on soil and surrounding ecosystems.
   - Film-forming and reflective antitranspirants, in particular, may leave residual compounds on leaves and in the soil, which could affect soil health and microbial activity. While these compounds are generally considered safe, excessive use over time may alter the soil ecosystem, particularly in orchards where soil health is essential for long-term productivity.
2. **Cost Considerations for Farmers**
   - The use of antitranspirants can add to operational costs, particularly if frequent applications

are required in areas with prolonged dry or hot conditions. This added expense can be challenging for small-scale farmers or those operating in low-margin markets, where cost efficiency is critical.

- For large-scale operations, the cost of antitranspirants may be offset by improved yield and fruit quality; however, the economic feasibility varies based on crop type, regional climate, and water availability. Careful cost-benefit analysis is essential to determine whether the benefits of antitranspirants justify the expense in each particular growing context.

C. Use of Plant Growth Regulators (PGRs) in Stress Management

Plant growth regulators (PGRs) are compounds that influence plant growth and development by modulating hormonal pathways. They play a critical role in stress management by enhancing a plant's tolerance to abiotic stressors, including drought, salinity, and temperature extremes. Common PGRs used in stress management include abscisic acid (ABA), gibberellins, cytokinins, and ethylene inhibitors.

1. Types of PGRs and Their Mode of Action

Plant growth regulators (PGRs) are substances that influence plant growth and development by modulating hormonal pathways. They play a crucial role in stress management by enhancing a plant's tolerance to abiotic stressors such as drought, salinity, and temperature extremes. The primary types of PGRs used in stress management include abscisic acid (ABA), cytokinins, gibberellins, and ethylene inhibitors. Each type has a unique mode of action that helps plants maintain physiological functions, support growth, and sustain productivity during stress conditions.

Abscisic Acid (ABA)

1. **Role in Drought and Salinity Tolerance**
   - Abscisic acid (ABA) is a central hormone in the plant's stress response, particularly in drought and salinity tolerance. ABA levels in plants increase significantly under drought conditions, initiating a series of physiological changes that help the plant conserve water.
   - One of ABA's primary roles is to regulate stomatal closure. Stomata are small openings on the leaf surface through which plants lose water through transpiration. By promoting stomatal closure, ABA helps plants reduce water loss, allowing them to retain moisture and survive in low-water environments. This mechanism is crucial during prolonged drought periods when water availability is limited.
   - Under saline conditions, ABA also supports osmotic adjustment, helping cells maintain water balance despite the high salt concentration in the soil. By enhancing root growth and modifying root architecture, ABA enables the plant to access deeper soil moisture, improving its resilience to water-deficit and saline environments.
2. **Enhanced Root Growth for Better Water Access**
   - ABA promotes root growth by stimulating root elongation and the formation of lateral roots, which increases the root system's ability to explore soil for moisture and nutrients. Deeper and more extensive root systems are essential for drought-tolerant plants, as they allow access to water stored in lower soil layers, beyond the reach of shallow-rooted plants.

- In crops like grapes, citrus, and tomatoes, ABA application during dry periods enhances root development, improving water uptake and reducing drought-induced stress. This increased rooting capacity supports fruit development and helps sustain yield under water-limited conditions.

Cytokinins

1. **Delay of Leaf Senescence and Promotion of Cell Division**
   - Cytokinins are essential PGRs that promote cell division and delay leaf senescence, making them valuable for maintaining plant vigor during stress. Cytokinins counteract the aging process in leaves by preserving chlorophyll content, which is crucial for photosynthesis. Under drought and heat stress, maintaining chlorophyll levels helps plants continue producing energy for growth and productivity.
   - By delaying senescence, cytokinins support the longevity and functionality of leaves, enabling the plant to sustain photosynthesis over a longer period. This effect is particularly beneficial during drought, when water scarcity could otherwise lead to premature leaf drop, reducing the plant's photosynthetic capacity.
2. **Support for Photosynthetic Activity and Chlorophyll Retention**
   - During heat and drought stress, cytokinins improve photosynthetic efficiency by reducing chlorophyll degradation. Chlorophyll is the pigment responsible for capturing light energy, and its degradation under stress reduces the plant's ability to convert light into energy.
   - In crops like apples and wheat, cytokinin treatments help maintain chlorophyll levels, which supports continued photosynthesis even under stressful conditions. This effect is particularly valuable in fruit crops, where sustained energy production is essential for fruit development and quality.
3. **Improvement of Ion Homeostasis in Saline Conditions**
   - Cytokinins also play a role in enhancing ion homeostasis under saline conditions. High soil salinity disrupts nutrient uptake by increasing the concentration of sodium (Na^+) and chloride (Cl^-) ions in the root zone, leading to ion toxicity and nutrient imbalances.
   - Cytokinins improve cellular integrity by promoting the compartmentalization of toxic ions, preventing their accumulation in sensitive tissues. In salt-sensitive crops like tomatoes and strawberries, cytokinins help maintain nutrient balance, reducing the negative impacts of salinity on growth and yield.

Gibberellins

1. **Promotion of Growth Processes and Stress Recovery**
   - Gibberellins are PGRs that regulate growth processes such as stem elongation, seed germination, and fruit development. Under stress conditions, gibberellins promote recovery by supporting cell expansion, allowing plants to continue growing despite adverse environmental factors.

- Gibberellins are particularly important for maintaining reproductive growth in fruit crops, as stress conditions often inhibit flowering and fruit set. By promoting cellular expansion and elongation, gibberellins help counteract the growth inhibition caused by stress, enabling the plant to produce fruits even under challenging conditions.

2. **Enhancement of Stem and Root Growth Under Stress**

- In addition to promoting recovery, gibberellins support the growth of stems and roots, which are vital for nutrient and water transport. During stress periods, when root function may be compromised due to soil conditions, gibberellins stimulate root elongation and branching, increasing the root system's ability to access water and nutrients.
- For instance, in crops like rice and maize, gibberellins have been shown to enhance root growth in saline and water-deficient soils, helping the plants adapt to stress. This improved root function supports better nutrient uptake and helps maintain the plant's productivity under adverse conditions.

3. **Reproductive Growth Support During Stress**

- Gibberellins play a crucial role in supporting reproductive growth by promoting flowering, fruit set, and fruit development. During stress conditions, when resources are limited, gibberellins enable plants to allocate resources to reproductive organs, ensuring that they can still produce viable fruits.
- In crops like grapes and citrus, gibberellin treatments during flowering and fruit set stages help maintain yield and improve fruit quality by enhancing cell division and expansion. This support for reproductive growth ensures that stress does not severely impact the plant's ability to produce a harvestable crop.

Ethylene Inhibitors

1. **Ethylene's Role in Ripening, Senescence, and Stress Responses**

- Ethylene is a hormone that regulates ripening, senescence, and stress responses in plants. Under stress, ethylene levels often rise, triggering senescence and accelerating leaf abscission, which can lead to premature fruit drop and reduced yield.
- High ethylene levels in fruit crops can cause premature ripening, reducing fruit quality, and post-harvest shelf life. To manage this, ethylene inhibitors are used to suppress ethylene synthesis or action, delaying senescence and reducing the stress-induced effects of ethylene on the plant.

2. **Aminoethoxyvinylglycine (AVG) and Its Role in Reducing Ethylene-Induced Stress**

- Aminoethoxyvinylglycine (AVG) is a commonly used ethylene inhibitor that slows down ethylene production in plants, helping reduce the impact of ethylene on growth and development during stress. AVG application is especially beneficial in fruit crops like apples, where stress-induced ethylene production can lead to early fruit drop and reduced quality.
- By inhibiting ethylene, AVG helps maintain the plant's productivity and prevents premature senescence, ensuring that fruits stay on the tree longer and develop to full

maturity. This is particularly useful during drought or heat stress, where high ethylene levels can otherwise accelerate fruit ripening and reduce marketable yield.

3. **Delaying Senescence and Preventing Fruit Drop**
   - Ethylene inhibitors delay senescence in leaves and fruits, which helps maintain photosynthetic activity and extends the productive period of the plant. By reducing ethylene-induced stress responses, these inhibitors help plants continue growing and producing despite challenging environmental conditions.
   - In fruit crops like tomatoes, citrus, and avocados, ethylene inhibitors prevent premature fruit drop, allowing the fruits to remain on the plant until they reach optimal maturity. This delay in senescence and fruit drop improves yield and ensures that the harvested fruits meet quality standards for size, flavor, and color.

2. Practical Use of PGRs in Stress Management

Plant growth regulators (PGRs) are valuable tools for managing abiotic stress in fruit crops, as they modulate physiological responses to environmental challenges. By applying PGRs such as abscisic acid (ABA), cytokinins, gibberellins, and ethylene inhibitors, growers can help fruit crops mitigate the effects of drought, salinity, heat stress, and other adverse conditions. These PGRs enhance water retention, ion balance, photosynthetic efficiency, and recovery potential, ensuring that crops remain productive even under stress.

Drought and Water Stress Management

1. **ABA Applications for Stomatal Closure**
   - In drought-prone areas, abscisic acid (ABA) is commonly applied to fruit crops like grapes and apples to reduce water loss. ABA plays a crucial role in regulating stomatal closure, which conserves water by minimizing transpiration. By reducing the opening of stomata, ABA applications help maintain the plant's internal water balance during periods of low rainfall or limited irrigation.
   - In grapes, ABA treatments are particularly beneficial during the dry season, as they prevent excessive water loss and help sustain vine health. This water conservation allows grapevines to allocate resources toward berry development, enhancing yield and quality despite limited water availability.
2. **Cytokinin Treatments to Delay Leaf Senescence**
   - Cytokinins are also valuable in drought management, as they delay leaf senescence, allowing plants to retain photosynthetic capacity during water stress. By preserving chlorophyll content, cytokinins enable plants to continue photosynthesis, ensuring energy production and growth even when water is scarce.
   - In apple orchards, cytokinin applications during drought conditions keep the leaves green and functional, allowing trees to photosynthesize efficiently. This delayed senescence supports better fruit set, size, and flavor, as the trees can sustain growth and fruit development without prematurely dropping leaves or reducing photosynthetic output.

- The combined use of ABA and cytokinins helps maximize water-use efficiency, allowing fruit crops to withstand prolonged drought and maintain productivity.

Salinity Tolerance

1. **Cytokinins for Ion Homeostasis and Root Health**
   - Saline conditions increase the concentration of sodium (Na^+) and chloride (Cl^-) ions in the soil, which can be toxic to plants and disrupt nutrient uptake. Cytokinins help manage salinity stress by promoting ion balance and maintaining root health. They aid in the compartmentalization of sodium and chloride ions, preventing these toxic ions from accumulating in sensitive tissues.
   - In salt-sensitive crops like tomatoes and pomegranates, cytokinin treatments improve the plant's ability to exclude sodium and maintain a favorable potassium-to-sodium ratio. This ion homeostasis reduces the detrimental effects of salinity, supporting steady growth and yield under salt stress conditions.
2. **Gibberellins for Enhanced Root Growth**
   - Gibberellins play a complementary role in salinity management by promoting root growth, which enables plants to access moisture and nutrients more effectively in saline soils. Gibberellins stimulate the growth of root systems, allowing plants to explore deeper soil layers and avoid high salt concentrations near the surface.
   - For crops like pomegranates, which are cultivated in arid regions where salinity can be a major concern, gibberellin treatments enhance root length and density, improving the plant's resilience to salinity stress. This deeper rooting capacity helps the plant avoid salt-affected soil zones, ensuring healthier growth and reduced salt toxicity.
3. **Combining Cytokinins and Gibberellins for Optimal Salinity Management**
   - Cytokinins and gibberellins together support the plant's ability to tolerate saline environments. While cytokinins manage ion balance and prevent sodium toxicity, gibberellins strengthen the root system, enabling the plant to absorb water and nutrients from less saline parts of the soil.
   - This combination is particularly effective for fruit crops in coastal or semi-arid regions, where soil salinity is a persistent issue. By improving root function and ion exclusion, these PGRs allow crops like tomatoes, grapes, and citrus to maintain productivity and withstand saline conditions without severe yield losses.

Heat Stress and Temperature Extremes

1. **Ethylene Inhibitors to Prevent Premature Ripening and Fruit Drop**
   - High temperatures can lead to elevated ethylene levels, accelerating ripening, fruit drop, and senescence. Ethylene inhibitors, such as aminoethoxyvinylglycine (AVG), are applied to crops like tomatoes and strawberries to prevent premature ripening and fruit drop under high temperatures.

- By reducing ethylene synthesis, AVG applications slow down the ripening process, ensuring that fruits reach full maturity before harvest. This delayed ripening helps maintain fruit size, texture, and quality, preventing the loss of marketable produce due to heat-induced ripening.
- In strawberry fields, ethylene inhibitors improve yield and reduce the likelihood of fruit loss during heat waves, allowing growers to maximize their harvest even under high-temperature conditions.

2. **Cytokinins for Chlorophyll Stability and Photosynthesis**
   - Cytokinins support plants under heat stress by stabilizing chlorophyll and enhancing photosynthetic efficiency. High temperatures often lead to chlorophyll degradation, reducing the plant's ability to capture sunlight and convert it into energy.
   - Cytokinin treatments maintain chlorophyll content in heat-stressed crops, enabling them to continue photosynthesis at optimal levels. In crops like mangoes and peppers, cytokinins help plants retain leaf greenness, promoting steady growth and reducing the risk of heat-induced wilting.
   - This enhanced photosynthetic stability is particularly valuable during prolonged heat periods, as it allows plants to produce the energy needed for growth, flowering, and fruit set, even under temperature extremes.
3. **Integrated Approach for Heat and Salinity Stress**
   - For crops exposed to both heat and salinity stress, a combination of cytokinins, gibberellins, and ethylene inhibitors can be used to improve resilience. Cytokinins and gibberellins address root and ion-related issues from salinity, while ethylene inhibitors mitigate the effects of high temperatures on fruit quality.
   - By managing both heat and salinity stress, this integrated approach supports balanced growth, water-use efficiency, and fruit development, ensuring that crops remain productive and marketable despite challenging environmental conditions.

Stress Recovery

1. **Gibberellins for Post-Stress Growth and Development**
   - Following stress events, gibberellins are applied to fruit crops to promote recovery and vigor. Gibberellins stimulate cell division, elongation, and expansion, processes that are essential for restoring plant health after stress conditions.
   - For example, in mango and citrus orchards, gibberellin applications following drought conditions support the resumption of growth, enhancing leaf production and promoting fruit development. This recovery ensures that trees regain their productive potential and can achieve a successful harvest even after experiencing water stress.
2. **Role of Gibberellins in Fruit Development and Yield Improvement**
   - During the recovery phase, gibberellins enhance fruit development by encouraging cell expansion within the fruit, leading to improved size and quality. This effect is

particularly valuable in high-value crops, where fruit size and appearance directly impact marketability and profit.

- In citrus, gibberellin treatments post-stress help ensure that the trees produce high-quality fruit with optimal size, texture, and juiciness, meeting market standards and enhancing profitability. For mango growers, gibberellins support fruit growth, enabling larger, well-formed fruits that appeal to consumers and increase market value.

3. **Cytokinins and ABA for Sustained Recovery and Resilience**

- In addition to gibberellins, cytokinins and ABA are used to support sustained recovery and resilience. Cytokinins delay senescence in stressed plants, allowing leaves to remain functional and contributing to energy production during the recovery phase. ABA assists in root recovery, promoting hydration and nutrient uptake as the plant resumes normal growth.
- This combination of PGRs aids in a holistic recovery process, where cytokinins sustain photosynthetic capacity, ABA improves water and nutrient uptake, and gibberellins enhance growth and fruit development. By using these PGRs strategically after a stress event, growers help plants return to full productivity, ensuring that stress does not lead to long-term yield losses.

3. Benefits and Challenges in the Use of PGRs

Plant growth regulators (PGRs) play an increasingly important role in modern agriculture by enhancing a plant's ability to cope with abiotic stress, improving yield and quality, and promoting sustainable production practices. However, the benefits of PGRs must be carefully weighed against potential challenges, as improper application can lead to unintended consequences for plant health and crop productivity. Both the advantages and limitations of PGRs are essential for maximizing their effectiveness in stress management.

Enhanced Stress Tolerance

1. **Improving Resilience Through Hormonal Modulation**

- PGRs enhance a plant's natural resilience by modulating hormonal pathways that govern stress responses. For instance, abscisic acid (ABA) is known to regulate stomatal closure, helping plants conserve water during drought. This modulation allows plants to adapt to environmental stresses in a controlled way, reducing the need for external interventions such as excessive irrigation.
- In crops like citrus, grapes, and apples, PGR applications improve drought and salinity tolerance, enabling the plants to maintain hydration and nutrient uptake. By supporting the plant's physiological functions, such as photosynthesis, respiration, and ion homeostasis, PGRs help crops survive and thrive even under challenging conditions.

2. **Proactive Stress Management for Long-Term Productivity**

- The use of PGRs offers a proactive approach to managing stress, as they can be applied before or during stress periods to prepare plants for environmental challenges. This preventive use of PGRs allows growers to optimize plant health and productivity in anticipation of stress events, rather than relying solely on reactive solutions after stress has already affected the crop.

- By supporting essential processes like root growth, leaf retention, and osmotic balance, PGRs help plants withstand multiple stress factors, including drought, heat, and salinity. In perennial fruit crops such as mangoes, peaches, and avocados, enhanced stress tolerance leads to healthier plants that are better able to recover from stress and remain productive over their lifespan.

3. **Reduction in Yield Losses Due to Abiotic Stress**

- Abiotic stresses, such as heat waves, drought, and soil salinity, can lead to significant yield losses if not managed effectively. PGRs help mitigate these losses by strengthening the plant's defense mechanisms, allowing it to continue producing fruits despite environmental challenges.
- In high-value crops like table grapes and apples, where even small yield losses can impact profitability, the use of PGRs provides a reliable method for minimizing the impact of stress on overall production. This resilience is crucial for ensuring consistent crop output and meeting market demand, even during adverse weather conditions.

Improved Yield and Quality

1. **Enhanced Fruit Set, Size, and Uniformity**

- The application of PGRs, such as gibberellins and cytokinins, promotes improved fruit set, size, and uniformity, directly benefiting yield and quality. Gibberellins stimulate cell expansion and elongation, leading to larger fruits, while cytokinins delay leaf senescence, preserving photosynthetic capacity and supporting fruit development.
- In crops like tomatoes, strawberries, and apples, improved fruit size and uniformity enhance market appeal, as consumers often prefer fruits that are consistent in size and appearance. This uniformity also simplifies post-harvest handling and packaging, making it easier for growers to meet quality standards.

2. **Increased Sugar Content and Nutritional Quality**

- PGRs influence the accumulation of sugars, acids, and other phytochemicals in fruits, which directly affects flavor, texture, and nutritional value. For instance, cytokinins maintain chlorophyll and photosynthetic activity, allowing the plant to produce and store more sugars in the fruit.
- In grapes, for example, the use of PGRs like cytokinins and gibberellins leads to higher sugar accumulation, improving flavor and enhancing the quality of both fresh and wine grapes. This increase in quality not only boosts consumer satisfaction but also raises the market value of the produce, contributing to higher profitability for growers.

3. **Extended Shelf Life and Reduced Post-Harvest Losses**

- Ethylene inhibitors, such as aminoethoxyvinylglycine (AVG), delay ripening and senescence, extending the shelf life of fruits and reducing post-harvest losses. This is especially valuable for fruits that are sensitive to premature ripening and spoilage, such as bananas, tomatoes, and avocados.

- Extended shelf life allows for longer storage and transportation periods, making it easier for growers to reach distant markets and reducing waste along the supply chain. For high-value crops where freshness is essential, the use of ethylene inhibitors improves marketability and minimizes economic losses.

Potential Limitations

1. **Risk of Overuse and Negative Impacts on Photosynthesis**
   - While PGRs offer significant benefits, overuse or improper application can interfere with plant metabolism. For instance, excessive use of ABA to induce stomatal closure can lead to prolonged reduction in CO_2 uptake, limiting photosynthesis and potentially slowing plant growth.
   - Stomatal closure restricts gas exchange, which is essential for photosynthesis. In crops with high photosynthetic demands, such as apples and tomatoes, a reduction in CO_2 uptake can result in lower carbohydrate production, ultimately impacting fruit growth and yield. This challenge highlights the importance of using ABA in precise doses and applying it only during critical stress periods.
2. **Excessive Vegetative Growth from Cytokinins**
   - Cytokinins promote cell division and delay senescence, which can be beneficial for stress management. However, excessive cytokinin application may lead to increased vegetative growth, diverting energy and resources away from fruit production.
   - In fruit crops, where the primary goal is to optimize yield rather than foliage growth, excess vegetative development can negatively impact productivity. For example, in citrus and apples, excessive foliage can create shading, reducing light penetration to the fruit and affecting fruit size, color, and quality. This potential side effect makes it essential to carefully monitor cytokinin dosage and avoid over-application.
3. **Balance Between Vegetative and Reproductive Growth**
   - Gibberellins, while effective in promoting fruit growth and recovery, can also promote vegetative growth when applied in high amounts. This can sometimes lead to imbalances between vegetative and reproductive growth, especially in young plants or during periods of rapid growth.
   - For example, in grapes, excessive gibberellin application may lead to elongated shoots and larger leaves, which compete with developing fruit for nutrients and energy. This competition can affect fruit set and size, potentially reducing yield if vegetative growth becomes too dominant. To avoid these issues, gibberellins should be applied at optimal stages of growth and in controlled amounts to ensure balanced plant development.

4. Economic and Practical Considerations for Growers

- While the benefits of PGRs are substantial, they come at an economic cost, which may be a limitation for some growers, particularly small-scale farmers. Regular PGR applications, especially when combined with other management practices, increase input costs, affecting overall profitability.
- Additionally, the effectiveness of PGRs can be influenced by environmental factors such as temperature, humidity, and soil conditions, which vary from one region to another. This variability makes it essential for growers to adapt PGR application rates and timing to local conditions, requiring knowledge and experience to maximize benefits while minimizing costs.
- To address these challenges, growers must conduct cost-benefit analyses to determine whether the potential gains in yield and quality justify the expense of PGR applications. In some cases, PGRs may be most feasible for high-value crops where the return on investment is substantial.

Chapter - 6

Stress Management Techniques

A. HSP Inducers in Stress Management

Heat shock proteins (HSPs) are critical in helping plants survive under abiotic stress conditions such as drought, heat, and salinity. When induced by environmental stressors, HSPs function as molecular chaperones that stabilize proteins, protect cellular structures, and ensure proper protein folding. The activation of HSPs through chemical inducers or stress conditions helps plants mitigate stress impacts, maintain cellular functions, and improve resilience. Alongside HSP inducers, other stress management techniques such as soil moisture conservation, mulching, hydrophilic polymers, rainwater harvesting, skimming technology, and contingency planning contribute to sustainable stress management. Together, these strategies enhance crop stability, sustainability, and resilience in the face of climate variability.

1. Role of HSPs in Plant Stress Tolerance in Fruit Crops

Heat shock proteins (HSPs) are essential molecular chaperones that play a pivotal role in protecting fruit crops against environmental stress-induced damage. As fruit crops are frequently exposed to abiotic stresses like high temperatures and drought, their cellular proteins and membranes are at risk of structural and functional damage. HSPs act to stabilize these cellular components by assisting in protein folding, preventing protein aggregation, and maintaining the integrity of cellular membranes. By reducing cellular damage and promoting metabolic stability, HSPs help fruit crops sustain growth, yield, and quality even in the face of adverse environmental conditions.

Protecting Plant Cells Against Stress-Induced Damage

1. **Assisting Protein Folding Under Stress Conditions**
   - Proteins must fold into specific three-dimensional shapes to function correctly. Environmental stresses, such as heat and drought, can cause proteins to misfold or denature, leading to loss of function and cellular instability. HSPs, particularly those in the HSP70 and HSP90 families, act as molecular chaperones that bind to partially unfolded proteins and help them regain their correct structure.
   - This protein-folding assistance is especially important in fruit crops like tomatoes and strawberries, where cellular proteins need to be functional to support growth, nutrient

DOI: 10.1201/9781003717409-6

transport, and fruit development. By aiding in proper protein folding, HSPs enable these crops to continue metabolic activities, which are crucial for producing high-quality fruits.

2. **Preventing Protein Aggregation Under High-Stress Conditions**

   - Under prolonged heat or drought stress, denatured proteins are prone to aggregation, forming clumps that interfere with cellular processes. Aggregated proteins are not only non-functional but can also create toxic conditions within cells, leading to accelerated cell damage and even cell death.
   - HSPs prevent protein aggregation by binding to misfolded proteins and maintaining them in a state that allows for eventual refolding or degradation. This anti-aggregation function is vital in fruit crops exposed to high temperatures, where protein stability directly affects cell viability and, consequently, crop yield. For example, in grapevines, where temperature stress can influence berry quality, HSPs mitigate the effects of protein aggregation, preserving cell function and supporting fruit development.

3. **Stabilizing Cellular Membranes**

   - Besides their role in protein stabilization, HSPs play a critical role in protecting cellular membranes. High temperatures and water stress can make cell membranes more fluid and prone to leakage, disrupting the transport of water, ions, and nutrients. In fruit crops, membrane stability is essential for maintaining turgor pressure, nutrient distribution, and overall cellular integrity.
   - Small HSPs, in particular, are known to stabilize membranes by binding to lipid bilayers, preventing them from becoming too permeable under stress. This membrane-stabilizing function is crucial in crops like apples and mangoes, where stable cellular compartments support fruit firmness, juiciness, and texture, qualities essential for marketability. By preserving membrane integrity, HSPs ensure that cells remain functional, sustaining the processes necessary for fruit quality and longevity.

Classification and Specific Roles of HSP Families in Fruit Crops

HSPs are classified into several families based on their molecular weight, each performing specialized functions in stress response and cellular protection. The primary HSP families include HSP100, HSP70, HSP60, and small HSPs, each contributing uniquely to stress tolerance in fruit crops.

1. **HSP100 Family**

   - The HSP100 family is essential for protein disaggregation and cellular recovery after prolonged stress. HSP100 proteins help dissolve aggregated proteins, bringing them back to a soluble state that allows for proper refolding or degradation.
   - In fruit crops like citrus, where severe heat stress can affect flowering and fruit set, HSP100 proteins are crucial for reversing protein aggregation, helping plants recover post-stress. By enabling the plant to re-establish cellular homeostasis, HSP100 proteins support ongoing fruit development and protect overall yield.

2. **HSP70 Family**

   - The HSP70 family is one of the most important groups of HSPs for stress tolerance, particularly in maintaining protein stability and preventing aggregation. HSP70 proteins bind to newly synthesized or stress-damaged proteins, creating a controlled environment that allows for correct folding and function.
   - In crops like tomatoes and peaches, where proteins are integral to growth, photosynthesis, and fruit formation, HSP70s play a vital role in preserving enzyme activity during stress. By ensuring proteins retain their functional shape, HSP70 proteins help fruit crops continue essential metabolic processes, even in hot and dry conditions, which is crucial for maintaining fruit yield and quality.

3. **HSP60 Family**

   - The HSP60 family, also known as chaperonins, primarily assists in protein folding within organelles like chloroplasts and mitochondria. These organelles are essential for energy production, photosynthesis, and respiration, making HSP60 proteins critical for maintaining metabolic activity under stress.
   - In fruit crops such as grapes, mangoes, and citrus, HSP60 proteins stabilize the machinery within chloroplasts, allowing continued photosynthetic activity even under high temperatures. This stability helps maintain energy levels needed for fruit growth and ripening, reducing the impact of heat on fruit size, color, and sugar content. HSP60 proteins thus ensure that critical energy-related processes remain efficient during environmental stress.

4. **Small HSPs**

   - Small HSPs (sHSPs) are highly adaptable and play a unique role in stabilizing membranes and other cellular structures under stress. Small HSPs bind to proteins and cellular membranes, preventing their denaturation and maintaining cell compartmentalization, which is crucial for sustaining cellular function.
   - In fruits like strawberries and pomegranates, where membrane integrity is tied directly to texture and juiciness, small HSPs are essential for protecting cell walls and membrane-bound compartments from dehydration. This protection is particularly valuable under drought stress, where cellular hydration and membrane fluidity are critical for preserving fruit quality, flavor, and shelf life.

Importance of HSPs in Heat and Drought Tolerance in Fruit Crops

1. **Enhancing Drought Tolerance by Supporting Cellular Hydration**

   - HSPs play a significant role in managing drought stress by maintaining cellular hydration and supporting osmotic balance. During drought conditions, cells face osmotic stress due to reduced water availability, which can lead to dehydration and impaired cellular activities.
   - By stabilizing proteins involved in osmotic regulation, HSPs enable cells to retain

water and maintain turgor pressure, a critical factor for sustaining growth and fruit development. In drought-sensitive crops like tomatoes and grapes, HSPs help prevent water loss, ensuring that the fruits maintain size, juiciness, and structural integrity even during prolonged dry periods.

2. **Mitigating Heat Stress for Stable Fruit Development**
   - High temperatures can disrupt cellular functions by causing proteins to unfold, membranes to become more fluid, and metabolic pathways to slow down. HSPs mitigate heat stress by refolding denatured proteins, preserving enzyme function, and stabilizing cellular membranes.
   - In crops like apples, pears, and grapes, where fruit quality depends on stable growth conditions, HSPs provide protection that supports consistent fruit development. By preventing heat-induced damage, HSPs help maintain growth, flavor, and texture, which are crucial for consumer appeal and market value.
3. **Supporting Recovery and Resilience After Stress Events**
   - Beyond immediate stress protection, HSPs are essential for the recovery process following periods of heat or drought. By restoring protein function and stabilizing cellular structures, HSPs help plants regain normal metabolic activity after the stress subsides.
   - This recovery function is particularly valuable for perennial fruit crops, such as citrus, stone fruits, and grapes, where sustained productivity depends on the plant's ability to recover from stress. By enhancing resilience and supporting rapid recovery, HSPs contribute to stable yields over multiple growing seasons, allowing fruit crops to produce consistently despite fluctuating environmental conditions.
4. **Preventing Long-Term Stress Damage for Improved Fruit Quality**
   - Heat and drought stress can lead to long-term cellular damage that affects the size, appearance, flavor, and nutritional quality of fruits. HSPs prevent this long-term damage by protecting proteins, membranes, and cellular compartments, enabling fruit crops to maintain high-quality produce.
   - In high-value crops like strawberries, apples, and mangoes, where fruit quality is directly tied to marketability, HSPs help ensure that the fruits develop to their full potential, free from stress-induced blemishes or structural degradation. This improved quality supports higher market value and consumer satisfaction, reinforcing the economic importance of HSPs in fruit production.

2. Induction of HSPs for Enhanced Stress Tolerance in Fruit Crops

The induction of heat shock proteins (HSPs) has become an essential tool in managing abiotic stress for fruit crops. HSP inducers, including naturally occurring compounds like salicylic acid and synthetic chemicals, initiate the production of HSPs within the plant even before exposure to environmental stress. By preemptively activating HSP pathways, these inducers prime fruit crops to respond more efficiently to upcoming stress, creating a state of "readiness" that allows plants to rapidly deploy protective measures when exposed to high temperatures or drought conditions. This proactive approach is especially beneficial during sensitive developmental phases like flowering, fruit set, and early fruit development when fruit crops are highly susceptible to temperature fluctuations and water scarcity.

Mechanism of Action of HSP Inducers in Fruit Crops

1. **Activating Plant Defense Pathways**
   - HSP inducers work by activating defense pathways in fruit crops, initiating physiological changes that mimic the plant's natural response to stress. Salicylic acid, a signaling molecule involved in stress responses, triggers the expression of HSP genes, even under non-stress conditions, effectively "priming" the plant to respond quickly when real stress occurs.
   - Upon induction, transcription factors activate the genes responsible for HSP production, leading to a buildup of HSPs in cells. This early accumulation of HSPs prepares the plant for sudden temperature changes or water shortages, enabling it to sustain cellular functions and protect vital structures during stress.
2. **Enhancing Rapid Response Mechanisms**
   - Priming fruit crops with HSP inducers enhances their rapid response capabilities, allowing them to deploy defenses faster and with greater efficiency than unprimed plants. This rapid response reduces the lag time between stress exposure and the activation of protective mechanisms, minimizing cellular damage.
   - When exposed to actual stress, pre-induced fruit crops can quickly utilize stored HSPs to prevent protein denaturation, maintain membrane integrity, and reduce oxidative stress. This rapid response is critical during periods of sudden environmental change, such as heatwaves, where every hour of delay in activating defenses could lead to irreversible damage to sensitive fruit tissues and significant yield loss.
3. **Supporting Protein and Membrane Stability in Fruit Cells**
   - HSPs play a central role in stabilizing proteins and membranes within fruit cells. They bind to partially unfolded proteins, refolding them into functional shapes and preventing the formation of toxic aggregates. This stabilizing effect is crucial in fruit crops like tomatoes and grapes, where high temperatures or drought can otherwise lead to protein degradation and impaired cellular functions.
   - In addition, HSPs help maintain membrane fluidity and integrity, reducing the risk of lipid peroxidation and other damage that can compromise cell structure. This protection is vital for fruit crops, where membrane stability directly influences water retention, nutrient transport, and photosynthetic efficiency—all of which are essential for fruit development and quality under stress.

Specific Applications in Fruit Crops

1. **Tomatoes**
   - Tomato plants are highly sensitive to heat, particularly during the flowering and fruit set stages, where elevated temperatures can lead to poor pollination, reduced fruit set, and smaller fruits. Applying HSP inducers such as salicylic acid before anticipated heat

events can significantly improve tomato resilience by enhancing protein stability and protecting cellular integrity.

- The pre-induction of HSPs in tomatoes allows the plants to maintain their photosynthetic activity and nutrient transport systems even under high temperatures. This results in a better fruit set and improved fruit size, flavor, and texture. In areas prone to heatwaves, the use of HSP inducers provides tomato growers with a tool to ensure stable yields and high-quality fruits despite challenging growing conditions.

2. **Grapes**
   - In grapevines, heat stress during flowering and berry development can lead to smaller berries, poor color, and compromised flavor—factors that are critical for the quality of table grapes and wine grapes alike. HSP inducers can protect grapevines from heat-induced stress by stabilizing proteins and cellular structures that are essential for berry development and sugar accumulation.
   - By applying HSP inducers before heatwaves, grapevines can retain essential compounds such as anthocyanins and flavonoids, which contribute to berry color, taste, and nutritional quality. The stabilization of these compounds not only supports the aesthetic and flavor qualities of the grapes but also improves their marketability. For wine producers, HSP induction helps ensure consistency in grape quality, maintaining flavor profiles essential for high-value wines.
3. **Citrus**
   - Citrus fruits, including oranges, lemons, and limes, are also highly vulnerable to heat and drought stress. High temperatures during flowering or fruit set can lead to excessive flower drop and reduced fruit quality. HSP inducers applied during early development stages can enhance the resilience of citrus plants, allowing them to retain more flowers and set higher-quality fruit even under thermal stress.
   - The induction of HSPs in citrus plants stabilizes enzymes involved in photosynthesis and respiration, ensuring continuous energy production and fruit development despite temperature fluctuations. This protection is especially beneficial for citrus growers in regions with hot summers, where maintaining fruit set and quality under high temperatures is essential for economic viability.
4. **Stone Fruits (e.g., Peaches and Plums)**
   - Stone fruits like peaches and plums are highly sensitive to both heat and drought stress, particularly during the flowering and early fruiting stages. HSP inducers applied at these stages prepare the plants for sudden heat stress, which can otherwise lead to fruit drop, smaller fruit size, and reduced sweetness.
   - By stabilizing cellular proteins and maintaining membrane integrity, HSP inducers help stone fruit plants sustain growth and fruit quality. This preemptive protection is particularly valuable for commercial peach and plum orchards, where fruit size, sweetness, and texture are critical quality parameters for consumer markets.

Benefits of HSP Inducers During Critical Growth Stages in Fruit Crops

1. **Flowering and Fruit Set Resilience**
   - Flowering and fruit set are among the most temperature-sensitive stages in fruit crop development. Even minor temperature increases during these phases can reduce pollination success and lead to poor fruit set, which directly impacts yield. HSP inducers prime plants to better withstand these stresses by stabilizing proteins and cellular structures critical to floral and reproductive tissues.
   - For crops like grapes, citrus, and tomatoes, HSP induction during flowering and fruit set can significantly improve reproductive success, resulting in higher fruit retention and better yield stability. This resilience during early development stages ensures that growers can achieve higher yields and consistent production, even in areas prone to heat spikes.
2. **Maintaining Fruit Quality Under Heat and Drought Stress**
   - Heat stress can accelerate ripening in fruit crops, leading to smaller, less flavorful fruits with reduced sugar content. By stabilizing proteins and protecting metabolic pathways, HSP inducers help fruit crops maintain their sugar and nutrient levels, contributing to better fruit quality.
   - In table grape production, HSP inducers help maintain berry size, sugar content, and flavor by preventing the heat-induced breakdown of sugars and other critical compounds. Similarly, in tomatoes and stone fruits, HSP induction helps maintain desirable flavor, size, and color, which are essential for market acceptance and consumer satisfaction.
3. **Extended Protective Effects for Continuous Stress Management**
   - One of the significant benefits of HSP inducers is their ability to provide extended protection, allowing fruit crops to endure prolonged or repeated stress events. For example, in citrus orchards facing recurring heat waves, pre-treatment with HSP inducers ensures that the trees retain their resilience across multiple high-temperature periods.
   - This continuous protection is beneficial for long-season crops or perennial fruit trees, as it allows growers to maintain productivity without having to reapply stress management treatments frequently. For crops like grapes and citrus, where stress events may recur throughout the season, HSP inducers offer a sustainable method for managing stress over time.
4. **Minimizing Yield Losses and Economic Impact**
   - By enhancing fruit crops' resilience to stress, HSP inducers reduce yield losses that would otherwise occur due to environmental fluctuations. This minimization of yield loss translates directly to economic stability for growers, particularly for high-value crops like grapes, citrus, and tomatoes, where slight reductions in yield or quality can significantly impact profitability.
   - The economic benefits are particularly important for growers in regions where unpredictable climate patterns pose frequent risks to crop output. By investing in HSP inducers, growers can ensure more consistent yields and higher-quality produce, which supports both market demand and farm profitability.

3. HSPs in Drought and Salinity Management in Fruit Crops

Drought and salinity are two of the most challenging abiotic stresses for fruit crops, often resulting in decreased growth, reduced yield, and compromised fruit quality. Heat shock proteins (HSPs) play a pivotal role in enabling fruit crops to withstand these stresses by helping maintain osmotic balance, stabilizing cellular membranes, and reducing oxidative damage. In saline conditions, HSPs are particularly beneficial for protecting root cells from ion toxicity, thereby improving water and nutrient uptake. By supporting cellular integrity and stress resilience, HSPs contribute to the overall productivity and quality of fruit crops even in harsh environmental conditions.

a. Maintaining Osmotic Balance, Stabilizing Cell Membranes, and Reducing Oxidative Damage

1. **Osmotic Balance and Cellular Hydration**
   - During drought and salinity stress, fruit crops experience osmotic stress due to a reduction in available water and an increase in external salt concentration. This osmotic imbalance can lead to cellular dehydration, disrupting metabolic activities essential for growth and fruit development.
   - HSPs assist in maintaining osmotic balance by stabilizing proteins and enzymes involved in osmotic regulation. By preserving the functionality of these proteins, HSPs allow fruit cells to retain water and manage solute concentrations, preventing excessive water loss and maintaining cellular hydration.
   - In fruit crops such as grapes and citrus, where osmotic balance is crucial for berry and fruit development, HSPs ensure that cells remain hydrated, enabling continuous growth and reducing the likelihood of fruit shrinkage or quality loss due to water stress.

2. **Stabilization of Cell Membranes**
   - Under drought and salinity stress, cell membranes are particularly vulnerable to damage due to dehydration and high ion concentrations. Membrane stability is critical in fruit crops because it affects cellular integrity, nutrient transport, and overall metabolic efficiency.
   - HSPs, particularly small HSPs, bind to cell membranes, preventing them from losing structure and becoming too permeable under stress conditions. By stabilizing the lipid bilayer, HSPs maintain membrane fluidity and prevent the leakage of essential ions, which is crucial for sustaining cellular function.
   - In fruits like tomatoes and pomegranates, where membrane integrity is directly tied to water retention and nutrient uptake, HSPs help ensure that cells maintain optimal conditions for fruit development, even in low water or high salinity environments.

3. **Reduction of Oxidative Damage**
   - Drought and salinity stress often result in the accumulation of reactive oxygen species (ROS) in cells, leading to oxidative damage. High levels of ROS can damage proteins, lipids, and DNA, resulting in impaired cellular functions and decreased plant vitality.
   - HSPs help reduce oxidative stress by stabilizing antioxidant enzymes, such as superoxide

dismutase (SOD) and catalase, that neutralize ROS. By supporting these enzymes, HSPs minimize cellular damage, ensuring that essential metabolic processes continue even under stress.

- For instance, in apples and strawberries, which are sensitive to oxidative stress, HSPs play an essential role in protecting cells from ROS accumulation. This protection helps maintain fruit quality, reduces premature aging, and supports sustained growth under drought and salinity conditions.

b. Application of HSP Inducers for Improved Drought Tolerance in Fruit Crops

The use of HSP inducers is a promising approach for enhancing drought tolerance in fruit crops. By pre-activating HSP pathways, these inducers prepare plants to handle water scarcity more effectively, supporting moisture retention, cellular hydration, and overall growth.

1. **Priming Fruit Crops for Drought Tolerance**
   - In drought-prone regions, applying HSP inducers like salicylic acid to fruit crops can significantly improve resilience to water stress. These inducers activate the HSP production pathways, leading to a buildup of HSPs in plant cells that primes the plant for efficient stress response.
   - For crops such as citrus and grapes, which require considerable water for fruit development, HSP inducers provide a preventive measure that enhances their ability to retain moisture and continue photosynthesis under drought conditions. This preparation reduces the impact of water scarcity, allowing the plants to support growth and fruit set despite limited water availability.
2. **Enhanced Water Retention and Cellular Hydration**
   - HSP inducers help fruit crops maintain cellular hydration by supporting osmotic balance and reducing water loss. This effect is particularly beneficial in crops grown in arid or semi-arid regions, where water availability is unpredictable and drought can severely impact productivity.
   - In crops like figs and olives, which are cultivated in water-limited areas, the induction of HSPs leads to better water retention in tissues, supporting steady fruit growth. By retaining cellular hydration, these fruit crops can withstand drought conditions for extended periods without experiencing significant reductions in yield or fruit quality.
3. **Sustained Growth and Yield Stability in Drought Conditions**
   - Drought stress often leads to stunted growth, reduced leaf area, and smaller fruit sizes, all of which can decrease the economic value of fruit crops. By inducing HSPs before drought conditions arise, growers can help plants maintain steady growth and avoid common drought-related setbacks.
   - In mango and avocado orchards, for instance, the use of HSP inducers supports the trees' growth processes, ensuring consistent leaf expansion, root development, and fruiting. This sustained growth contributes to yield stability and fruit quality, which are critical for profitability in regions where drought is a recurring challenge.

4. **Improved Nutrient Uptake and Ion Management Under Salinity Stress**
   - In saline environments, excessive sodium (Na^+) and chloride (Cl^-) ions in the soil interfere with nutrient uptake, leading to ion toxicity in root cells. This condition is particularly challenging for salt-sensitive fruit crops like strawberries, citrus, and pomegranates, which suffer from osmotic stress and nutrient imbalances under high salinity.
   - HSPs protect root cells from ion toxicity by stabilizing the cellular machinery responsible for ion transport and compartmentalization. By managing the influx and efflux of ions, HSPs help these crops maintain favorable potassium-to-sodium ratios, which is essential for nutrient balance and cellular function.
   - In citrus, for example, the induction of HSPs under salinity stress supports root cell health, enhancing the tree's ability to exclude sodium while retaining essential nutrients like potassium and calcium. This improved ion homeostasis allows salt-sensitive fruit crops to better withstand salinity stress, preserving growth and productivity in soils affected by salt buildup.
5. **Supporting Fruit Quality and Marketability**
   - Drought and salinity stress often impact fruit quality parameters, such as size, flavor, texture, and nutritional content. By stabilizing proteins and cellular structures, HSPs help maintain the internal conditions necessary for high-quality fruit production, even under challenging conditions.
   - In crops like tomatoes and grapes, which are valued for their flavor and nutritional quality, HSPs ensure that sugars, acids, and phenolic compounds are synthesized and stored efficiently. This protection supports the development of well-formed, flavorful fruits that meet market standards, contributing to better marketability and consumer satisfaction.
6. **Economic Benefits of HSP Induction in Drought-Prone Regions**
   - For fruit growers in drought-prone regions, the use of HSP inducers is an economically viable approach to managing water stress. By reducing the need for frequent irrigation, HSP inducers help conserve water resources and lower operational costs.
   - In addition, the yield stability and quality improvements supported by HSPs enhance profitability for farmers. For high-value fruit crops, such as table grapes, strawberries, and citrus, maintaining yield and quality under drought conditions ensures that growers can meet market demands and sustain income despite environmental challenges.

B. Soil Moisture Conservation Techniques for fruit crops

Soil moisture conservation is essential for sustainable fruit production, particularly in regions that experience seasonal drought or irregular rainfall. By conserving soil moisture, growers can optimize water use, improve crop resilience to dry spells and support consistent fruit development. Mulching is one of the most effective techniques for conserving soil moisture, and it offers additional benefits, including soil temperature moderation, weed suppression, and soil health improvement.

1. Mulching

Mulching involves covering the soil surface around fruit crops with various organic or synthetic materials to create a protective layer. This practice is widely used in fruit orchards and other perennial cropping systems to retain moisture, reduce evaporation, and regulate soil temperature. Mulching can be particularly beneficial for fruit crops like citrus, apples, strawberries, and grapes, which have high water demands and are sensitive to soil moisture fluctuations.

Types of Mulching Materials and Their Mechanisms for Moisture Conservation

1. **Organic Mulches**
   - Organic mulches include natural materials like straw, grass clippings, leaves, bark, compost, and wood chips. These mulches gradually decompose, adding organic matter to the soil, which enhances its water-holding capacity over time. Organic mulches are especially beneficial in fruit orchards, where soil structure and health are important for long-term productivity.
   - By covering the soil surface, organic mulches reduce direct sunlight exposure, lowering soil temperature and reducing evaporation. This effect helps maintain a cooler and more stable environment around the root zone, which is crucial for fruit crops during dry seasons.
   - Additionally, organic mulches suppress weed growth by blocking sunlight from reaching weed seeds. This weed suppression is important because weeds compete with fruit crops for soil moisture, nutrients, and sunlight. By minimizing competition, mulching ensures that more water and nutrients are available for the fruit plants, enhancing their growth and fruiting potential.
2. **Synthetic Mulches**
   - Synthetic mulches, such as plastic films and biodegradable sheets, are also widely used for moisture conservation in fruit crops. Plastic films, typically black or reflective, create an impermeable barrier that minimizes soil evaporation by preventing direct exposure to sunlight and air.
   - Reflective mulches are commonly used in orchards to moderate soil temperature by reflecting sunlight, which reduces heat stress on the root zone during hot seasons. In crops like strawberries and melons, reflective mulches also deter certain pests by making it difficult for them to locate the plants, offering an additional protective benefit.
   - Biodegradable plastic mulches combine the benefits of plastic with the advantages of organic decomposition, as they break down over time, reducing the need for removal and disposal. These mulches are increasingly popular in fruit farms because they conserve soil moisture effectively while eliminating the environmental concerns associated with non-biodegradable plastics.
3. **Living Mulches**
   - Living mulches involve planting ground-cover crops, such as clover or legumes, around fruit trees or between rows of fruit crops. These living covers reduce soil erosion, enhance

soil structure, and provide habitat for beneficial insects. By shading the soil surface, living mulches also reduce evaporation, contributing to moisture conservation.

- In addition, legumes as living mulches offer the added benefit of nitrogen fixation, which naturally enriches the soil, reducing the need for synthetic fertilizers. In apple and peach orchards, for example, living mulches can improve soil fertility while conserving moisture, creating a symbiotic environment that supports both crop growth and soil health.

Benefits of Mulching in Fruit Orchards

1. **Improved Water Availability**
   - Mulching significantly improves water availability for fruit crops by conserving soil moisture. By reducing the rate of evaporation, mulch keeps the soil moist for longer periods, ensuring that water is readily accessible to plant roots. This is especially beneficial in fruit orchards, where deep-rooted trees and shrubs require consistent moisture for fruit development.
   - In grape and citrus orchards, mulching during dry seasons helps maintain soil moisture levels, reducing the need for frequent irrigation. This not only conserves water but also minimizes the risk of drought stress, which can lead to fruit drop, smaller fruits, and lower yields.
2. **Soil Temperature Regulation**
   - Mulching moderates soil temperature by insulating the soil from extreme fluctuations. During hot weather, mulch keeps the soil cooler by blocking direct sunlight, protecting roots from heat stress. This temperature moderation is essential for fruit crops, as high soil temperatures can inhibit root growth and reduce nutrient uptake.
   - In strawberries, for example, mulching prevents soil temperatures from rising too high, ensuring optimal conditions for root function and fruit growth. The same principle applies in apple and peach orchards, where stable soil temperatures support steady fruit development and prevent stress-related issues such as wilting or stunted growth.
3. **Weed Suppression and Reduced Competition**
   - Weeds compete with fruit crops for soil moisture, nutrients, and sunlight, which can reduce the growth and yield potential of the main crop. Mulching creates a physical barrier that suppresses weed growth by preventing weed seeds from germinating or blocking sunlight from reaching the weeds.
   - In orchards with high-value crops like apples, grapes, and berries, effective weed suppression ensures that the trees and shrubs have access to sufficient moisture and nutrients. By reducing weed pressure, mulching supports healthier and more vigorous growth, leading to larger and higher-quality fruits.
4. **Enhanced Soil Structure and Health**
   - Organic mulches decompose over time, adding organic matter to the soil and improving its structure. This organic matter enhances soil aeration, increases water infiltration, and

improves moisture-holding capacity, all of which contribute to long-term soil health and productivity.

- The addition of organic matter also supports beneficial microbial activity. Microorganisms in the soil help break down organic material, releasing nutrients that are essential for plant growth. In fruit orchards, this improved soil health results in better root growth, higher nutrient availability, and increased resilience to stress.
- For example, in mango and citrus orchards, the use of organic mulches such as compost and wood chips promotes soil fertility, supporting continuous fruit production and improving fruit size and flavor over time.

5. **Reduction of Soil Erosion**

- In sloped orchards or areas with heavy rainfall, mulching helps prevent soil erosion by absorbing the impact of raindrops and reducing surface runoff. Mulch holds the soil in place, preventing the loss of topsoil and preserving soil structure.
- For vineyards located on hillsides, where soil erosion can be a major concern, mulching provides an effective way to protect the soil and prevent nutrient loss. By retaining topsoil, mulching contributes to healthier vine growth and supports the production of high-quality grapes.

6. **Environmental and Economic Benefits**

- Mulching is an environmentally friendly practice that conserves water, reduces the need for chemical herbicides, and improves soil health. By reducing evaporation and improving soil structure, mulching minimizes the need for frequent irrigation, which is particularly beneficial in regions facing water scarcity.
- Economically, mulching reduces production costs associated with weed control, irrigation, and soil amendments. For growers of high-value fruit crops, such as strawberries, blueberries, and apples, these cost savings enhance profitability while promoting sustainable production practices.

Mulching in Different Fruit Crop Systems

1. **Citrus and Apple Orchards**

- In citrus and apple orchards, mulching helps maintain consistent moisture levels around the root zone, which is essential for fruit size and juiciness. Organic mulches such as wood chips and compost are commonly used to conserve soil moisture, enhance nutrient cycling, and improve soil structure.
- For citrus, mulching also minimizes weed growth around the base of the trees, reducing competition for water and nutrients. In apple orchards, mulching contributes to improved soil health, which supports root growth and increases fruit yield and quality.

2. **Grape Vineyards**

- Mulching in grape vineyards provides insulation to the soil, which protects roots from temperature extremes and conserves moisture, especially in dry regions. Reflective plastic mulches are sometimes used to reduce soil temperatures while enhancing light reflection, which can improve grape ripening and color.

- Organic mulches like straw and compost are also beneficial for soil health in vineyards, supporting microbial activity and adding organic matter. This added organic matter helps improve soil structure and water-holding capacity, which is crucial for grapevine productivity and resilience to drought.

3. **Berry Crops (e.g., Strawberries and Blueberries)**
   - In strawberry and blueberry production, mulching is essential for moisture conservation and weed control. Straw and black plastic mulches are commonly used to protect berries from soil contact, which reduces fruit rot and maintains fruit cleanliness.
   - For blueberries, which prefer acidic soils, pine bark and wood chip mulches provide added benefits by lowering soil pH and enriching the soil with organic matter. These mulches improve water retention, support root health, and enhance berry quality, making them a key component of sustainable berry production.

2. Hydrophilic Polymers (Superabsorbent Polymers - SAPs)

Hydrophilic polymers, commonly known as superabsorbent polymers (SAPs), are water-retentive soil amendments that can significantly enhance soil moisture levels, particularly in water-scarce environments. These polymers can absorb and hold large amounts of water relative to their weight, acting as a reservoir that gradually releases moisture to plant roots as needed. By improving soil water retention and reducing the frequency of irrigation, SAPs support healthier and more resilient growth in fruit crops, especially under drought-prone conditions. SAPs are particularly valuable for fruit growers aiming to optimize water use and improve crop productivity with limited water resources.

Mechanism of SAPs in Soil Moisture Conservation

1. **Water Absorption and Release Properties**
   - Hydrophilic polymers are capable of absorbing water hundreds of times their own weight. When added to the soil, SAPs expand as they absorb moisture from rainfall, irrigation, or natural groundwater. This absorbed water is stored within the polymer's structure, where it remains until soil moisture levels begin to drop.
   - As the surrounding soil dries, the SAPs gradually release the stored water, making it available to plant roots. This controlled release of moisture ensures a sustained water supply for fruit crops, even during periods of drought or reduced irrigation, allowing the plants to access moisture when it is most needed.
2. **Extended Water Availability**
   - The unique water-holding ability of SAPs extends the availability of water in the soil, which is particularly beneficial for deep-rooted fruit crops. By reducing the rate of water loss to the atmosphere, SAPs maintain soil moisture levels for longer periods. This extended water availability allows fruit crops to sustain growth without the immediate need for additional irrigation.

- In fruit orchards where frequent irrigation may be challenging or costly, SAPs provide a practical solution by reducing the reliance on external water sources. This is especially valuable in regions prone to seasonal droughts, where maintaining consistent soil moisture is critical for fruit quality and yield.

Benefits of SAPs for Fruit Crop Production

1. **Enhanced Water-Use Efficiency**
   - SAPs improve water-use efficiency by optimizing the distribution of moisture within the root zone, allowing fruit crops to absorb water more effectively. With SAPs in the soil, less water is lost to evaporation or deep drainage, which maximizes the efficiency of each irrigation event.
   - In crops like strawberries, grapes, and blueberries, where optimal soil moisture is essential for flavor and quality, SAPs help plants use available water more effectively. This efficiency not only conserves water but also reduces the cost and labor associated with frequent irrigation, making it economically beneficial for fruit growers.
2. **Improved Root Development and Drought Resilience**
   - By providing a consistent moisture supply, SAPs support healthy root development, enabling fruit crops to establish deep and extensive root systems. A well-developed root system enhances a plant's ability to explore the soil for nutrients and water, increasing its resilience to periods of drought.
   - For crops like apples, citrus, and peaches, where drought stress can lead to early fruit drop and reduced fruit size, SAPs help maintain adequate moisture around the root zone, reducing the stress on the plant. This resilience allows the crop to continue fruiting under water-limited conditions, resulting in higher yields and better-quality fruits.
3. **Reduction in Irrigation Frequency and Water Conservation**
   - SAPs reduce the need for frequent irrigation, which is particularly advantageous in fruit orchards where water availability may be restricted. By holding water within the soil, SAPs minimize the rapid loss of moisture, allowing longer intervals between irrigation events.
   - In drought-prone regions, SAPs help fruit growers conserve water by reducing the volume and frequency of irrigation required to maintain soil moisture levels. This water-saving capacity is critical for sustainable fruit production, as it enables growers to allocate limited water resources more effectively across their orchards.
4. **Stabilized Soil Moisture for Uniform Fruit Development**
   - Consistent soil moisture is essential for uniform fruit development in crops like tomatoes, melons, and berries, where fluctuating moisture levels can lead to issues like fruit cracking or uneven ripening. By stabilizing soil moisture, SAPs help maintain an even water supply that supports balanced growth and high-quality fruit development.
   - For strawberries and blueberries, SAPs ensure that the roots receive a steady water supply, which is critical for achieving the desired size, flavor, and texture of the fruit.

This uniformity in fruit quality is particularly valuable for market-focused growers who aim to produce high-grade fruit that meets consumer expectations.

Specific Applications of SAPs in Fruit Crop Systems

1. **Strawberry Production**
   - Strawberries are highly sensitive to soil moisture fluctuations, which can impact fruit size, sweetness, and shelf life. SAPs applied to strawberry fields help maintain a consistent level of soil moisture, ensuring optimal conditions for flowering, fruit set, and fruit enlargement.
   - The gradual release of water from SAPs reduces the risk of water stress, which is especially beneficial during flowering and fruiting stages, when water demand is high. This stability helps strawberries develop uniformly, reducing the risk of small or deformed fruits and improving overall crop quality.
2. **Grape Vineyards**
   - In grapevines, maintaining soil moisture is crucial for both yield and quality, as water stress during critical stages of berry development can impact sugar accumulation and berry size. SAPs are often used in vineyards to retain moisture in the root zone, particularly in dry regions where irrigation resources are limited.
   - By extending water availability, SAPs support consistent vine growth and berry development, enhancing the grape's flavor profile, color, and sugar content. This is particularly valuable for wine grape production, where grape quality directly affects wine characteristics and market value.
3. **Citrus Orchards**
   - Citrus trees have a high water requirement, particularly during fruit enlargement and ripening stages. SAPs applied to citrus orchards help maintain moisture levels in the root zone, allowing the trees to access water during dry spells without additional irrigation.
   - In regions with limited rainfall, SAPs are especially beneficial in citrus production, as they allow trees to sustain fruit growth under water-scarce conditions. This moisture stability prevents fruit drop, ensures larger fruit size, and improves juice content, contributing to higher-quality harvests.
4. **Blueberries and Other Berry Crops**
   - Blueberries and other berry crops, such as raspberries and blackberries, have shallow root systems and are sensitive to dry soil conditions. SAPs improve soil moisture retention around the roots, ensuring a stable water supply that supports healthy berry development.
   - Consistent soil moisture is essential for achieving the optimal sweetness and firmness of blueberries, which are highly valued in the fresh market. By using SAPs, berry growers can reduce the risk of water stress, support steady fruit development, and produce berries that meet high quality standards.

5. **Pome and Stone Fruits (e.g., Apples, Peaches, and Plums)**
    - In pome and stone fruits like apples, peaches, and plums, SAPs help mitigate the effects of drought by stabilizing soil moisture during critical growth periods. For apples, maintaining soil moisture during fruit enlargement is essential for achieving the desired size, flavor, and texture.
    - In peach and plum orchards, SAPs reduce the risk of fruit shrinkage, enhance juiciness, and contribute to improved fruit firmness, all of which are important for marketable quality. The use of SAPs also prevents drought-induced stress that can lead to reduced yields or early fruit drop, ensuring a more profitable harvest for orchardists.

Environmental and Economic Benefits of Using SAPs in Fruit Production

1. **Water Conservation and Sustainability**
    - SAPs support sustainable water management by reducing irrigation requirements and conserving water in water-limited regions. This water-saving capacity is particularly advantageous for fruit growers facing increasing water costs and restrictions.
    - By minimizing water usage, SAPs contribute to the long-term sustainability of fruit production, helping growers optimize resources and reduce their environmental footprint.
2. **Reduced Production Costs**
    - Although SAPs require an initial investment, their ability to extend irrigation intervals and conserve water leads to cost savings over time. Fewer irrigation events reduce labor, energy, and water expenses, improving the economic efficiency of fruit production.
    - For high-value fruit crops, the consistent moisture provided by SAPs also improves yield and quality, contributing to higher market returns. This combination of reduced costs and enhanced quality makes SAPs a valuable addition to fruit production systems.
3. **Improved Soil Health and Reduced Soil Erosion**
    - By retaining moisture in the root zone, SAPs help maintain soil structure and reduce the risk of soil erosion, particularly on sloped or exposed orchard soils. This soil stability supports healthier root systems and prevents nutrient leaching, which is beneficial for long-term soil fertility and crop health.
    - In addition, SAPs can improve soil aeration and reduce compaction, creating an environment that supports beneficial microbial activity and nutrient cycling. Healthier soil promotes better root growth, leading to stronger and more resilient fruit crops.

3. Hydrophilic Polymers (Superabsorbent Polymers - SAPs)

Hydrophilic polymers, commonly known as superabsorbent polymers (SAPs), are water-retentive soil amendments that can significantly enhance soil moisture levels, particularly in water-scarce environments. These polymers can absorb and hold large amounts of water relative to their weight, acting as a reservoir that gradually releases moisture to plant roots as needed. By improving soil water retention and reducing the frequency of irrigation, SAPs support healthier and more resilient growth in fruit crops, especially under drought-prone conditions. SAPs are particularly valuable for fruit growers aiming to optimize water use and improve crop productivity with limited water resources.

Mechanism of SAPs in Soil Moisture Conservation

1. **Water Absorption and Release Properties**
   - Hydrophilic polymers are capable of absorbing water hundreds of times their own weight. When added to the soil, SAPs expand as they absorb moisture from rainfall, irrigation, or natural groundwater. This absorbed water is stored within the polymer's structure, where it remains until soil moisture levels begin to drop.
   - As the surrounding soil dries, the SAPs gradually release the stored water, making it available to plant roots. This controlled release of moisture ensures a sustained water supply for fruit crops, even during periods of drought or reduced irrigation, allowing the plants to access moisture when it is most needed.
2. **Extended Water Availability**
   - The unique water-holding ability of SAPs extends the availability of water in the soil, which is particularly beneficial for deep-rooted fruit crops. By reducing the rate of water loss to the atmosphere, SAPs maintain soil moisture levels for longer periods. This extended water availability allows fruit crops to sustain growth without the immediate need for additional irrigation.
   - In fruit orchards where frequent irrigation may be challenging or costly, SAPs provide a practical solution by reducing the reliance on external water sources. This is especially valuable in regions prone to seasonal droughts, where maintaining consistent soil moisture is critical for fruit quality and yield.

Benefits of SAPs for Fruit Crop Production

1. **Enhanced Water-Use Efficiency**
   - SAPs improve water-use efficiency by optimizing the distribution of moisture within the root zone, allowing fruit crops to absorb water more effectively. With SAPs in the soil, less water is lost to evaporation or deep drainage, which maximizes the efficiency of each irrigation event.
   - In crops like strawberries, grapes, and blueberries, where optimal soil moisture is essential for flavor and quality, SAPs help plants use available water more effectively. This efficiency not only conserves water but also reduces the cost and labor associated with frequent irrigation, making it economically beneficial for fruit growers.
2. **Improved Root Development and Drought Resilience**
   - By providing a consistent moisture supply, SAPs support healthy root development, enabling fruit crops to establish deep and extensive root systems. A well-developed root system enhances a plant's ability to explore the soil for nutrients and water, increasing its resilience to periods of drought.

- For crops like apples, citrus, and peaches, where drought stress can lead to early fruit drop and reduced fruit size, SAPs help maintain adequate moisture around the root zone, reducing the stress on the plant. This resilience allows the crop to continue fruiting under water-limited conditions, resulting in higher yields and better-quality fruits.

3. **Reduction in Irrigation Frequency and Water Conservation**
   - SAPs reduce the need for frequent irrigation, which is particularly advantageous in fruit orchards where water availability may be restricted. By holding water within the soil, SAPs minimize the rapid loss of moisture, allowing longer intervals between irrigation events.
   - In drought-prone regions, SAPs help fruit growers conserve water by reducing the volume and frequency of irrigation required to maintain soil moisture levels. This water-saving capacity is critical for sustainable fruit production, as it enables growers to allocate limited water resources more effectively across their orchards.

4. **Stabilized Soil Moisture for Uniform Fruit Development**
   - Consistent soil moisture is essential for uniform fruit development in crops like tomatoes, melons, and berries, where fluctuating moisture levels can lead to issues like fruit cracking or uneven ripening. By stabilizing soil moisture, SAPs help maintain an even water supply that supports balanced growth and high-quality fruit development.
   - For strawberries and blueberries, SAPs ensure that the roots receive a steady water supply, which is critical for achieving the desired size, flavor, and texture of the fruit. This uniformity in fruit quality is particularly valuable for market-focused growers who aim to produce high-grade fruit that meets consumer expectations.

Specific Applications of SAPs in Fruit Crop Systems

1. **Strawberry Production**
   - Strawberries are highly sensitive to soil moisture fluctuations, which can impact fruit size, sweetness, and shelf life. SAPs applied to strawberry fields help maintain a consistent level of soil moisture, ensuring optimal conditions for flowering, fruit set, and fruit enlargement.
   - The gradual release of water from SAPs reduces the risk of water stress, which is especially beneficial during flowering and fruiting stages, when water demand is high. This stability helps strawberries develop uniformly, reducing the risk of small or deformed fruits and improving overall crop quality.

2. **Grape Vineyards**
   - In grapevines, maintaining soil moisture is crucial for both yield and quality, as water stress during critical stages of berry development can impact sugar accumulation and berry size. SAPs are often used in vineyards to retain moisture in the root zone, particularly in dry regions where irrigation resources are limited.
   - By extending water availability, SAPs support consistent vine growth and berry development, enhancing the grape's flavor profile, color, and sugar content. This is

particularly valuable for wine grape production, where grape quality directly affects wine characteristics and market value.

3. **Citrus Orchards**
   - Citrus trees have a high water requirement, particularly during fruit enlargement and ripening stages. SAPs applied to citrus orchards help maintain moisture levels in the root zone, allowing the trees to access water during dry spells without additional irrigation.
   - In regions with limited rainfall, SAPs are especially beneficial in citrus production, as they allow trees to sustain fruit growth under water-scarce conditions. This moisture stability prevents fruit drop, ensures larger fruit size, and improves juice content, contributing to higher-quality harvests.
4. **Blueberries and Other Berry Crops**
   - Blueberries and other berry crops, such as raspberries and blackberries, have shallow root systems and are sensitive to dry soil conditions. SAPs improve soil moisture retention around the roots, ensuring a stable water supply that supports healthy berry development.
   - Consistent soil moisture is essential for achieving the optimal sweetness and firmness of blueberries, which are highly valued in the fresh market. By using SAPs, berry growers can reduce the risk of water stress, support steady fruit development, and produce berries that meet high quality standards.
5. **Pome and Stone Fruits (e.g., Apples, Peaches, and Plums)**
   - In pome and stone fruits like apples, peaches, and plums, SAPs help mitigate the effects of drought by stabilizing soil moisture during critical growth periods. For apples, maintaining soil moisture during fruit enlargement is essential for achieving the desired size, flavor, and texture.
   - In peach and plum orchards, SAPs reduce the risk of fruit shrinkage, enhance juiciness, and contribute to improved fruit firmness, all of which are important for marketable quality. The use of SAPs also prevents drought-induced stress that can lead to reduced yields or early fruit drop, ensuring a more profitable harvest for orchardists.

Environmental and Economic Benefits of Using SAPs in Fruit Production

1. **Water Conservation and Sustainability**
   - SAPs support sustainable water management by reducing irrigation requirements and conserving water in water-limited regions. This water-saving capacity is particularly advantageous for fruit growers facing increasing water costs and restrictions.
   - By minimizing water usage, SAPs contribute to the long-term sustainability of fruit production, helping growers optimize resources and reduce their environmental footprint.
2. **Reduced Production Costs**
   - Although SAPs require an initial investment, their ability to extend irrigation intervals and conserve water leads to cost savings over time. Fewer irrigation events reduce labor,

energy, and water expenses, improving the economic efficiency of fruit production.

- For high-value fruit crops, the consistent moisture provided by SAPs also improves yield and quality, contributing to higher market returns. This combination of reduced costs and enhanced quality makes SAPs a valuable addition to fruit production systems.

3. **Improved Soil Health and Reduced Soil Erosion**

   - By retaining moisture in the root zone, SAPs help maintain soil structure and reduce the risk of soil erosion, particularly on sloped or exposed orchard soils. This soil stability supports healthier root systems and prevents nutrient leaching, which is beneficial for long-term soil fertility and crop health.
   - In addition, SAPs can improve soil aeration and reduce compaction, creating an environment that supports beneficial microbial activity and nutrient cycling. Healthier soil promotes better root growth, leading to stronger and more resilient fruit crops.

4. Increasing Water Use Efficiency (WUE) in Fruit Crops

Water use efficiency (WUE) is a measure of the yield or biomass produced per unit of water used. In fruit crops, where high water demands and variable climate conditions often intersect, enhancing WUE is crucial for achieving optimal yields while conserving water. Increasing WUE not only supports sustainable agriculture in water-scarce regions but also helps maintain fruit quality, reduces irrigation costs, and minimizes environmental impact. Advanced irrigation techniques such as drip irrigation, deficit irrigation, and precise scheduling based on crop water needs play a key role in optimizing WUE for fruit crops.

Importance of Water Use Efficiency in Fruit Crop Production

1. **Sustainable Water Management**

   - Increasing WUE in fruit crops ensures that water resources are used efficiently, reducing wastage and preserving water availability for future agricultural needs. In areas where water is scarce, improving WUE enables farmers to meet the demands of high-water-requirement crops without exhausting local water supplies.
   - For example, in semi-arid regions where citrus and grape production is common, efficient water use helps conserve limited resources, allowing growers to sustain production over multiple seasons while minimizing stress on regional water supplies.

2. **Enhancing Crop Resilience to Water Scarcity**

   - Improved WUE helps fruit crops withstand periods of water scarcity or irregular rainfall by ensuring that plants receive water in precise amounts, exactly when needed. This resilience is critical for fruit crops that are sensitive to drought stress, as it supports steady growth and fruit development even under fluctuating moisture conditions.
   - In fruit crops such as tomatoes, strawberries, and melons, maintaining high WUE helps reduce the impact of water shortages on fruit size, flavor, and quality, allowing growers to produce marketable crops despite challenging conditions.

3. **Economic and Environmental Benefits**

   - Efficient water use reduces the costs associated with frequent irrigation, conserving resources like labor, energy, and fuel used in pumping and distributing water. For high-value fruit crops, where irrigation expenses can be substantial, maximizing WUE contributes to higher profitability.
   - Environmentally, increased WUE minimizes water runoff and deep percolation, which can carry fertilizers and pesticides into groundwater or nearby water bodies. By enhancing WUE, growers reduce the environmental impact of their irrigation practices, contributing to sustainable and eco-friendly fruit production.

Techniques for Increasing Water Use Efficiency in Fruit Crops

1. **Drip Irrigation**

   - Drip irrigation is a highly efficient system that delivers water directly to the root zone of each plant through emitters placed along irrigation lines. By targeting the root zone, drip irrigation minimizes water loss from evaporation and runoff, ensuring that a greater proportion of water reaches the crop roots where it is needed.
   - In high-value crops like citrus, grapes, and pomegranates, drip irrigation helps maintain consistent moisture levels, supporting optimal growth and fruit quality. For instance, in grapevines, drip irrigation promotes balanced growth and berry development, which are critical for wine production, as consistent moisture is linked to grape flavor and sugar concentration.
   - The precision of drip irrigation also enables growers to adjust water application according to the specific needs of each crop and soil type, further enhancing WUE by delivering only as much water as required.

2. **Deficit Irrigation**

   - Deficit irrigation is a water-saving technique that involves applying less water than the full crop water requirement, with the goal of controlling plant growth and focusing water usage on critical growth stages. In fruit crops, deficit irrigation is typically used during periods when slight water stress does not negatively affect yield or quality, such as after fruit set or during early growth.
   - In stone fruits like peaches and cherries, deficit irrigation can help control excessive vegetative growth, redirecting more resources towards fruit development. This practice not only conserves water but also improves fruit size, sweetness, and firmness, as mild water stress can enhance sugar accumulation and concentration.
   - In citrus orchards, regulated deficit irrigation applied during non-critical growth stages can reduce water use without compromising fruit quality, allowing growers to maintain high yields while saving water.

3. **Irrigation Scheduling Based on Crop Water Requirements**

   - Irrigation scheduling involves planning irrigation events based on the specific water needs of each crop at different growth stages. By matching irrigation with crop demand, growers can optimize water usage, providing just the right amount of moisture when plants need it most.

- Advanced tools like soil moisture sensors, tensiometers, and weather-based irrigation models are used to determine optimal irrigation timing. In apple and pear orchards, for example, sensors monitor soil moisture levels, ensuring that irrigation is applied only when the soil begins to dry, which prevents over-irrigation and promotes efficient water use.
- Precise scheduling is particularly beneficial for fruit crops with specific moisture requirements, such as blueberries and raspberries, where consistent soil moisture is essential for flavor, texture, and berry size. By scheduling irrigation accurately, growers avoid water stress that can compromise fruit quality and yield.

4. Mulching for Reduced Evaporation

- Mulching, the practice of covering the soil surface with organic or synthetic materials, enhances WUE by reducing water loss from evaporation and improving soil moisture retention. Organic mulches like straw, compost, and wood chips help insulate the soil, keeping it cool and moist, which benefits fruit crops that are sensitive to moisture fluctuations.
- In citrus and mango orchards, mulching reduces evaporation, allowing the soil to retain moisture longer, and reducing the need for frequent irrigation. This extended soil moisture supports steady root hydration, which is critical for fruit enlargement and quality during dry seasons.
- Mulching also contributes to better soil structure and health, which in turn improves water infiltration and retention, creating a stable environment for roots and supporting higher WUE.

5. Improving Soil Structure and Organic Matter Content

- Enhancing soil structure and organic matter content improves WUE by increasing the soil's capacity to retain and store water. Organic matter improves soil porosity and water-holding capacity, allowing more moisture to be stored in the root zone and made available to plants.
- For fruit orchards, incorporating compost or other organic materials into the soil helps create a more favorable rooting environment, which enhances the ability of plants to access water. In crops like avocado and apple trees, healthier soil structure supports extensive root development, allowing the plants to use available water more effectively and reducing the frequency of irrigation.

6. Precision Agriculture and Sensor-Based Irrigation

- Precision agriculture techniques, including the use of soil moisture sensors, evapotranspiration (ET) models, and remote sensing, provide real-time data on crop water needs. This information allows growers to apply irrigation precisely where and when it is needed, maximizing WUE.

- In high-value fruit crops like almonds and pistachios, where water availability can affect kernel size and quality, sensor-based irrigation systems ensure that each tree receives optimal moisture without over-irrigating. This precision reduces water waste, enhances tree health, and supports consistent yields and quality, making it a valuable tool for managing water use in commercial orchards.

Benefits of Improved Water Use Efficiency for Fruit Crops

1. **Increased Yield and Fruit Quality**
   - Enhanced WUE directly contributes to increased yield and improved fruit quality by ensuring that plants receive consistent moisture. For example, in grapes, consistent soil moisture levels support berry growth, sugar accumulation, and flavor development, which are critical for wine quality.
   - In crops like tomatoes and watermelons, improved WUE prevents moisture stress that can lead to issues such as fruit cracking or irregular ripening. By maintaining optimal moisture levels, growers can produce high-quality fruits that meet market standards for size, taste, and appearance.
2. **Reduced Irrigation Costs and Resource Conservation**
   - Efficient water use lowers the costs associated with irrigation, including water expenses, labor, and energy for pumping and distribution. For fruit producers, these cost savings contribute to higher profitability, especially for water-intensive crops like citrus, apples, and avocados.
   - Improved WUE also conserves resources, as less water is drawn from local sources. This conservation is particularly important in arid regions, where water scarcity is a concern, and supports the sustainable use of water resources over the long term.
3. **Enhanced Resilience to Drought and Climate Variability**
   - Fruit crops grown with high WUE are better able to withstand drought conditions, as efficient water use strategies ensure that plants receive the necessary moisture, even under water-limited conditions. This resilience is essential for crops like stone fruits and pomegranates, where drought stress can impact fruit quality and yield.
   - By adapting water management practices to improve WUE, growers can protect their crops from climate-induced water shortages, helping them maintain productivity and quality in an unpredictable climate.
4. **Reduced Environmental Impact**
   - Increasing WUE minimizes water runoff, deep percolation, and the risk of nutrient leaching, reducing the environmental impact of irrigation practices. By using water more efficiently, growers lower the risk of groundwater contamination and reduce their overall footprint.
   - This environmental benefit is especially valuable in regions with fragile ecosystems, where water conservation helps protect biodiversity and maintain ecosystem health.

Sustainable water use practices in fruit production contribute to the preservation of natural resources, supporting both agricultural and ecological resilience.

C. Skimming Technology

Skimming technology is an innovative approach to water extraction that enables farmers to access freshwater in regions where groundwater is predominantly saline. By extracting water from shallow aquifers, this technology provides a reliable source of low-salinity water that is suitable for irrigation without disturbing the underlying saline layers. In coastal and arid regions, where saltwater intrusion and limited freshwater resources make agriculture challenging, skimming technology offers a sustainable solution that supports crop health and productivity.

1. Concept and Application of Skimming Technology

1. **Extracting Fresh Water from Shallow Aquifers**
   - Skimming technology is designed to extract freshwater from shallow aquifers, typically located within the upper layers of groundwater. In areas where deeper aquifers are saline or brackish, shallow aquifers often contain a layer of fresher water that can be accessed for agricultural use.
   - To minimize the risk of drawing up saline water, skimming wells are designed with perforations only in the upper sections, allowing the well to selectively extract water from the fresher, shallow layers. This precision prevents the mixing of saline water from deeper layers, ensuring that the extracted water remains low in salinity and suitable for irrigation.
2. **Application in Coastal and Arid Regions**
   - In coastal areas, saltwater intrusion is a common issue due to the proximity of saline seawater and the potential for seawater to seep into freshwater aquifers. Skimming technology is particularly useful in these regions, as it allows farmers to access freshwater from shallow sources without risking contamination from deeper, saline water.
   - In arid and semi-arid regions, where natural freshwater resources are scarce and groundwater quality is often compromised by high salt content, skimming technology provides a sustainable alternative for irrigation. By tapping into shallow freshwater layers, this technology enables agriculture in areas that might otherwise be unsuitable for crop production due to salinity challenges.
3. **Design and Structure of Skimming Wells**
   - Skimming wells are constructed with a specific design that enables selective extraction from shallow aquifers. Unlike conventional wells that penetrate deep into the groundwater table, skimming wells are shallow, with perforations placed only in the upper portion to access the freshest water.
   - These wells are often lined with filters to further prevent the entry of saline water and sediment. In some cases, additional barriers or impermeable layers may be used to separate the skimming well from deeper saline aquifers, ensuring the quality and consistency of the extracted water.

Benefits of Skimming Technology for Sustainable Agriculture

1. Reducing the Impact of Salinity on Crop Health

- Salinity is a major stress factor for crops, as it affects water uptake, nutrient absorption, and overall plant health. Excess salt in irrigation water can lead to osmotic stress, ion toxicity, and reduced crop yields. By accessing freshwater through skimming technology, farmers can reduce the salinity levels in irrigation water, preventing salt accumulation in the root zone and promoting healthier plant growth.
- In salt-sensitive fruit crops, such as strawberries, citrus, and grapes, skimming technology supports improved root health and nutrient uptake, resulting in higher-quality fruits and better crop resilience to salinity stress.

2. Supporting Sustainable Irrigation Practices

- Skimming technology is aligned with sustainable water management practices, as it enables the use of local freshwater resources without over-extracting or contaminating deeper aquifers. By accessing only the top layer of groundwater, skimming wells reduce the risk of groundwater depletion and help maintain the integrity of saline aquifers.
- This selective extraction contributes to the long-term sustainability of water resources, particularly in regions where overuse of groundwater can lead to ecological imbalance or aquifer collapse. In water-scarce areas, skimming technology allows growers to manage limited water resources more responsibly, supporting both agricultural productivity and environmental health.

3. Enabling Agriculture in Salinity-Prone Regions

- Skimming technology opens up agricultural opportunities in regions where high salinity has traditionally limited crop production. By providing access to freshwater in areas with saline groundwater, skimming technology enables farmers to cultivate high-value fruit crops that would otherwise struggle to thrive.
- In coastal areas with salt-sensitive crops, such as citrus or pomegranates, skimming technology ensures a steady supply of fresh water, reducing dependence on external water sources and increasing agricultural viability in salinity-prone regions.

4. Economic Advantages and Cost Savings

- Compared to desalination or long-distance water transport, skimming technology is a cost-effective solution for accessing freshwater. The construction of skimming wells requires less infrastructure and lower maintenance costs, making it a financially viable option for small and medium-sized farms.
- For farmers facing high irrigation costs due to the need to manage salinity, skimming technology offers an economical alternative that reduces the expenses associated with salinity mitigation. By lowering salinity levels in irrigation water, skimming technology helps prevent yield losses, improving profitability and productivity in regions with challenging water conditions.

5. **Reducing Soil Degradation and Improving Soil Health**
   - Saline irrigation water contributes to soil salinization, which can degrade soil structure, reduce fertility, and inhibit plant growth. Skimming technology prevents soil degradation by providing low-salinity water that reduces salt buildup in the soil profile.
   - For crops like mangoes, olives, and grapes, which are cultivated in arid or semi-arid regions, maintaining soil health is essential for long-term productivity. By ensuring that irrigation water is free from excessive salts, skimming technology supports better soil structure, nutrient availability, and root health, fostering a productive growing environment.

Practical Applications of Skimming Technology in Fruit Crop Systems

1. **Citrus and Pomegranate Orchards**
   - Citrus and pomegranate crops are sensitive to saline conditions, particularly during critical growth stages like flowering and fruit set. In coastal regions where saltwater intrusion can impact groundwater quality, skimming technology provides a reliable source of freshwater, supporting consistent growth and high-quality fruit production.
   - By reducing the salt concentration in irrigation water, skimming technology improves the health and productivity of citrus and pomegranate trees, helping growers produce marketable fruit with optimal size, flavor, and nutrient content.
2. **Grape Vineyards**
   - Salinity can adversely affect grapevine growth, berry quality, and yield, especially in coastal vineyards where groundwater salinity is a concern. Skimming technology provides a sustainable irrigation source for grapevines, reducing salt-induced stress and promoting balanced vine growth and berry ripening.
   - With access to low-salinity water, vineyards are able to support grapevines with consistent moisture and nutrient uptake, leading to better flavor development, sugar concentration, and color in grapes—qualities essential for both table and wine grapes.
3. **Olive Groves in Arid Regions**
   - Olive trees are relatively drought-tolerant but can be affected by high salinity levels in irrigation water, which can impact olive size, oil quality, and yield. In arid regions where freshwater is scarce, skimming technology allows olive growers to access freshwater for irrigation without drawing up saline water from deeper layers.
   - By maintaining low-salinity irrigation, skimming technology supports healthy root function, better nutrient absorption, and improved fruit quality in olive groves, which is critical for producing premium-grade olive oil.
4. **Date Palm Cultivation in Desert Areas**
   - Date palms are often grown in desert regions where groundwater is the primary irrigation source. However, high salinity levels can limit date palm productivity and fruit quality. Skimming technology enables date farmers to access low-salinity water from shallow

aquifers, supporting optimal growth and fruit production even in challenging conditions.

- By preventing salt accumulation in the soil and root zone, skimming technology enhances the resilience of date palms to salinity, enabling consistent yields and high-quality dates that are valued in local and export markets.

5. **Strawberries and Other Berry Crops**

   - Strawberries and other berries are highly sensitive to salinity, as salt stress can reduce fruit size, flavor, and overall crop quality. In coastal farms where groundwater salinity poses a challenge, skimming technology provides a dependable source of freshwater, supporting the production of high-quality berries.
   - Low-salinity water obtained through skimming wells helps maintain ideal soil conditions for berry crops, ensuring that fruits develop with the desired sweetness, texture, and size, which are crucial for marketability.

Environmental Benefits of Skimming Technology

1. **Protecting Groundwater Quality**

   - Skimming technology reduces the need for extensive groundwater extraction from deeper layers, helping protect the quality of both shallow and deep aquifers. By focusing water extraction on shallow freshwater layers, skimming technology minimizes the risk of drawing up saline water, preserving the integrity of groundwater resources.
   - This selective extraction helps maintain aquifer balance and prevents over-salinization, which is particularly important for protecting groundwater ecosystems and maintaining water quality in coastal and arid regions.

2. **Supporting Biodiversity and Soil Health**

   - By providing low-salinity water for irrigation, skimming technology minimizes the risk of soil salinization, which is beneficial for both crop health and the surrounding environment. Reduced salinity levels in the soil support soil biodiversity, promoting beneficial microbial activity and improving soil structure.
   - Healthier soil enhances root growth, nutrient cycling, and water retention, which not only benefits the crops but also contributes to the sustainability of the ecosystem. Skimming technology thus promotes a balanced, eco-friendly approach to water management, benefiting both agriculture and the environment.

Benefits of Skimming Technology for Salinity Management in Fruit Crops

Salinity poses a significant challenge to fruit crop production, particularly in coastal and arid regions where saline groundwater can lead to soil degradation and reduced crop yields. Skimming technology addresses this challenge by selectively extracting low-salinity water from shallow aquifers, providing a sustainable source of irrigation water that minimizes salt buildup in the soil. This approach is particularly beneficial for salt-sensitive fruit crops, as it reduces the risk of salinity stress, supports soil health, and promotes long-term productivity.

1. Reducing Soil Salinization

1. Minimizing Salt Buildup in the Root Zone

- One of the primary benefits of skimming technology is its ability to reduce salt accumulation in the root zone by providing low-salinity irrigation water. For salt-sensitive crops, such as strawberries, tomatoes, and citrus, exposure to saline water can lead to osmotic stress, ion toxicity, and impaired root function, all of which negatively impact growth and yield.
- By using skimming technology to access fresh or low-salinity water from shallow aquifers, growers can irrigate their crops without introducing excessive salts into the soil. This reduction in salt input helps maintain a favorable root environment, ensuring that plants can absorb water and nutrients efficiently, which is crucial for healthy growth and fruit development.

2. Preventing Soil Salinization Over Time

- Salinization is a gradual process that can lead to irreversible soil degradation if not properly managed. Continuous use of saline irrigation water leads to the accumulation of salts in the soil, which can eventually render the land unproductive for agriculture.
- Skimming technology offers a sustainable solution by allowing farmers to irrigate with low-salinity water, slowing down or even preventing the salinization process. This is especially important for perennial fruit crops, such as citrus trees and grapevines, which require stable soil conditions for long-term productivity. By reducing salt buildup, skimming technology ensures that orchards and vineyards can continue to thrive without suffering the damaging effects of salinity.

3. Maintaining Optimal Soil Moisture Balance

- Salinity affects soil moisture retention by reducing the soil's ability to hold water, which can exacerbate drought stress. By providing low-salinity water, skimming technology helps maintain a healthier moisture balance in the soil, reducing the risk of dehydration in fruit crops.
- This improved moisture balance is beneficial for crops like berries and tomatoes, which are sensitive to fluctuations in soil moisture. With skimming technology, growers can ensure a steady supply of low-salinity water, which supports consistent soil moisture levels and promotes uniform fruit development.

2. Supporting Salt-Sensitive Crops

1. Enhanced Growth and Yield in Salt-Sensitive Crops

- Salt-sensitive crops, such as strawberries, tomatoes, and citrus, are particularly vulnerable to salinity stress, as high salt concentrations interfere with root water uptake and nutrient absorption. By providing low-salinity water, skimming technology enables these crops to grow in environments where saline groundwater might otherwise restrict productivity.
- In strawberry farms, where soil salinity can reduce fruit size, flavor, and yield, skimming

technology provides an effective way to manage water quality, ensuring that plants receive fresh water that supports optimal growth. Similarly, in citrus orchards, low-salinity irrigation promotes healthier root systems, which enhances the tree's ability to produce high-quality, marketable fruit.

2. **Improving Fruit Quality and Marketability**

   - Salinity affects not only the yield but also the quality of fruit, impacting characteristics like size, sweetness, and texture. High salinity levels can cause smaller, less flavorful fruits, which may not meet market standards. By using skimming technology to irrigate with fresh water, growers can maintain the quality parameters that are essential for consumer satisfaction.
   - For example, in grape production, skimming technology helps avoid salt-induced flavor changes, which are critical for wine grapes. By reducing salinity stress, skimming technology supports the production of grapes with balanced sugar and acidity levels, contributing to better wine quality and higher market value.

3. **Protecting Salt-Sensitive Root Systems**

   - Salt stress can lead to root damage by creating osmotic pressure that prevents roots from absorbing water. In crops like tomatoes and blueberries, which have delicate root systems, excessive salt exposure can cause root dehydration, reducing nutrient uptake and impairing plant growth.
   - By providing low-salinity water, skimming technology protects these sensitive root systems, ensuring that plants can access the water and nutrients they need. This protection supports healthy root development, which is essential for consistent growth and fruit production in salt-sensitive crops.

3. Preserving Soil Structure and Fertility

1. **Maintaining Soil Structure and Porosity**

   - Saline water can alter soil structure by causing clay particles to disperse, leading to soil compaction and reduced porosity. Compacted soils are less able to retain water and nutrients, creating challenging conditions for root growth and plant health.
   - Skimming technology prevents these issues by supplying fresh or low-salinity water that preserves soil structure. This practice supports soil porosity and aeration, which are crucial for root expansion and healthy soil biology. For fruit crops like apples, pears, and cherries, a well-structured soil environment is essential for strong root systems that can support high yields and quality fruit.

2. **Protecting Long-Term Soil Fertility**

   - Saline irrigation water contributes to nutrient imbalances in the soil, as high salt concentrations interfere with the availability of essential elements like potassium, calcium, and magnesium. This imbalance affects soil fertility, reducing the soil's ability to support productive crops over time.

- By providing low-salinity water, skimming technology helps maintain nutrient availability and balance, supporting the soil's long-term fertility. For perennial crops, such as stone fruits and citrus trees, this preservation of soil fertility is critical for sustaining productivity over many years.

3. **Encouraging Beneficial Soil Microbial Activity**

- Salinity can disrupt soil microbial communities, as high salt concentrations create an inhospitable environment for many beneficial microorganisms. These microorganisms play a vital role in nutrient cycling, organic matter decomposition, and disease suppression, all of which contribute to soil health.
- Skimming technology supports a favorable soil environment by reducing salinity levels, allowing beneficial microbes to thrive. Healthy microbial activity enhances nutrient availability and helps create a resilient soil ecosystem that supports strong crop growth. For fruit crops like grapes, where soil health is closely tied to vine vigor and fruit quality, maintaining a balanced microbial community is essential.

4. Enhancing Crop Resilience in Coastal and Semi-Arid Regions

1. **Mitigating the Effects of Saltwater Intrusion**

- In coastal regions, saltwater intrusion is a significant issue, as seawater seeps into freshwater aquifers, increasing groundwater salinity. Skimming technology allows farmers to access low-salinity water from shallow aquifers, minimizing the impact of saltwater intrusion on crop irrigation.
- For salt-sensitive crops grown near coastal areas, such as strawberries and blueberries, skimming technology provides a sustainable solution that protects crops from salinity stress. This reliable access to low-salinity water supports year-round production and helps coastal farms remain productive even as saltwater intrusion poses a growing challenge.

2. **Supporting Agriculture in Semi-Arid Zones**

- In semi-arid and arid regions, natural freshwater resources are often limited, and saline groundwater can make agriculture difficult. Skimming technology enables farmers in these areas to tap into shallow freshwater layers, supporting sustainable agriculture in regions where water quality is a concern.
- For semi-arid fruit orchards, such as pomegranates and olives, skimming technology offers a way to access reliable irrigation water without exacerbating soil salinity. This access to freshwater enhances crop resilience to drought and salinity, enabling growers to produce high-quality fruit in challenging environments.

3. **Reducing Dependency on External Water Sources**

- Skimming technology reduces the need to import freshwater from external sources, which can be costly and unsustainable in the long term. By providing a local source of low-salinity water, skimming technology supports self-sufficiency in water management, particularly in regions where access to freshwater is limited.

- In regions that experience seasonal drought or water scarcity, having a dependable, low-salinity water source allows growers to maintain production levels without relying heavily on alternative sources. This independence in water management promotes resilience and supports sustainable agriculture in salinity-affected areas.

D. Contingency Planning for Stress Situations

Contingency planning is a proactive approach in agriculture that involves developing strategies to manage and mitigate the impact of unpredictable environmental stresses. These stressors, including drought, floods, extreme temperatures, and other climate-induced challenges, can significantly affect crop productivity and quality. By creating contingency plans, growers can prepare for various stress scenarios, minimizing potential losses and enhancing the resilience of their operations. For fruit crop production, where yield and quality are sensitive to environmental changes, contingency planning is essential to ensure stable and sustainable production.

1. Importance of Contingency Planning in Agriculture

1. **Mitigating the Impact of Climate Variability**
    - Climate variability poses a significant threat to agricultural productivity, especially in regions prone to erratic weather patterns. Unpredictable droughts, heavy rainfall, heatwaves, and cold snaps can cause immediate and long-term damage to crops, affecting yield, quality, and profitability. Contingency planning enables growers to anticipate these events and implement strategies to minimize their impact on fruit crops.
    - For example, in regions susceptible to drought, a contingency plan may include a water management strategy with drought-resistant crop varieties, water-saving irrigation methods, and soil moisture conservation practices. These measures help reduce the vulnerability of crops to water scarcity and allow growers to continue producing high-quality fruits even during prolonged dry periods.
2. **Ensuring Agricultural Productivity and Food Security**
    - In many regions, agriculture is a primary source of income and food security. Environmental stresses can disrupt food supply chains, causing economic losses and food shortages. Contingency planning plays a critical role in maintaining productivity and ensuring food security by providing farmers with the tools to adapt to changing conditions.
    - By incorporating adaptive practices such as crop rotation, alternative cropping systems, and soil management techniques, contingency plans help sustain productivity even under adverse conditions. This resilience is especially important for fruit crops, which often have longer growing cycles and require stable environmental conditions for optimal yield and quality.
3. **Reducing Crop Losses and Financial Risks**
    - Unpredictable weather events can lead to significant crop losses and financial setbacks for growers. Extreme weather conditions may result in poor fruit set, reduced fruit quality, or even complete crop failure. Contingency planning helps mitigate these risks

by preparing farmers to implement responsive measures that protect their crops and investments.

- For instance, in the case of an unexpected frost event, a contingency plan might include protective measures such as frost blankets or wind machines, which help prevent frost damage to sensitive fruit crops like apples, peaches, and grapes. These preventive actions reduce the risk of yield loss, allowing growers to protect their income and maintain market supply.

Adaptive Strategies in Contingency Planning for Agriculture

1. Diversified Crop Rotation and Alternative Cropping Systems

- Crop rotation and alternative cropping systems are effective strategies for building resilience to environmental stress. By diversifying crops and rotating between different fruit varieties or including cover crops, farmers can improve soil health, reduce pest pressure, and mitigate the effects of stress on any single crop.
- For example, in citrus orchards, planting nitrogen-fixing cover crops between tree rows improves soil fertility and helps retain soil moisture, which is particularly valuable during droughts. In apple and peach orchards, rotating with crops that require less water or are more resistant to pests can reduce resource demands and improve overall system resilience.

2. Soil Amendments and Soil Health Management

- Soil health is critical for supporting resilient crops that can withstand stress conditions. Soil amendments, such as compost, organic matter, and biochar, improve soil structure, water retention, and nutrient availability, creating a favorable environment for crop growth.
- In drought-prone regions, soil amendments increase the soil's water-holding capacity, reducing the need for frequent irrigation and helping crops endure water scarcity. For instance, the use of organic mulches in berry farms helps conserve soil moisture, reduces evaporation, and enhances soil fertility, enabling plants to thrive even under dry conditions.

3. Water-Saving Practices and Efficient Irrigation

- Water is a precious resource, especially in areas with limited rainfall. Efficient irrigation methods, such as drip irrigation and micro-sprinklers, deliver water directly to the root zone, minimizing wastage and ensuring that crops receive adequate moisture even during dry spells.
- Contingency plans often incorporate water-saving practices to maximize WUE. For high-value fruit crops like grapes, citrus, and avocados, drip irrigation systems are particularly effective, as they ensure that water is used optimally and reduce the risk of drought stress. Additionally, rainwater harvesting systems may be part of the plan, collecting and storing rainwater to supplement irrigation during dry periods.

4. **Implementing Early Warning Systems and Monitoring**

   - Early warning systems that use climate data, weather forecasts, and remote sensing technologies enable farmers to monitor potential stress events and respond quickly. By receiving timely alerts, growers can implement protective measures in advance, minimizing the potential damage to their crops.
   - In frost-sensitive crops like grapes, peaches, and citrus, monitoring temperature fluctuations allows farmers to take preventive action, such as deploying frost protection equipment or adjusting irrigation schedules. Early warning systems help improve decision-making, allowing growers to protect their crops from unexpected temperature drops, heatwaves, or other environmental stressors.

5. **Selecting Drought-Resistant and Resilient Crop Varieties**

   - Choosing crop varieties that are inherently resilient to specific environmental stresses, such as drought, heat, or salinity, is an essential component of contingency planning. These varieties are bred or selected for their ability to withstand adverse conditions without significant loss of yield or quality.
   - For instance, drought-tolerant grape varieties are preferred in regions with limited water availability, as they can produce quality fruit under water-limited conditions. Similarly, salt-tolerant citrus varieties may be chosen for coastal areas where saltwater intrusion is a concern. By selecting resilient varieties, growers reduce the risk of crop failure and improve the adaptability of their orchards to varying environmental conditions.

6. **Protective Infrastructure and Measures**

   - Physical protective measures, such as shade nets, windbreaks, and frost blankets, help shield crops from extreme weather. Shade nets are useful for fruit crops that are sensitive to high temperatures and sunburn, such as apples and tomatoes, as they reduce the intensity of sunlight and keep temperatures within an optimal range.
   - Windbreaks provide protection from strong winds, which can damage fruit trees, cause fruit drop, and increase water loss from soil. Frost blankets and wind machines are also effective for managing frost risk in cold-sensitive crops. These protective measures create a more stable growing environment, allowing fruit crops to withstand environmental stresses more effectively.

7. **Financial Risk Management and Crop Insurance**

   - Crop insurance is an essential tool in contingency planning that provides financial protection in case of crop losses due to environmental stress events. By insuring their crops, farmers can reduce the economic impact of unpredictable weather events, ensuring that they have financial support to recover and reinvest in the following growing season.
   - For high-value fruit crops, insurance coverage offers peace of mind and stability, as it allows growers to recover from losses due to events like drought, floods, frost, or heatwaves. Contingency plans that incorporate crop insurance provide a financial safety net that contributes to the long-term resilience of agricultural operations.

Integrating Contingency Planning into Agricultural Operations

1. **Customized Plans for Region-Specific Stressors**
    - Effective contingency planning involves customizing strategies to address the unique stressors in each region. For example, coastal areas may focus on managing saltwater intrusion, while arid regions prioritize drought resilience. Developing region-specific contingency plans ensures that growers are equipped with the most relevant tools and strategies for their local conditions.
    - In tropical regions, contingency plans may include flood management practices, such as raised beds and drainage systems, to prevent waterlogging in fruit crops like bananas and pineapples. By tailoring plans to regional needs, growers can improve the effectiveness of their adaptive measures and achieve greater resilience to environmental stress.
2. **Training and Capacity Building for Farmers**
    - Providing training on contingency planning and adaptive techniques is essential to empower farmers to manage stress effectively. Workshops, extension programs, and on-farm demonstrations help farmers understand the benefits of adaptive practices, familiarize them with new technologies, and build their capacity to respond to stress events.
    - For fruit growers, training in efficient irrigation, soil health management, and climate monitoring improves their ability to implement effective contingency measures. This education enhances the resilience of farming communities, promoting sustainable practices that support productivity in the face of climate variability.
3. **Collaboration with Research Institutions and Extension Services**
    - Collaborating with agricultural research institutions and extension services provides growers with access to the latest advancements in stress management. These partnerships offer insights into improved crop varieties, advanced irrigation techniques, and innovative soil management practices, which can be incorporated into contingency plans.
    - For example, research institutions may offer guidance on selecting fruit varieties suited to specific climates, such as drought-tolerant apple varieties or heat-resistant tomatoes. By integrating research-based knowledge, contingency plans become more robust, supporting growers with science-backed solutions to environmental challenges.

Implementation of Adaptive Cropping Strategies

Adaptive cropping strategies are a crucial aspect of contingency planning in agriculture, helping growers adjust to changing environmental conditions and mitigate the risks associated with climate variability, drought, and extreme weather. By selecting resilient crop varieties, modifying planting dates, and incorporating cover crops, farmers can create more sustainable and productive systems. These adaptive strategies are essential for fruit crops, which are often sensitive to water availability, temperature fluctuations, and soil health.

1. Selecting Drought-Tolerant and Resilient Crop Varieties

1. **Drought-Tolerant Varieties for Water-Limited Environments**
   - Selecting drought-tolerant varieties is a foundational strategy in adaptive cropping, especially in regions prone to water scarcity. These varieties are bred to withstand low moisture conditions, maintaining productivity and quality despite reduced water availability. In fruit crops, drought-tolerant varieties can help ensure consistent yields and minimize the risk of crop failure during dry periods.
   - For example, drought-tolerant grape varieties have been developed for arid regions, where water is scarce and high temperatures can intensify moisture stress. These grape varieties exhibit traits such as deeper root systems, smaller leaf size, and improved water-use efficiency, all of which contribute to their resilience under drought conditions.
2. **Salt-Tolerant Varieties for Coastal and Saline-Prone Areas**
   - In coastal regions or areas with high salinity, salt-tolerant varieties can reduce the impact of saline irrigation water on crop health. Salt-tolerant fruit varieties, such as certain citrus and pomegranate types, have been developed to cope with elevated salt levels, ensuring better growth, yield, and fruit quality under salinity stress.
   - These varieties are particularly valuable for managing soil and water salinity, as they can grow in challenging environments without suffering the effects of salt-induced stress. By choosing salt-tolerant crops, growers can continue to produce marketable fruit in regions where salinity limits the viability of conventional varieties.
3. **Early-Maturing and Short-Season Varieties**
   - Early-maturing and short-season varieties are ideal for regions with shorter growing windows or areas prone to late-season drought. By selecting varieties with shorter growth cycles, farmers can harvest their crops before the onset of severe water shortages or temperature extremes.
   - For instance, early-maturing strawberry and melon varieties allow growers in drought-prone regions to complete the growing season before water resources become critically low. This approach not only reduces the overall water demand but also increases the likelihood of achieving a successful harvest in variable climates.

2. Adjusting Planting Dates to Avoid Stress Periods

1. **Planting to Avoid Peak Stress Seasons**
   - Adjusting planting dates is an effective strategy for minimizing exposure to stressful environmental conditions. By aligning planting schedules with favorable weather patterns, growers can avoid periods of extreme temperature, drought, or excessive rainfall that can negatively impact crop growth and yield.
   - In crops like apples, where late frost events can damage blossoms, shifting planting dates slightly earlier or later helps avoid frost damage. For summer crops, adjusting planting dates to align with cooler periods or the rainy season can reduce water stress, ensuring that plants establish well before high temperatures or dry spells set in.

2. **Staggered Planting for Continuous Production**
   - Staggered planting involves dividing the planting of a crop into multiple stages, with each stage separated by a few weeks. This technique provides flexibility in case of unexpected weather events and ensures that not all crops are at the same developmental stage if a stress event occurs.
   - In crops like tomatoes and berries, staggered planting extends the harvest season and reduces the risk of losing an entire crop to an unexpected drought or heatwave. This approach also allows growers to adjust their irrigation and fertilization schedules based on the needs of each planting, optimizing resource use and reducing waste.

3. Crop Rotation and Diversification

1. **Rotating Crops to Improve Soil Health and Water Retention**
   - Crop rotation is the practice of growing different types of crops sequentially on the same plot of land. In fruit crop production, rotating with non-fruit crops, such as legumes, helps improve soil health, enhance nutrient cycling, and reduce pest pressure. Rotating crops also supports soil water retention, as different plants have unique rooting structures and water uptake patterns.
   - For example, rotating fruit crops with cover crops like clover or alfalfa improves soil structure and water-holding capacity, as these crops add organic matter and nitrogen to the soil. This improvement in soil health enhances the resilience of fruit crops to drought stress and reduces the need for frequent irrigation.
2. **Integrating Annual and Perennial Crops**
   - Integrating annual crops, like leafy greens or legumes, with perennial fruit trees helps build system resilience by diversifying resource use and reducing soil erosion. Annual crops can be rotated in the spaces between trees, which minimizes open soil surfaces and enhances water retention.
   - In orchards, intercropping with legumes provides an added benefit by fixing nitrogen, which supports the nutrient needs of fruit trees. This approach not only improves soil fertility but also helps create a more balanced ecosystem, enhancing the orchard's ability to withstand environmental stressors.
3. **Diversified Cropping Systems for Pest and Disease Management**
   - Diversifying crops within a production system helps reduce the risk of pest and disease outbreaks, as it disrupts pest life cycles and limits the spread of diseases. In fruit production, this diversification supports crop resilience by maintaining a balanced ecosystem, where beneficial organisms can thrive and naturally control pests.
   - For example, intercropping blueberries with flowering plants attracts pollinators and beneficial insects, which support pollination and control of pest populations. This biodiversity reduces the need for chemical inputs, enhancing soil and crop health while improving the system's adaptability to environmental stress.

4. Cover Crops for Soil Structure and Water Retention

1. **Using Cover Crops to Enhance Soil Structure**

   - Cover crops, such as legumes, grasses, and brassicas, are grown between main crop cycles to improve soil structure and prevent erosion. The root systems of cover crops create channels in the soil, increasing porosity, promoting water infiltration, and reducing surface runoff.
   - In fruit orchards, cover crops like clover and rye grass reduce soil compaction and provide organic matter, which improves water retention and supports a healthy root environment. For deep-rooted fruit crops, such as apple trees, these improvements in soil structure support better root growth and nutrient uptake, which are essential for maintaining productivity under stress.

2. **Cover Crops for Water Retention and Drought Resilience**

   - By adding organic matter to the soil, cover crops enhance the soil's ability to retain water, which is particularly valuable in drought-prone regions. Organic matter acts as a sponge, holding moisture near the root zone and reducing the need for frequent irrigation.
   - For instance, in vineyards, cover crops help maintain soil moisture and protect grapevines from drought stress, which can impact berry size, sugar concentration, and flavor. This additional moisture buffering provided by cover crops is crucial for maintaining consistent fruit quality, even during dry spells.

3. **Erosion Control and Soil Protection During Extreme Weather Events**

   - Cover crops protect the soil from erosion, particularly in areas susceptible to heavy rains or strong winds. By covering the soil surface, these crops prevent topsoil from being washed or blown away, which preserves soil fertility and stability.
   - In hilly or sloped orchards, such as olive and almond farms, cover crops provide a protective barrier against soil erosion, preventing nutrient loss and maintaining soil depth. This protection is essential for sustaining productivity in regions where extreme weather events can compromise soil health and structure.

Benefits of Adaptive Cropping Strategies in Stress Management

1. **Enhanced Resilience to Climate Variability**

   - Adaptive cropping strategies, including drought-tolerant varieties, adjusted planting schedules, crop rotation, and cover cropping, create a more resilient agricultural system that can better withstand the effects of climate variability. These strategies help fruit growers maintain productivity, minimize crop losses, and ensure consistent supply despite fluctuations in weather patterns.
   - For example, selecting drought-tolerant varieties and planting them early in the season allows growers to mitigate the impact of late-season drought, ensuring that crops reach maturity and produce high-quality fruit even under reduced water availability.

2. **Improved Soil Health and Productivity**
   - Adaptive cropping strategies improve soil health by enhancing soil structure, increasing organic matter, and supporting beneficial microbial activity. Healthy soil is more resilient to stress, as it retains moisture, provides a balanced nutrient profile, and supports vigorous root growth.
   - In fruit production, these soil health benefits translate into better root development, improved nutrient uptake, and higher yields. Crops like apples, cherries, and blueberries benefit from healthier soil, which supports consistent growth, improves fruit size and quality, and reduces susceptibility to disease.
3. **Reduction in Input Costs and Environmental Impact**
   - Adaptive strategies reduce the need for chemical inputs, such as synthetic fertilizers and pesticides, by promoting natural soil fertility and pest control. Cover crops, crop rotation, and resilient varieties contribute to a balanced ecosystem that requires fewer inputs, reducing costs and minimizing the environmental footprint of fruit production.
   - By reducing reliance on chemical inputs, adaptive cropping strategies support sustainable practices that protect water quality, enhance biodiversity, and reduce the carbon footprint associated with synthetic inputs.
4. **Enhanced Crop Quality and Marketability**
   - Adaptive cropping strategies contribute to improved crop quality by promoting stable growth and reducing stress-related damage. For fruit crops, where size, flavor, and appearance are critical for market value, these strategies ensure that fruits reach market standards and maintain consumer appeal.
   - For instance, drought-tolerant varieties produce fruit that maintains size and sweetness even under water-limited conditions, which is essential for high-value crops like grapes, berries, and citrus. This consistency in quality supports profitability and helps growers remain competitive in the market.

E. Stability and Sustainability Indices

Stability and sustainability indices are crucial tools in agriculture for evaluating crop performance and the long-term viability of agricultural practices. Stability indices measure a crop's ability to perform consistently under variable environmental conditions, providing insights into its resilience and adaptability. Sustainability indices, on the other hand, assess the environmental, economic, and social impacts of agricultural systems, helping farmers and researchers make decisions that promote resource efficiency and ecological balance. Together, these indices contribute to developing resilient, productive, and sustainable farming systems, especially important for fruit crop production in diverse and changing climates.

1. Stability Indices for Assessing Crop Performance

Stability indices are metrics that help quantify how well a crop performs under different environmental conditions. These indices provide insights into a crop's resilience to stress factors, yield stability, and adaptability to various soil and climate conditions. Stability indices are valuable in breeding programs and crop selection, as they help identify varieties that can maintain high yields and quality across multiple environments, making them ideal for regions prone to climatic variability.

1. **Yield Stability Index (YSI)**
   - The Yield Stability Index (YSI) measures the consistency of a crop's yield across diverse environments and varying weather conditions. A high YSI indicates that the crop performs well regardless of environmental changes, making it suitable for regions with variable rainfall, temperature, or soil quality.
   - For example, in fruit crops like grapes, apples, and tomatoes, a high YSI is desirable because it reflects the crop's ability to produce stable yields even under stressful conditions, such as drought or unexpected temperature fluctuations. Breeders and researchers use the YSI to identify varieties that can perform reliably across different growing seasons, supporting steady production and reducing the risk of crop failure.
2. **Environmental Stability Index (ESI)**
   - The Environmental Stability Index (ESI) evaluates a crop's ability to adapt to specific environmental factors, such as temperature, rainfall, and soil type. The ESI provides a measure of how well a crop performs under a range of environmental conditions, indicating its resilience to stress factors like drought, salinity, or heat.
   - For fruit crops grown in areas with challenging environments, such as arid or coastal regions, the ESI is particularly useful. In citrus and pomegranate orchards, for instance, a high ESI would indicate that the crop can tolerate saline soils or variable moisture levels, reducing the need for intensive water and soil management. This adaptability is crucial for sustaining production in regions with limited or fluctuating resources.
3. **Coefficient of Variation (CV) for Yield Stability**
   - The Coefficient of Variation (CV) is another stability index that measures yield variability relative to the average yield across multiple environments. A lower CV suggests that the crop has more stable yields and is less affected by environmental changes, making it suitable for regions with inconsistent growing conditions.
   - In perennial crops like apple and cherry orchards, where stable annual yields are essential for profitability, a low CV reflects the crop's reliability and ability to produce consistent yields. This stability allows growers to plan more effectively, ensuring predictable harvest volumes and market supply despite seasonal variations in weather.
4. **Adaptability Index (AI)**
   - The Adaptability Index (AI) assesses a crop's ability to perform well across a range of environmental conditions, identifying varieties that are not only stable but also adaptable. Crops with a high AI can adjust to both favorable and unfavorable conditions, making them valuable for regions with extreme weather fluctuations.
   - For example, in grape production, varieties with a high AI can adapt to both wet and dry seasons, ensuring stable berry quality and yield. This adaptability is particularly

important for high-value fruit crops, where quality consistency directly impacts market value. Breeding programs use the AI to develop resilient varieties that can thrive across diverse climates, enhancing agricultural sustainability and reducing the impact of climate-related risks.

2. Sustainability Indices for Long-Term Agricultural Viability

Sustainability indices evaluate the environmental, economic, and social impacts of agricultural practices, helping farmers and policymakers make informed decisions that promote long-term resource efficiency and ecological health. In fruit crop production, where inputs like water, nutrients, and energy are essential, sustainability indices are valuable for tracking the overall impact of farming practices and ensuring that they support a balanced ecosystem.

1. **Water Use Efficiency (WUE) Index**
    - The Water Use Efficiency (WUE) Index measures the yield produced per unit of water used. This index is essential for assessing how efficiently a crop utilizes water, particularly in water-scarce regions where efficient irrigation is critical for sustainability.
    - For fruit crops, a high WUE indicates that the plant produces substantial yield with minimal water, which is beneficial for reducing irrigation costs and conserving water resources. In drought-prone areas, WUE is a key sustainability metric for crops like grapes, citrus, and strawberries, where water-efficient practices are crucial for maintaining productivity and profitability.
2. **Nutrient Use Efficiency (NUE) Index**
    - The Nutrient Use Efficiency (NUE) Index evaluates how effectively a crop uses nutrients, such as nitrogen and phosphorus, to produce yield. High NUE values indicate that the crop can maximize yield with minimal nutrient inputs, reducing the need for synthetic fertilizers and minimizing the risk of nutrient leaching.
    - In fruit crops like apples, peaches, and blueberries, maintaining a high NUE is essential for producing quality fruit without excessive fertilizer use, which can impact soil health and water quality. The NUE index guides sustainable fertilization practices, supporting soil fertility and reducing the environmental impact of nutrient runoff.
3. **Carbon Footprint (CF) Index**
    - The Carbon Footprint (CF) Index measures the greenhouse gas emissions associated with crop production, including emissions from fertilizers, machinery, irrigation, and transport. A lower CF index indicates that the farming practices have a smaller carbon footprint, contributing to climate-friendly agriculture.
    - For sustainable fruit production, especially in high-value crops like avocados, apples, and grapes, minimizing carbon emissions is increasingly important due to consumer demand for eco-friendly products. The CF index helps growers track their carbon impact and implement practices, such as reduced tillage and renewable energy use, to lower their carbon footprint.

4. **Soil Health Index (SHI)**

   - The Soil Health Index (SHI) is a comprehensive metric that evaluates soil physical, chemical, and biological properties to assess its overall health and capacity to support crop growth. Indicators include soil organic matter, nutrient levels, pH, and microbial activity, all of which influence crop resilience and productivity.
   - A high SHI is indicative of a healthy soil ecosystem that promotes root growth, nutrient cycling, and water retention, all of which are crucial for fruit crop production. In orchards and vineyards, maintaining a high SHI supports sustainable yields, improves fruit quality, and reduces the need for soil amendments. The SHI is essential for long-term soil management, ensuring that fruit crops can thrive without compromising soil fertility.

5. **Biodiversity Index (BI)**

   - The Biodiversity Index (BI) measures the diversity of plant and animal species within and around agricultural fields. A high BI reflects a balanced ecosystem that supports beneficial insects, pollinators, and soil organisms, all of which contribute to natural pest control and pollination.
   - In fruit orchards, a high BI reduces reliance on chemical pest control, as natural predators help manage pest populations. In apple and cherry orchards, promoting biodiversity attracts pollinators and beneficial insects, enhancing fruit set and yield. The BI supports sustainable crop management by reducing pesticide use and fostering ecological stability.

6. **Economic Sustainability Index (ESI)**

   - The Economic Sustainability Index (ESI) evaluates the long-term profitability of agricultural practices, factoring in input costs, yield stability, and market value. A high ESI indicates that the farming system is financially sustainable, allowing growers to maintain production without excessive costs or financial risk.
   - For high-value fruit crops like citrus, blueberries, and avocados, achieving a high ESI requires balancing input costs with market returns. Sustainable practices, such as efficient water and nutrient use, contribute to economic viability by reducing expenses while supporting yield and quality. The ESI helps farmers make informed choices that enhance both profitability and resource conservation.

Benefits of Stability and Sustainability Indices in Fruit Crop Production

1. **Supporting Resilient and Adaptable Crop Systems**

   - Stability indices allow breeders and researchers to identify crop varieties that are resilient to environmental stress and adaptable to diverse climates. For fruit crop production, stable and adaptable varieties reduce the risk of yield loss due to climate variability, ensuring a reliable supply and supporting growers' livelihoods.
   - In vineyards, for instance, varieties with high stability indices ensure consistent grape quality across different growing seasons, supporting the wine industry's need for predictable flavor profiles and production volumes.

2. **Promoting Sustainable Resource Use and Environmental Health**
    - Sustainability indices guide farmers in adopting practices that optimize water, nutrient, and energy use, minimizing the environmental impact of fruit crop production. These indices encourage practices that enhance soil health, reduce greenhouse gas emissions, and promote biodiversity, supporting an ecologically balanced farming system.
    - For example, by tracking the Water Use Efficiency (WUE) index, citrus growers in arid regions can adopt irrigation strategies that conserve water while maintaining productivity, ensuring the sustainability of their water resources.
3. **Guiding Breeding Programs for Future-Proof Varieties**
    - Stability indices play an essential role in breeding programs by identifying varieties that perform consistently under stress, making them ideal for future climates with unpredictable weather patterns. Breeders use stability data to develop new varieties that combine yield stability with resilience to drought, salinity, and other stressors.
    - In breeding programs for apples, cherries, and peaches, stability indices help develop varieties that maintain quality and yield under climate variability, supporting the long-term productivity and adaptability of these fruit crops.
4. **Providing Insights for Informed Decision-Making**
    - Stability and sustainability indices provide data-driven insights that help farmers make informed decisions about crop selection, resource management, and farming practices. By using these indices, farmers can optimize their operations for both short-term productivity and long-term sustainability.
    - In high-value fruit crops, where resource use efficiency directly impacts profitability, these indices support strategies that balance productivity with environmental responsibility, ensuring that production remains economically viable and ecologically sound.

2. Sustainability Indices for Long-Term Agricultural Health

Sustainability indices are critical tools in evaluating the overall health and long-term viability of agricultural systems. They provide insights into how farming practices impact environmental resources, economic stability, and social well-being, offering a comprehensive view of sustainability in agriculture. For fruit crop production, where input use, environmental sensitivity, and consumer demand for eco-friendly products are increasingly significant, sustainability indices enable farmers, researchers, and policymakers to make informed decisions that align with environmental goals and promote resilience. By considering key factors like soil health, water use efficiency, carbon footprint, and biodiversity, sustainability indices help ensure that agricultural practices support food security for future generations while protecting natural resources.

1. Importance of Sustainability Indices in Agriculture

Sustainability indices provide a framework for assessing whether agricultural practices contribute to or detract from the long-term health of ecosystems and communities. They take into account a wide range of factors—environmental, economic, and social—that collectively determine the viability of farming systems over time. By using these indices, farmers can implement practices that not only improve yield and quality but also enhance resource efficiency, reduce environmental impact, and support social and economic resilience.

1. **Holistic Assessment of Agricultural Practices**
    - Sustainability indices evaluate the impact of agriculture on soil, water, air, and biodiversity, as well as economic and social factors. This comprehensive view allows farmers to understand how each component of their operation affects the broader ecosystem and society.
    - For example, a fruit orchard may use sustainability indices to assess the combined impact of irrigation, fertilization, and pest management practices on water quality, soil health, and carbon emissions. By tracking these indices, the orchard manager can make adjustments to minimize negative environmental impacts while maintaining productivity.
2. **Guiding Resource Efficiency and Conservation**
    - Sustainability indices encourage efficient use of resources, including water, nutrients, and energy, by highlighting practices that conserve these inputs. In regions with limited resources, sustainability indices are essential for managing inputs responsibly, ensuring that agricultural systems remain viable without depleting or polluting natural resources.
    - For high-water-demand crops like citrus and avocados, water sustainability indices help farmers track irrigation efficiency, reducing waste and conserving local water supplies. This efficiency contributes to both environmental sustainability and cost savings, supporting the economic viability of water-intensive crops.
3. **Supporting Long-Term Food Security**
    - By promoting resource-efficient and environmentally responsible practices, sustainability indices contribute to food security by ensuring that agricultural land remains productive over the long term. Practices that deplete or degrade soil, water, and biodiversity compromise the ability of future generations to grow food, underscoring the importance of sustainable approaches.
    - In fruit production, sustainability indices encourage practices that maintain soil fertility and water availability, ensuring that orchards and vineyards can continue to provide food and income for decades. This continuity is particularly crucial for perennial crops like apples and grapes, where productivity depends on maintaining a healthy growing environment year after year.

2. Key Sustainability Indices in Agriculture

Sustainability indices encompass various metrics, each addressing a specific aspect of agricultural health. By measuring factors such as soil health, water use, carbon emissions, and biodiversity, these indices provide actionable insights that help farmers balance productivity with ecological responsibility.

1. **Soil Health Index (SHI)**
    - The Soil Health Index (SHI) assesses soil quality based on physical, chemical, and biological factors, including organic matter, nutrient availability, pH, and microbial

diversity. A high SHI reflects a healthy, productive soil ecosystem that supports plant growth, nutrient cycling, and water retention.

- In fruit crop production, a robust SHI indicates that soil management practices, such as the use of organic amendments and reduced tillage, are maintaining or improving soil fertility. Healthier soils support stronger root systems, reduce erosion, and improve water retention, all of which contribute to higher yields and quality in crops like apples, strawberries, and citrus.

2. **Water Sustainability Index (WSI)**

- The Water Sustainability Index (WSI) measures the efficiency of water use in agricultural practices, evaluating factors like irrigation methods, water quality, and conservation efforts. A high WSI indicates that water resources are being used efficiently, with minimal waste and minimal impact on local water availability.
- For high-value fruit crops, such as grapes and avocados, water efficiency is essential for both productivity and environmental sustainability. Using WSI, growers can assess their irrigation practices, make necessary adjustments to conserve water, and reduce the impact of agriculture on local water supplies.

3. **Carbon Footprint (CF) Index**

- The Carbon Footprint (CF) Index measures the greenhouse gas emissions associated with various agricultural activities, such as fertilizer application, fuel use, and soil management. A lower CF index indicates that farming practices are generating fewer emissions, contributing to climate-friendly agriculture.
- In fruit orchards, where mechanized activities like pruning, harvesting, and irrigation can contribute to emissions, tracking the CF index helps identify areas where carbon reduction strategies can be implemented. For example, using solar-powered irrigation systems or minimizing tractor usage reduces emissions, supporting sustainable production while responding to consumer demand for eco-friendly practices.

4. **Biodiversity Index (BI)**

- The Biodiversity Index (BI) measures the diversity of plant, animal, and microbial life within agricultural ecosystems. A high BI reflects a balanced and thriving ecosystem where beneficial organisms, such as pollinators and natural pest predators, play a role in supporting crop health and productivity.
- In fruit orchards, enhancing biodiversity through practices like intercropping, planting cover crops, and maintaining natural habitats for beneficial insects reduces dependence on chemical inputs, supports pollination, and promotes soil health. For example, in apple orchards, high biodiversity attracts pollinators and pest predators, reducing pesticide use and improving crop quality.

5. **Economic Sustainability Index (ESI)**

- The Economic Sustainability Index (ESI) evaluates the financial viability of agricultural practices by measuring factors like input costs, yield stability, and market profitability. A high ESI suggests that practices are economically sustainable, providing a stable income for farmers without excessive dependence on costly inputs.

- In fruit crop production, achieving a high ESI requires efficient use of resources and resilient crop varieties that perform well under variable conditions. For example, drip irrigation in high-value crops like blueberries and avocados enhances water-use efficiency and reduces irrigation costs, contributing to higher profitability and economic resilience.

6. **Social Sustainability Index (SSI)**
   - The Social Sustainability Index (SSI) evaluates the social impact of agricultural practices, including labor conditions, community engagement, and access to resources. A high SSI reflects practices that promote fair labor, improve local food security, and support community well-being.
   - In fruit farming, particularly in regions reliant on seasonal labor, the SSI highlights practices that ensure fair wages, safe working conditions, and community involvement. For example, investing in fair labor practices in apple orchards fosters a more resilient workforce and strengthens community relationships, supporting the overall sustainability of the farming operation.

Benefits of Sustainability Indices for Long-Term Agricultural Health

1. **Promoting Resource Conservation and Environmental Protection**
   - Sustainability indices encourage practices that conserve resources and reduce environmental impacts, such as efficient water use, nutrient management, and emissions reduction. By minimizing inputs and managing resources responsibly, farmers can protect natural resources and reduce the ecological footprint of agriculture.
   - For instance, a high WSI ensures that irrigation practices in orchards are conserving water, preserving local water sources for other agricultural and community needs. This conservation is crucial in arid regions, where water scarcity threatens both agriculture and community resilience.
2. **Improving Resilience to Climate Change**
   - By tracking and managing sustainability indices, farmers can adopt practices that enhance resilience to climate change. Improved soil health, biodiversity, and water efficiency reduce the vulnerability of crops to drought, extreme temperatures, and other climate-related stresses.
   - For example, by maintaining high soil organic matter through sustainable practices, growers increase the soil's water-holding capacity, which helps buffer fruit crops like grapes and berries from drought stress. These practices enhance the resilience of farms, ensuring that productivity and quality are maintained despite climate variability.
3. **Supporting Economic Viability and Long-Term Profitability**
   - Sustainability indices provide insights into cost-effective practices that reduce input expenses and increase yield stability, supporting economic viability. By balancing input

costs with productivity, farmers can achieve long-term profitability while minimizing environmental impact.

- For high-value crops, such as avocados and citrus, sustainability indices help optimize practices that enhance yield and quality while reducing resource costs. For instance, efficient nutrient use supports steady fruit production without over-reliance on fertilizers, improving profitability and reducing financial risks.

4. **Enhancing Consumer Confidence and Market Access**
   - Sustainability indices support the adoption of eco-friendly practices, which align with consumer preferences for sustainably produced food. For fruit crops, where consumers are increasingly concerned about environmental impact, high sustainability scores enhance product appeal and marketability.
   - For example, fruit growers using organic fertilizers and reducing pesticide use can attract consumers seeking organic or eco-friendly products. Meeting sustainability standards also opens opportunities in certified organic and fair-trade markets, increasing market access and boosting profitability.
5. **Guiding Policy and Research for Sustainable Agriculture**
   - Sustainability indices provide valuable data for policymakers and researchers, helping guide investments in sustainable agriculture, conservation programs, and breeding research. By identifying areas of improvement and best practices, sustainability indices inform policy decisions that support resource conservation and food security.
   - In fruit crop production, sustainability indices highlight the need for research on drought-resistant varieties, water-efficient technologies, and regenerative soil practices. This information guides funding and research priorities, ensuring that resources are directed toward sustainable solutions that benefit both growers and ecosystems.

References

Altaf MA, Shahid R, Ren M, Altaf MM, Jahan MS, et al. 2021. Melatonin mitigates nickel toxicity by improving nutrient uptake fluxes, root architecture system, photosynthesis, and antioxidant potential in tomato seedlings. Journal of Soil Science and Plant Nutrition 21:1842–55

Altaf MA, Shahid R, Altaf MM, Kumar R, Naz S, et al. 2022. Melatonin: first-line soldier in tomato under abiotic stress current and future perspective. Plant Physiology and Biochemistry 185:188–97

Andersen, P.C.; Lombard, P.B. and Westwood, M.N. 1984a. Effect of root anaerobiosis on the water relations of several *Pyrus* species. *Physiol. Plant*, 62: 245-252.

Andersen, P.C.; Lombard, P.B. and Westwood, M.N. 1984b. Leaf conductance, growth, and survival of willow and deciduous fruit tree species under flooded soil conditions. *J.Am. Soc. Hort. Sci.* 109: 132-138.

Arkhipova T, Martynenko E, Sharipova G, Kuzmina L, Ivanov I, et al. 2020. Effects of plant growth promoting rhizobacteria on the content of abscisic acid and salt resistance of wheat plants. Plants 9:1429

Arif Y, Sami F, Siddiqui H, Bajguz A, Hayat S. 2020. Salicylic acid in relation to other phytohormones in plant: a study towards physiology and signal transduction under challenging environment. Environmental and Experimental Botany 175:104040

Artlip, T. S., Wisniewski, M. E., Macarisin, D., and Norelli, J. L. 2009. Ectopic expression of a spinach *SOD* gene in young apple trees enhances abiotic stress resistance. *Acta Horticulturae*, 839: 645-650.

Arun Kumar Shanker and B. Venkateswarlu Abiotic Stress in Plants – Mechanisms and Adaptations InTech, Croatia

Ashraf M, Foolad MR. 2007. Roles of glycine betaine and proline in improving plant abiotic stress resistance. Environmental and Experimental Botany 59:206–16

Ballard, J.K. and Proebsting, E.L. 1978. Frost and frost control in Washington orchards. *Wash. State Univ. Coop. Ext. Ser. Bull.* 634.

Bradford, K.J. & Hsiao, T.C. (1982). Physiological responses to moderate water stress. In: Physiological plant ecology, O.L. Lange; P.S. Nobel; C.B. Osmond; H. Ziegler, (Ed.), pp. 263-324, Springer-Verlag, Berlin

Bartolini, G., Mazuelos, C. and Troncoso, A. 1991. Influence of Na2 SO4 and NaCl salts on survival, growth, and mineral composition of young olive plants in inert sand culture. *Adv. Hort. Sci.* 5: 73-76.

Bartolozzi, F., and Fontanazza, G.1999. Assessment of frost tolerance in Olive (*Olea europaea* L.). *Scientia Horticulturae*, 81: 309-319.

Bernstein, L. 1964. Effects of salinity on mineral composition and growth of plants. *Proc. 4th Int. Coll. Plant Analysis and Fertilizer Problems*, Bruxelles, Belgium, 4: 25-45.

Blanc, M.L., Geslin, H., Holzberg, I.A. and Mason, B. 1963. Protection against frost damage. *World Meteorol. Org. Tech. Note*. 51.

Blumm A. 1988. Plant Breeding for Stress Environments. CRC Publication, USA. Christiansen, MN and Lewis CF. 1982. Breeding Plants for Less Favourable Environments. Wiley International Science, USA.

Bongi, G. and Palliotti, A. 1994. Olive. P. 165-187. In: B. Schaffer and P.c. Andersen (eds.) Handbook environmental physiology of fruit crops. CRC Press, Boca Raton, FL.

Brewer, R.F.; Burns, R.M. and Opitz, K.W. 1973. Evaluation of man-made fog for frost protection. *Citrograph*, 59: 9-10.

Bulgari R, Franzoni G, Ferrante A. 2019. Biostimulants application in horticultural crops under abiotic stress conditions. Agronomy 9:306

Burns, R.M. 1970. Testing foliar sprays for frost protection of young citrus. *Proc. Fla. State Hort. Soc.* 83: 92-95.

Ciura J, Kruk J. 2018. Phytohormones as targets for improving plant productivity and stress tolerance. Journal of Plant Physiology 229:32–40

Chartzoulakis, K., Patakas, A., Kofidis, G., Bosabalidis, A. and Nastou, A. 2002. Water stress affects leaf anatomy, gas exchange, water relations, and growth of two avocado cultivars. *Scientia Horticulturae*, 95: 39-50.

Cornforth JW, Milborrow BV, Ryback G, Wareing PF. 1965. Chemistry and physiology of 'dormins' in sycamore: Identity of sycamore 'dormin' with abscisin II. Nature 205:1269–70

Cresti, M.; Ciampolini, F.; Tattini, M. and Cimato, A. 1994. Effect of salinity on productivity and oil quality of olive (*Olea europeae* L.) plants. *Adv. Hort. Sci.* 8: 211-214.

Danquah A, de Zelicourt A, Colcombet J, Hirt H. 2014. The role of ABA and MAPK signaling pathways in plant abiotic stress responses. Biotechnology Advances 32:40–52

Dar NA, Amin I, Wani W, Wani SA, Shikari AB, et al. 2017. Abscisic acid: a key regulator of abiotic stress tolerance in plants. Plant Gene 11:106–11

Davies, F.S. Buchanan, D.W. and Anderson, J.A. 1981. Water stress and cold hardiness in field-grown citrus. *J. Am. Soc. Hort. Sci.* 106: 197-200.

Depuydt S, Hardtke CS. 2011. Hormone signaling crosstalk in plant growth regulation. Current Biology 21:R365–R373

Devi R, Behera B, Raza B, Mangal V, Altaf MA, et al. 2022. An insight into microbes mediated heavy metal detoxification in plants: a review. Journal of Soil Science and Plant Nutrition 22:914–36

Dhillon BS, Tyagi RK, Saxena S and Randhawa GJ. 2005. Plant genetic resources: horticultural crops. Narosa Publ. House.

Dwivedi P and Dwivedi RS. 2005. Physiology of abiotic stress in plants. Agrobios. Janick JJ. 1986. Horticultural science. 4th Ed. WH Freeman and Co.

Edgerton, L.J. 1966. Some effects of gibberellin and growth retardants on bud development and cold hardiness of peach. *Proc. Am. Soc. Hort. Sci.* 88: 197-203.

Feller U, Vaseva II. 2014. Extreme climatic events: impacts of drought and high temperature on physiological processes in agronomically important plants. Frontiers in Environmental Science 2:39

Finch, A.H. 1977. Use of artificial fog for low-temperature control and its possible value in climate control to improve citrus fruiting in a desert climate. *Proc. Intl. Soc. Citriculture*, 1: 209-210.

Flore, J.A. and Howell, G.S. 1987. Environmental and physiological factors that influence cold hardiness and frost resistance in perennial crops, p. 139-150. In: F. Prodi, F. Rossi, G. Cristoferi (eds.). *Intl. Conf. on Agrometeorology*. Comune di Cesena, Cesena, Italy.

Freeman, J. A. and Carne, I.C. 1970. Use of succinic acid 2,2-dimethylhydrazide (Alar) to reduce winter injury in strawberries. *Can. J. Plant Sci.* 50: 189-190.

Funt, R.C. and Ferree, D.C. 1985. Influence of ethephon on bloom delay, hardiness, and yield of Redhaven peaches, Ohio, USA. *Proc. 5th Intl. Symp.* Growth Regulators Fruit Production, Bologna-Rimini, Italy, p.17 [Abstr.]

Glenn, D.M. Prado, E. Erez, A. Mc Ferson, J. and Puterka, G.J. 2002. A reflective processed - kaolin particle film affects fruit temperature, radiation, reflection, and solar injury in apples. *J. Am. Soc. Hort. Sci.* 127: 188-193.

Gucci, R. and Tattini, M. 1997. Salinity tolerance in Olive. *Horticultural Reviews*, 21: 177-214.

Guschina IA, Harwood JL, Smith M, Beckett RP. 2002. Abscisic acid modifies the changes in lipids brought about by water stress in the moss *Atrichum ndrogynum*. New Phytologist 156:255–64

Hedden P, Thomas SG. 2012. Gibberellin biosynthesis and its regulation. *Biochemical Journal* 444:11–25

He M, He C, Ding N. 2018. Abiotic stresses: general defenses of land plants and chances for engineering multistress tolerance. Frontiers in Plant Science 9:1771

Holmberg N and Bülow L. (1998) Improving stress tolerance in plants by gene transfer. *Trends inPlant Science* 3:61–66.

Holubowicz, T. and Boe, A.A. 1969. Development of cold hardiness in apple seedlings treated with gibberellic acid and abscisic acid. *J. Am. Soc. Hort. Sci.* 94: 661-664.

Hoque TS, Hossain MA, Mostofa MG, Burritt DJ, Fujita M, et al. 2016. Methylglyoxal: an emerging signaling molecule in plant abiotic stress responses and tolerance. *Frontiers in Plant Science* 7:1341

Islam MS, Hasan K, Islam B, Renu NA, Hakim MA, et al. 2021. Responses of water and pigments status, dry matter partitioning, seed production, and traits of yield and quality to foliar application of GA3 in mung bean (*Vigna radiata* L.). *Frontiers in Agronomy* 2:596850

Jaglo-Ottosen, K.R., Gilmour, G.S.J., Zarka, D.G., Schabenberger, O. and Thomashow, M.F. 1998.

Arabidopsis CBF1 overexpression induces *Cor* genes and enhances freezing tolerance. *Science*, 280: 104-106.

Jeyakumar, P. Kavino M. Kumar, N. and Soorianathasundaram, K. 2007. Physiological performance of papaya cultivars under abiotic stress conditions. *Acta Horticulturae* 740: 450-58.

Jiang K, Asami T. 2018. Chemical regulators of plant hormones and their applications in basic research and agriculture. Bioscience, Biotechnology, and Biochemistry 82:1265–300

Kanayama Y and Kochetor. 2015. Abiotic Stress Biology in Horticultural Plants. Springer.

Karlidag, H., E. Yildirim and M. Turan. 2009. Salicylic acid ameliorates the adverse effect of salt stress on strawberries. *J. Agric. Sci.*, 66: 271-278.

Kohli A, Sreenivasulu N, Lakshmanan P, Kumar PP. 2013. The phytohormone crosstalk paradigm takes center stage in understanding how plants respond to abiotic stresses. Plant Cell Reports 32:945–57

Koo, R.J.C. 1981. The effect of fall irrigation of freeze damage to citrus. *Proc. Fla. State. Hort. Soc.* 94: 37-39.

Koshita, Y. and Takahara, T. 2004. Effect of water stress on flower bud formation and plant hormone content of Satsuma mandarin (*Citrus unshiu* Marc.). *Scientia Horticulturae*, 99: 301-307.

Kramer PJ. 1980. Drought Stress and the Origin of Adaptation. In: Adaptation of Plants to Water and High Temperature Stress. John Wiley & Sons, USA.

Kuiper, P.J.C. 1964. Inducing resistance to freezing and desiccation in plants by decenylsuccinic acid. *Science*, 146: 544 – 546.

Ku Y, Sintaha M, Cheung M, Lam H. 2018. Plant hormone signaling crosstalks between biotic and abiotic stress responses. International Journal of Molecular Sciences 19:3206

Lee K, Lee HG, Yoon S, Kim HU, Seo PJ. 2015. The Arabidopsis MYB96 transcription factor is a positive regulator of Abscisic Acid-Insensitive 4 in the control of seed germination. *Plant Physiology* 168:677–89

Lee, S.H. Choi, J.H. Kim, W.S. Han, T.H. Park, Y.S. and Gemma, H. 2006. Effect of soil water stress on the development of stone cells in pear (*Pyrus pyrifolia* cv. 'Niitaka') flesh. *Scientia Horticulturae*, 110: 247-253.

Lee, H.J.; Kim, T.C; Kim, S.J.; and Park, S.J. 2005. Bruising injury of persimmon (*Diospyros kaki* cv. Fuyu) fruits. *Scientia Horticulturae*, 103: 179-185.

Lin, L.N. and Chang, Y.S. 2009. Effects of salicylic acid on the plant responses to heat stress. *Journal of the Taiwan Society for Horticultural Science*, 55(4):193-206

Liu Y, Huang W, Xian Z, Hu N, Lin D, et al. 2017. Overexpression of SlGRAS40 in tomato enhances tolerance to abiotic stresses and influences auxin and gibberellin signaling. *Frontiers in Plant Science* 8:1659

Llanes A, Masciarelli O, Ordóñez R, Isla MI, Luna V. 2014. Differential growth responses to sodium salts involve different abscisic acid metabolism and transport in Prosopis strombulifera. Biologia Plantarum 58:80–88

Lo SF, Ho THD, Liu YL, Jiang MJ, Hsieh KT, et al. 2017. Ectopic expression of specific GA2 oxidase mutants promotes yield and stress tolerance in rice. *Plant Biotechnology Journal* 15:850–64

Maloo SR. 2003. Abiotic Stress and Crop Productivity. Agrotech Publ. Academy, India.

Mangal V, Lal MK, Tiwari RK, Altaf MA, Sood S, et al. 2023. Molecular insights into the role of reactive oxygen, nitrogen, and sulphur species in conferring salinity stress tolerance in plants. Journal of Plant Growth Regulation 42:554–74

Mark C, Zór K, Heiskanen A, Dufva M, Emnéus J, et al. 2016. Monitoring intra-and extracellular redox capacity of intact barley aleurone layers responding to phytohormones. *Analytical Biochemistry* 515:1–8

Marin, L.; Benlloch, M. and Fernandez-Escobar, R. 1995. Screening for olive cultivars for salt tolerance. *Scientia Hort.* 64: 113-116.

Matthew A. Jenks and Paul M. Hasegawa. Abiotic Stress. Blackwell Publishing Ltd. UK.

Melgarejo, P., Martinez, J.J., Hernandez, F., Font, R.M., Barrows, P. and Erez, A. 2004. Kaolin treatment to reduce pomegranate sunburn. *Scientia Horticulturae*, 100: 349-353.

Mitchell, P.D.; Van Den Ende, B.; Jerie, P.H. and Chalmers, D.J. 1989. Response of Bartlett pear to withholding irrigation, regulated deficit irrigation, and tree spacing. *J.Am.Soc. Hort. Sci.* 114: 15-19.

Modlibowska, I. 1968. Effects of some growth regulators on frost damage. *Cryobiology*, 5:175- 187.

Nambara E, Okamoto M, Tatematsu K, Yano R, Seo M, et al. 2010. Abscisic acid and the control of seed dormancy and germination. Seed Science Research 20:55–67

Naor, A.; Stern, R.A., Gal, Y., Peres, M. and Flashman, M. 2005. Timing and severity of postharvest water stress affect following year productivity and fruit quality of field grown 'Snow Queen' nectarine. *J.Am.Soc. Hort. Sci.* 130: 806-812.

Nickell LG. 1983. Plant Growth Regulating Chemicals. CRC Publication, USA.

Nicolas, J.L., Richard-Forget, F.C., Goupy, P.M., Amiot, M.J., and Aubert, S.Y. 1994. Enzymatic browning reactions in apple and apple products. *CRC Crit. Rev. Food Sci. Nutr.* 34: 109-157.

Pareek, O.P., Vishal Nath and Mishra, A.L. 1996. Water harvesting and rainwater management in fruit crops. *Proceedings of Silver Jublee National Symposium on Arid Horticulture*, December 5-6, CCSHAU, Hisar: 49-61.

Pearce S, Huttly AK, Prosser IM, Li Y, Vaughan SP, et al. 2015. Heterologous expression and transcript analysis of gibberellins biosynthetic genes of grasses reveals novel functionality in the GA3 ox family. *BMC Plant Biology* 15:130

Peleg Z, Blumwald E. 2011. Hormone balance and abiotic stress tolerance in crop plants. Current Opinion in Plant Biology 14:290–95

Pellet, H.M. and Carter, J.V. 1981. Effect of nutritional factors on cold hardiness of plants. *Hort. Rev.* 3: 144-171.

Peter M.A. Toivonen and D. Mark Hodges. Abiotic Stress in Harvested Fruits and Vegetables. Intechopen .com.

Pretorius, J.J.B. and Wand, S.J.E. 2003. Late season stomatal sensitivity to microclimate is influenced by sink strength and soil moisture stress in 'Braester' apple trees in South Africa. *Scientia Horticulturae*, 98: 157-171.

Proebsting, E.L. and Mills, H.H. 1973. Bloom delay and frost survival in ethephon-treated sweet cherry. *Hort Science*, 8: 46-47.

Proebsting, E.L. and Mills, H.H. 1974. Time of gibberellin application determines the hardiness response of 'Bing' cherry buds and wood. *J. Am. Soc. Hort. Sci.* 99: 464-466.

Proebsting, E.L. and Mills, H.H. 1985. Cold resistance in peach, apricot and cherry as influenced by soil-applied paclobutrazol. *Hort Science*, 20: 88-90.

Raese, J.T. 1977. Induction of hardiness in apple tree shoots with ethephon, NAA, and growth retardants. *J. Am. Soc. Hort. Sci.* 102: 789-792.

Raese, J.T. 1983. Conductivity tests to screen fall-applied growth regulators to induce cold hardiness in young 'Delicious' apple trees. *J. Am. Soc. Hort. Sci.* 108: 172-176.

Rao NKS, Shivashankar KS, and Laxman RH. 2016. Abiotic Stress Physiology of Horticultural Crops. Springer.

Rao YR, Ansari MW, Singh AK, Bharti N, Rani V, et al. 2020. Ethylene mediated physiological response for in vitro development of salinity tolerant tomato. Journal of Plant Interactions 15:406–16

RuizSanchez, M.C., Domingo, R., Save, R., Biel, C. and Torrecillas, A. 1997. Effects of water stress and re-watering on leaf water relations of the lemon plant. *Biol. Plantarum*, 39: 623-631.

Sachs T. 2005. Auxin's role as an example of the mechanisms of shoot/root relations. *Plant and Soil* 268:13–19

Saini S, Kaur N, Pati PK. 2021. Phytohormones: key players in the modulation of heavy metal stress tolerance in plants. Ecotoxicol- ogy and Environmental Safety 223:112578

Salvi P, Manna M, Kaur H, Thakur T, Gandass N, et al. 2021. Phyto-hormone signaling and crosstalk in regulating drought stress response in plants. Plant Cell Reports 40:1305–29

Schaffer, B. Andersen, P.C. and Ploetz, R.C. 1992. Responses of fruit crops to flooding. *Horticultural Reviews*, 13: 257-313.

Singhal RK, Saha D, Skalicky M, Mishra UN, Chauhan J, et al. 2021. Crucial cell signaling crosstalk and integrative multi- omics techniques for salinity stress tolerance plants. *Frontiers in Plant Science* 12:670369

Stewart, I. and Leonard, C.D. 1960. Increased winter hardiness in citrus from maleic hydrazide sprays. *Proc. Am. Soc. Hort. Sci.* 75: 253-256.

Southwick, S.M., and Devenport, T.L. 1986. Characterization of water stress and low temperature effects on flower induction in citrus. *Plant Physiol.* 81: 26-29.

Tattini, M. Gucci, R. Coradeschi, M.A. Ponzio, C. And Everard, J.D. 1995. Growth, gas exchange, and ion content in *Olea europeae* plants during salinity stress and subsequent relief. *Physiol. Plant.* 95: 203-210.

Thomashow, M.F.1998. Role of cold-responsive genes in plant freezing tolerance. *Plant Physiology*, 118: 1-8.

Thomashow, M. F. 1999. Plant cold acclimation: freezing tolerance genes and regulatory mechanisms. *Annual Review of Plant Physiology and Plant Molecular Biology*, 50: 571- 599.

Turner NC and Kramer PJ. 1980. Adaptation of Plants to Water and High Temperature Stress. John Wiley & Sons, USA.

Wani SH, Kumar V, Shriram V, Sah SK. 2016. Phytohormones and their metabolic engineering for abiotic stress tolerance in crop plants. The Crop Journal 4:162–76

Westwood, M.N. and Bjornstad, H.O. 1981. Winter injury to apple cultivars is affected by growth regulators, weed control methods, and rootstocks. *J. Am. Soc. Hort. Sci.* 106: 430-432.

White, D.G. and Kennard, W.C. 1955. A preliminary report on the use of malic hydrazide to delay the blossoming of fruits. *Proc. Am. Soc. Hort. Sci.* 50: 147-151.

Xu ZY, Kim SY, Hyeon DY, Kim DH, Dong T, et al. 2013. The Arabidopsis NAC transcription factor ANAC096 cooperates with bZIP-type transcription factors in dehydration and osmotic stress responses. The Plant Cell 25:4708–24

Yamaguchi S. 2008. Gibberellin metabolism and its regulation. *Annual Review of Plant Biology* 59:225–51

Yin Y, Jiang X, Ren M, Xue M, Nan D, et al. 2018. AmDREB2C, from *Ammopiptanthus mongolicus*, enhances abiotic stress tolerance and regulates fatty acid composition in transgenic Arabidopsis. *Plant Physiology and Biochemistry* 130:517–28

For Product Safety Concerns and Information please contact our EU representative GPSR@taylorandfrancis.com Taylor & Francis Verlag GmbH, Kaufingerstraße 24, 80331 München, Germany